CAMBRIDGE LIBRARY COLLECTION

Books of enduring scholarly value

Darwin

Two hundred years after his birth and 150 years after the publication of 'On the Origin of Species', Charles Darwin and his theories are still the focus of worldwide attention. This series offers not only works by Darwin, but also the writings of his mentors in Cambridge and elsewhere, and a survey of the impassioned scientific, philosophical and theological debates sparked by his 'dangerous idea'.

Essays upon Some Controverted Questions

Thomas Henry Huxley (1825–95) became known as 'Darwin's bulldog' because of his forceful and energetic support for Darwin's theory, most famously at the legendary British Association meeting in Oxford in 1860. In fact, Huxley had some reservations about aspects of the theory, especially the element of gradual, continuous progress, but in public he was unwavering in his allegiance, saying in a letter to Darwin 'As for your doctrines I am prepared to go to the Stake if requisite'. In his 1892 Essays upon Some Controverted Questions, Huxley collected some of his previously published writings, of which the titles alone give some flavour of his pugnacious stance in debate: 'The interpreters of Genesis and the interpreters of Nature'; 'Science and pseudo-science'; 'Agnosticism and Christianity'. The passion for scientific truth which underlies everything he writes is well demonstrated in this lively and still-relevant collection.

Cambridge University Press has long been a pioneer in the reissuing of out-of-print titles from its own backlist, producing digital reprints of books that are still sought after by scholars and students but could not be reprinted economically using traditional technology. The Cambridge Library Collection extends this activity to a wider range of books which are still of importance to researchers and professionals, either for the source material they contain, or as landmarks in the history of their academic discipline.

Drawing from the world-renowned collections in the Cambridge University Library, and guided by the advice of experts in each subject area, Cambridge University Press is using state-of-the-art scanning machines in its own Printing House to capture the content of each book selected for inclusion. The files are processed to give a consistently clear, crisp image, and the books finished to the high quality standard for which the Press is recognised around the world. The latest print-on-demand technology ensures that the books will remain available indefinitely, and that orders for single or multiple copies can quickly be supplied.

The Cambridge Library Collection will bring back to life books of enduring scholarly value (including out-of-copyright works originally issued by other publishers) across a wide range of disciplines in the humanities and social sciences and in science and technology.

Essays upon Some Controverted Questions

THOMAS HENRY HUXLEY

CAMBRIDGE
UNIVERSITY PRESS

CAMBRIDGE UNIVERSITY PRESS

Cambridge, New York, Melbourne, Madrid, Cape Town, Singapore,
São Paolo, Delhi, Dubai, Tokyo

Published in the United States of America by Cambridge University Press, New York

www.cambridge.org
Information on this title: www.cambridge.org/9781108001557

This edition first published 1892
This digitally printed version 2009

ISBN 978-1-108-00155-7 Paperback

ESSAYS UPON SOME CONTROVERTED QUESTIONS

ESSAYS

UPON SOME

CONTROVERTED QUESTIONS

BY

THOMAS H. HUXLEY, F.R.S.

London

MACMILLAN AND CO.

AND NEW YORK

1892

CONTENTS

I AM indebted to the Editors of the *Nineteenth Century* and of the *Fortnightly Review* for permission to reprint such of the following Essays as have appeared in the pages of those periodicals : and so large a proportion of the papers has been published in the *Nineteenth Century* that my acknowledgments are especially due to Mr. Knowles.

T. H. H.

May 4, 1892.

PROLOGUE

Le plus grand service qu'on puisse rendre à la science est d'y faire place nette avant d'y rien construire.—CUVIER.

MOST of the Essays comprised in the present volume have been written during the last six or seven years, without premeditated purpose or intentional connection, in reply to attacks upon doctrines which I hold to be well founded; or in refutation of allegations respecting matters lying within the province of natural knowledge, which I believe to be erroneous; and they bear the mark of their origin in the controversial tone which pervades them.

Of polemical writing, as of other kinds of warfare, I think it may be said, that it is often useful, sometimes necessary, and always more or less of an evil. It is useful, when it attracts attention to topics which might otherwise be neglected; and when, as does sometimes happen, those who come to see a contest remain to think. It is necessary, when the interests of truth and of justice are at stake. It is an evil, in so far as controversy always tends to degenerate into quarrelling, to swerve from the great issue of what is right and what is wrong to the very small question

B

of who is right and who is wrong. I venture to hope that the useful and the necessary were more conspicuous than the evil attributes of literary militancy, when these papers were first published; but I have had some hesitation about reprinting them. If I may judge by my own taste, few literary dishes are less appetising than cold controversy; moreover, there is an air of unfairness about the presentation of only one side of a discussion, and a flavour of unkindness in the reproduction of "winged words," which, however appropriate at the time of their utterance, would find a still more appropriate place in oblivion. Yet, since I could hardly ask those who have honoured me by their polemical attentions to confer lustre on this collection, by permitting me to present their lucubrations along with my own; and since it would be a manifest wrong to them to deprive their, by no means rare, vivacities of language of such justification as they may derive from similar freedoms on my part; I came to the conclusion that my best course was to leave the essays just as they were written;[1] assuring my honourable adversaries that any heat of which signs may remain was generated, in accordance with the law of the conservation of energy, by the force of their own blows, and has long since been dissipated into space.

But, however the polemical concomitants of these discussions may be regarded—or better, disregarded

[1] With a few exceptions, which are duly noted when they amount to more than verbal corrections.

—there is no doubt either about the importance of the topics of which they treat, or as to the public interest in the " Controverted Questions " with which they deal. Or rather, the Controverted Question ; for disconnected as these pieces may, perhaps, appear to be, they are, in fact, concerned only with different aspects of a single problem, with which thinking men have been occupied, ever since they began seriously to consider the wonderful frame of things in which their lives are set, and to seek for trustworthy guidance among its intricacies.

Experience speedily taught them that the shifting scenes of the world's stage have a permanent background; that there is order amidst the seeming confusion, and that many events take place according to unchanging rules. To this region of familiar steadiness and customary regularity they gave the name of Nature. But, at the same time, their infantile and untutored reason, little more, as yet, than the playfellow of the imagination, led them to believe that this tangible, commonplace, orderly world of Nature was surrounded and interpenetrated by another intangible and mysterious world, no more bound by fixed rules than, as they fancied, were the thoughts and passions which coursed through their minds and seemed to exercise an intermittent and capricious rule over their bodies. They attributed to the entities, with which they peopled this dim and dreadful region, an unlimited amount of that power of modifying the course of events of which they themselves possessed a small share, and thus came to regard them as not merely beyond, but above, Nature.

Hence arose the conception of a "Supernature" antithetic to "Nature"—the primitive dualism of a natural world "fixed in fate" and a supernatural, left to the free play of volition—which has pervaded all later speculation and, for thousands of years, has exercised a profound influence on practice. For it is obvious that, on this theory of the Universe, the successful conduct of life must demand careful attention to both worlds; and, if either is to be neglected, it may be safer that it should be Nature. In any given contingency, it must doubtless be desirable to know what may be expected to happen in the ordinary course of things; but it must be quite as necessary to have some inkling of the line likely to be taken by supernatural agencies able, and possibly willing, to suspend or reverse that course. Indeed, logically developed, the dualistic theory must needs end in almost exclusive attention to Supernature, and in trust that its over-ruling strength will be exerted in favour of those who stand well with its denizens. On the other hand, the lessons of the great schoolmaster, experience, have hardly seemed to accord with this conclusion. They have taught, with considerable emphasis, that it does not answer to neglect Nature; and that, on the whole, the more attention paid to her dictates the better men fare.

Thus the theoretical antithesis brought about a practical antagonism. From the earliest times of which we have any knowledge, Naturalism and

Supernaturalism have consciously, or unconsciously, competed and struggled with one another; and the varying fortunes of the contest are written in the records of the course of civilisation, from those of Egypt and Babylonia, six thousand years ago, down to those of our own time and people.

These records inform us that, so far as men have paid attention to Nature, they have been rewarded for their pains. They have developed the Arts which have furnished the conditions of civilised existence; and the Sciences, which have been a progressive revelation of reality and have afforded the best discipline of the mind in the methods of discovering truth. They have accumulated a vast body of universally accepted knowledge; and the conceptions of man and of society, of morals and of law, based upon that knowledge, are every day more and more, either openly or tacitly, acknowledged to be the foundations of right action.

History also tells us that the field of the supernatural has rewarded its cultivators with a harvest, perhaps not less luxuriant, but of a different character. It has produced an almost infinite diversity of Religions. These, if we set aside the ethical concomitants upon which natural knowledge also has a claim, are composed of information about Supernature; they tell us of the attributes of supernatural beings, of their relations with Nature, and of the operations by which their interference with the ordinary course of events can be secured or averted. It does not appear, however, that supernaturalists have attained to any agreement about these matters, or that history indicates a widening of the influence of supernaturalism on practice, with the onward flow of time. On the contrary, the various religions are, to a great extent, mutually exclusive; and their adherents delight in charging each other, not merely with error, but with criminality, deserving and ensuing punishment of infinite severity. In singular contrast with natural knowledge, again, the acquaintance of mankind with the supernatural appears the more extensive and the more exact, and the influence of supernatural doctrines upon conduct the greater, the further back we go in time and the lower the stage of civilisation submitted to investigation. Historically, indeed, there would seem to be an inverse relation between supernatural and natural knowledge. As the latter has widened, gained in precision and in trustworthiness, so has the former shrunk, grown vague and questionable; as the one has more and more filled the sphere of action, so has the other retreated into the region of meditation, or vanished behind the screen of mere verbal recognition.

Whether this difference of the fortunes of Naturalism and of Supernaturalism is an indication of the progress, or of the regress, of humanity; of a fall from, or an advance towards, the higher life; is a matter of opinion. The point to which I wish to direct attention is that the difference exists and is making itself felt. Men are growing to be seriously alive to the fact that the historical evolution of humanity, which is generally, and I venture to think not unreasonably, regarded as progress, has been, and is being, accompanied by a co-ordinate elimination of the supernatural from its originally large occupation of men's thoughts. The question —How far is this process to go ?—is, in my apprehension, the Controverted Question of our time.

Controversy on this matter — prolonged, bitter, and fought out with the weapons of the flesh, as well as with those of the spirit—is no new thing to Englishmen. We have been more or less occupied with it these five hundred years. And, during that time, we have made attempts to establish a *modus vivendi* between the antagonists, some of which have had a world-wide influence ; though, unfortunately, none have proved universally and permanently satisfactory.

In the fourteenth century, the controverted question among us was, whether certain portions of the Supernaturalism of mediæval Christianity were well-founded. John Wicliff proposed a solution of the problem which, in the course of the following two hundred years, acquired wide popularity and vast historical importance : Lollards, Hussites, Lutherans, Calvinists, Zwinglians, Socinians, and Anabaptists, whatever their disagreements, concurred in the proposal to reduce the Supernaturalism of Christianity within the limits sanctioned by the Scriptures. None of the chiefs of Protestantism called in question either the supernatural origin and infallible authority of the Bible, or the exactitude of the account of the supernatural world given in its pages. In fact, they could not afford to entertain any doubt about these points, since the infallible Bible was the fulcrum of the lever with which they were endeavouring to upset the Chair of St. Peter. The "freedom of private judgment" which they proclaimed, meant no more, in practice, than permission to themselves to make free with the public judgment of the Roman Church, in respect of the canon and of the meaning to be attached to the words of the canonical books. Private judgment — that is to say, reason —was (theoretically, at any rate) at liberty to decide what books were and what were not to take the rank of "Scripture"; and to determine the sense of any passage in such books. But this sense, once ascertained to the mind of the sectary, was to be taken for pure truth—for the very word of God. The controversial efficiency of the principle of biblical infallibility lay in the fact that the conservative adversaries of the Reformers were not in a position to contravene it without entangling themselves in serious difficulties; while, since both Papists and Protestants agreed in taking efficient measures to stop the mouths of any more radical critics, these did not count.

The impotence of their adversaries, however, did not remove the inherent weakness of the position of the Protestants. The dogma of the infallibility of the Bible is no more self-evident than is that of the infallibility of the Pope. If the former is held by "faith," then the latter may be. If the latter is to be accepted, or rejected, by private judgment, why not the former? Even if the Bible could be proved anywhere to assert its own infallibility, the value of that self-assertion to those who dispute the point is not obvious. On the other hand, if the infallibility of the Bible was rested on that of a " primitive Church," the admission that the " Church " was formerly infallible was awkward in the extreme for those who denied its present infallibility. Moreover, no sooner was the Protestant principle applied to practice, than it became evident that even an infallible text, when manipulated by private judgment, will impartially countenance contradictory deductions; and furnish forth creeds and confessions as diverse as the quality and the information of the intellects which exercise, and the prejudices and passions which sway, such judgments. Every sect, confident in the derivative infallibility of its wire-drawing of infallible materials, was ready to supply its contingent of martyrs ; and to enable history, once more, to illustrate the truth, that steadfastness under persecution says much for the sincerity and still more for the tenacity, of the believer, but very little for the objective truth of that which he believes. No martyrs have sealed their faith with their blood more steadfastly than the Anabaptists.

Last, but not least, the Protestant principle con-tained within itself the germs of the destruction of the finality, which the Lutheran, Calvinistic, and other Protestant Churches fondly imagined they had reached. Since their creeds were professedly based on the canonical Scriptures, it followed that, in the long run, whoso settled the canon defined the creed. If the private judgment of Luther might legitimately conclude that the epistle of James was contemptible, while the epistles of Paul contained the very essence of Christianity, it must be permissible for some other private judgment, on as good or as bad grounds, to reverse these conclusions; the critical process which excluded the Apocrypha could not be barred, at any rate by people who rejected the authority of the Church, from extending its operations to Daniel, the Canticles, and Ecclesiastes; nor, having got so far, was it easy to allege any good ground for staying the further progress of criticism. In fact, the logical development of Protestantism could not fail to lay the authority of the Scriptures at the feet of Reason; and, in the hands of latitudinarian and rationalistic theologians, the despotism of the Bible was rapidly converted into an extremely limited monarchy. Treated with as much respect as ever, the sphere of its practical authority was minimised; and its decrees were valid only so far as they were countersigned by common sense, the responsible minister.

The champions of Protestantism are much given to glorify the Reformation of the sixteenth century as the emancipation of Reason; but it may be doubted if their contention has any solid ground; while there is a good deal of evidence to show, that aspirations after intellectual freedom had nothing whatever to do with the movement. Dante, who struck the Papacy as hard blows as Wicliff; Wicliff himself and Luther himself, when they began their work; were far enough from any intention of meddling with even the most irrational of the dogmas of mediæval Supernaturalism. From Wicliff to Socinus, or even to Münzer, Rothmann, and John of Leyden, I fail to find a trace of any desire to set reason free. The most that can be discovered is a proposal to change masters. From being the slave of the Papacy the intellect was to become the serf of the Bible; or, to speak more accurately, of somebody's interpretation of the Bible, which, rapidly shifting its attitude from the humility of a private judgment to the arrogant Cæsaro-papistry of a state-enforced creed, had no more hesitation about forcibly extinguishing opponent private judgments and judges, than had the old-fashioned Pontiff-papistry.

It was the iniquities, and not the irrationalities, of the Papal system that lay at the bottom of the revolt of the laity; which was, essentially, an attempt to shake off the intolerable burden of certain practical deductions from a Supernaturalism in which everybody, in principle, acquiesced. What was the gain to intellectual freedom of abolishing transubstantiation, image worship, indulgences, ecclesiastical infallibility; if consubstantiation, real-unreal presence mystifications, the bibliolatry, the "inner-light" pretensions, and the demonology, which are fruits of the same supernaturalistic tree, remained in enjoyment of the spiritual and temporal support of a new infallibility? One does not free a prisoner by merely scraping away the rust from his shackles.

It will be asked, perhaps, was not the Reformation one of the products of that great outbreak of many-sided free mental activity included under the general head of the Renascence? Melanchthon, Ulrich von Hutten, Beza, were they not all humanists? Was not the arch-humanist, Erasmus, fautor-in-chief of the Reformation, until he got frightened and basely deserted it?

From the language of Protestant historians, it would seem that they often forget that Reformation and Protestantism are by no means convertible terms. There were plenty of sincere and indeed zealous reformers, before, during, and after the birth and growth of Protestantism, who would have nothing to do with it. Assuredly, the rejuvenescence of science and of art; the widening of the field of Nature by geographical and astronomical discovery; the revelation of the noble ideals of antique literature by the revival of classical learning; the stir of thought, throughout all classes of society, by the printers' work, loosened traditional bonds and weakened the hold of mediæval Supernaturalism. In the interests of liberal culture and of national welfare, the humanists were eager to lend a hand to anything which tended to the discomfiture of their sworn enemies, the monks, and they willingly supported every movement in the direction of weakening ecclesiastical inter-ference with civil life. But the bond of a common enemy was the only real tie between the humanist

and the protestant; their alliance was bound to
be of short duration, and, sooner or later, to be
replaced by internecine warfare. The goal of the
humanists, whether they were aware of it or not,
was the attainment of the complete intellectual freedom of the antique philosopher, than which nothing
could be more abhorrent to a Luther, a Calvin, a Beza,
or a Zwingli.

The key to the comprehension of the conduct of
Erasmus, seems to me to lie in the clear apprehension of this fact. That he was a man of many
weaknesses may be true; in fact, he was quite
aware of them and professed himself no hero.
But he never deserted that reformatory movement
which he originally contemplated; and it was
impossible he should have deserted the specifically
Protestant reformation in which he never took part.
He was essentially a theological whig, to whom
radicalism was as hateful as it is to all whigs; or,
to borrow a still more appropriate comparison from
modern times, a broad churchman who refused to
enlist with either the High Church or the Low
Church zealots, and paid the penalty of being called
coward, time-server and traitor, by both. Yet really
there is a good deal in his pathetic remonstrance that
he does not see why he is bound to become a martyr
for that in which he does not believe; and a fair consideration of the circumstances and the consequences
of the Protestant reformation seems to me to go a long
way towards justifying the course he adopted.

Few men had better means of being acquainted

with the condition of Europe; none could be more competent to gauge the intellectual shallowness and self-contradiction of the protestant criticism of catholic doctrine; and to estimate, at its proper value, the fond imagination that the waters let out by the Renascence would come to rest amidst the blind alleys of the new ecclesiasticism. The bastard, whilom poor student and monk, become the familiar of bishops and princes, at home in all grades of society, could not fail to be aware of the gravity of the social position, of the dangers imminent from the profligacy and indifference of the ruling classes, no less than from the anarchical tendencies of the people who groaned under their oppression. The wanderer who had lived in Germany, in France, in England, in Italy, and who counted many of the best and most influential men in each country among his friends, was not likely to estimate wrongly the enormous forces which were still at the command of the Papacy. Bad as the churchmen might be, the statesmen were worse; and a person of far more sanguine temperament than Erasmus might have seen no hope for the future, except in gradually freeing the ubiquitous organisation of the Church from the corruptions which alone, as he imagined, prevented it from being as beneficent as it was powerful. The broad tolerance of the scholar and man of the world might well be revolted by the ruffianism, however genial, of one great light of Protestantism, and the narrow fanaticism, however learned and logical, of others; and to a cautious thinker, by whom, whatever his shortcomings, the ethical ideal of the Christian evangel was sincerely prized, it really was a fair question, whether it was worth while to bring about a political and social deluge, the end of which no mortal could foresee, for the purpose of setting up Lutheran, Zwinglian, and other Peterkins, in the place of the actual claimant to the reversion of the spiritual wealth of the Galilean fisherman.

Let us suppose that, at the beginning of the Lutheran and Zwinglian movement, a vision of its immediate consequences had been granted to Erasmus; imagine that to the spectre of the fierce outbreak of Anabaptist communism, which opened the apocalypse, had succeeded, in shadowy procession, the reign of terror and of spoliation in England, with the judicial murders of his friends, More and Fisher; the bitter tyranny of evangelistic clericalism in Geneva and in Scotland; the long agony of religious wars, persecutions, and massacres, which devastated France and reduced Germany almost to savagery; finishing with the spectacle of Lutheranism in its native country sunk into mere dead Erastian formalism, before it was a century old; while Jesuitry triumphed over Protestantism in three-fourths of Europe, bringing in its train a recrudescence of all the corruptions Erasmus and his friends sought to abolish; might not he have quite honestly thought this a somewhat too heavy price to pay for Protestantism; more especially, since no one was in a better position than himself to know how little the dogmatic foundation of the new confessions was able to bear the light which the inevitable progress of humanistic criticism would throw upon them? As the wiser of his contemporaries saw, Erasmus was, at heart, neither Protestant nor Papist, but an "Independent Christian"; and, as the wiser of his modern biographers have discerned, he was the precursor, not of sixteenth century reform, but of eighteenth century "enlightenment"; a sort of broad-church Voltaire, who held by his "Independent Christianity" as stoutly as Voltaire by his Deism.

In fact, the stream of the Renascence, which bore Erasmus along, left Protestantism stranded amidst the mudbanks of its articles and creeds : while its true course became visible to all men, two centuries later. By this time, those in whom the movement of the Renascence was incarnate became aware what spirit they were of; and they attacked Supernaturalism in its Biblical stronghold, defended by Protestants and Romanists with equal zeal. In the eyes of the "Patriarch," Ultramontanism, Jansenism, and Calvinism were merely three persons of the one "Infâme" which it was the object of his life to crush. If he hated one more than another, it was probably the last; while D'Holbach, and the extreme left of the free-thinking host, were disposed to show no more mercy to Deism and Pantheism.

The sceptical insurrection of the eighteenth century made a terrific noise and frightened not a few worthy people out of their wits ; but cool judges might have foreseen, at the outset, that the efforts of the later rebels were no more likely

than those of the earlier, to furnish permanent
resting-places for the spirit of scientific inquiry.
However worthy of admiration may be the acuteness, the common-sense, the wit, the broad humanity,
which abound in the writings of the best of the
free-thinkers ; there is rarely much to be said
for their work as an example of the adequate
treatment of a grave and difficult investigation.
I do not think any impartial judge will assert that,
from this point of view, they are much better than
their adversaries. It must be admitted that they
share to the full the fatal weakness of *à priori*
philosophising, no less than the moral frivolity
common to their age ; while a singular want of
appreciation of history, as the record of the moral and
social evolution of the human race, permitted them to
resort to preposterous theories of imposture, in order
to account for the religious phenomena which are
natural products of that evolution.

For the most part, the Romanist and Protestant
adversaries of the free-thinkers met them with arguments no better than their own ; and with vitupera-
tion, so far inferior that it lacked the wit. But one
great Christian apologist fairly captured the guns of
the free-thinking array, and turned their batteries
upon themselves. Speculative "infidelity" of the
eighteenth century type was mortally wounded by
the *Analogy* ; while the progress of the historical
and psychological sciences brought to light the
important part played by the mythopœic faculty ;
and, by demonstrating the extreme readiness of men

C

to impose upon themselves, rendered the calling in of sacerdotal co-operation, in most cases, a superfluity.

Again, as in the fourteenth and the sixteenth centuries, social and political influences came into play. The free-thinking *philosophes*, who objected to Rousseau's sentimental religiosity almost as much as they did to *L'Infâme*, were credited with the responsibility for all the evil deeds of Rousseau's Jacobin disciples, with about as much justification as Wicliff was held responsible for the Peasants' revolt, or Luther for the *Bauern-krieg*. In England, though our *ancien régime* was not altogether lovely, the social edifice was never in such a bad way as in France; it was still capable of being repaired; and our forefathers, very wisely, preferred to wait until that operation could be safely performed, rather than pull it all down about their ears, in order to build a philosophically planned house on brand-new speculative foundations. Under these circumstances, it is not wonderful that, in this country, practical men preferred the gospel of Wesley and Whitfield to that of Jean Jacques; while enough of the old leaven of Puritanism remained to ensure the favour and support of a large number of religious men to a revival of evangelical supernaturalism. Thus, by degrees, the free-thinking, or the indifference, prevalent among us in the first half of the eighteenth century, was replaced by a strong supernaturalistic reaction, which submerged the work of the free-thinkers; and even seemed, for a time, to have arrested the naturalistic movement of

which that work was an imperfect indication. Yet,
like Lollardry, four centuries earlier, free-thought
merely took to running underground, safe, sooner or
later, to return to the surface.

My memory, unfortunately, carries me back to the
fourth decade of the nineteenth century, when the
evangelical flood had a little abated and the tops of
certain mountains were soon to appear, chiefly in the
neighbourhood of Oxford; but when, nevertheless,
bibliolatry was rampant; when church and chapel
alike proclaimed, as the oracles of God, the crude
assumptions of the worst informed and, in natural
sequence, the most presumptuously bigoted, of all
theological schools.

In accordance with promises made on my behalf,
but certainly without my authorisation, I was very
early taken to hear " sermons in the vulgar tongue."
And vulgar enough often was the tongue in which
some preacher, ignorant alike of literature, of history,
of science, and even of theology, outside that
patronised by his own narrow school, poured forth,
from the safe entrenchment of the pulpit, invect-
ives against those who deviated from his notion
of orthodoxy. From dark allusions to " sceptics "
and " infidels," I became aware of the existence
of people who trusted in carnal reason; who
audaciously doubted that the world was made in six
natural days, or that the deluge was universal;
perhaps even went so far as to question the literal
accuracy of the story of Eve's temptation, or of

Balaam's ass; and, from the horror of the tones in which they were mentioned, I should have been justified in drawing the conclusion that these rash men belonged to the criminal classes. At the same time, those who were more directly responsible for providing me with the knowledge essential to the right guidance of life (and who sincerely desired to do so), imagined they were discharging that most sacred duty by impressing upon my childish mind the necessity, on pain of reprobation in this world and damnation in the next, of accepting, in the strict and literal sense, every statement contained in the protestant Bible. I was told to believe, and I did believe, that doubt about any of them was a sin, not less reprehensible than a moral delict. I suppose that, out of a thousand of my contemporaries, nine hundred, at least, had their minds systematically warped and poisoned, in the name of the God of truth, by like discipline. I am sure that, even a score of years later, those who ventured to question the exact historical accuracy of any part of the Old Testament and *à fortiori* of the Gospels, had to expect a pitiless shower of verbal missiles, to say nothing of the other disagreeable consequences which visit those who, in any way, run counter to that chaos of prejudices called public opinion.

My recollections of this time have recently been revived by the perusal of a remarkable document,[1]

[1] *Declaration on the Truth of Holy Scripture.* The *Times,* 18th December 1891.

signed by as many as thirty-eight out of the twenty
odd thousand clergymen of the Established Church.
It does not appear that the signataries are officially
accredited spokesmen of the ecclesiastical corporation
to which they belong ; but I feel bound to take their
word for it, that they are " stewards of the Lord,
who have received the Holy Ghost," and, therefore,
to accept this memorial as evidence that, though the
Evangelicism of my early days may be deposed from
its place of power, though so many of the colleagues
of the thirty - eight even repudiate the title of
Protestants, yet the green bay tree of bibliolatry
flourishes as it did sixty years ago. And, as in
those good old times, whoso refuses to offer incense
to the idol is held to be guilty of " a dishonour
to God," imperilling his salvation.

It is to the credit of the perspicacity of the
memorialists that they discern the real nature of the
Controverted Question of the age. They are awake
to the unquestionable fact that, if Scripture has
been discovered " not to be worthy of unques-
tioning belief," faith " in the supernatural itself " is,
so far, undermined. And I may congratulate myself
upon such weighty confirmation of an opinion in
which I have had the fortune to anticipate them.
But whether it is more to the credit of the courage,
than to the intelligence, of the thirty-eight that they
should go on to proclaim that the canonical scriptures
of the Old and New Testaments " declare incontrovertibly the actual historical truth in all records,
both of past events and of the delivery of predictions

to be thereafter fulfilled," must be left to the coming
generation to decide.

The interest which attaches to this singular document will, I think, be based by most thinking men,
not upon what it is, but upon that of which it is a
sign. It is an open secret, that the memorial is put
forth as a counterblast to a manifestation of opinion
of a contrary character, on the part of certain
members of the same ecclesiastical body, who
therefore have, as I suppose, an equal right to declare
themselves "stewards of the Lord and recipients of
the Holy Ghost." In fact, the stream of tendency
towards Naturalism, the course of which I have
briefly traced, has, of late years, flowed so strongly,
that even the Churches have begun, I dare not say
to drift, but, at any rate, to swing at their moorings.
Within the pale of the Anglican establishment, I
venture to doubt, whether, at this moment, there
are as many thorough-going defenders of "plenary
inspiration" as there were timid questioners of
that doctrine, half a century ago. Commentaries,
sanctioned by the highest authority, give up the
"actual historical truth" of the cosmogonical
and diluvial narratives. University professors of
deservedly high repute accept the critical decision
that the Hexateuch is a compilation, in which the
share of Moses, either as author or as editor, is not
quite so clearly demonstrable as it might be; highly
placed Divines tell us that the pre-Abrahamic
Scripture narratives may be ignored; that the book
of Daniel may be regarded as a patriotic romance of

the second century B.C. ; that the words of the writer
of the fourth Gospel are not always to be distinguished from those which he puts into the mouth
of Jesus. Conservative, but conscientious, revisers
decide that whole passages, some of dogmatic and
some of ethical importance, are interpolations. An
uneasy sense of the weakness of the dogma of
Biblical infallibility seems to be at the bottom of
a prevailing tendency once more to substitute the
authority of the " Church " for that of the Bible.
In my old age, it has happened to me to be taken to
task for regarding Christianity as a " religion of a
book " as gravely as, in my youth, I should have
been reprehended for doubting that proposition. It
is a no less interesting symptom that the State
Church seems more and more anxious to repudiate
all complicity with the principles of the Protestant
Reformation and to call itself " Anglo-Catholic."
Inspiration, deprived of its old intelligible sense, is
watered down into a mystification. The Scriptures
are, indeed, inspired; but they contain a wholly
undefined and indefinable " human element"; and
this unfortunate intruder is converted into a sort
of biblical whipping boy. Whatsoever scientific
investigation, historical or physical, proves to be
erroneous, the " human element" bears the blame;
while the divine inspiration of such statements, as by
their nature are out of reach of proof or disproof, is
still asserted with all the vigour inspired by conscious
safety from attack. Though the proposal to treat the
Bible " like any other book " which caused so much

scandal, forty years ago, may not yet be generally accepted, and though Bishop Colenso's criticisms may still lie, formally, under ecclesiastical ban, yet the Church has not wholly turned a deaf ear to the voice of the scientific tempter; and many a coy divine, while " crying I will ne'er consent," has consented to the proposals of that scientific criticism which the memorialists renounce and denounce.

A humble layman, to whom it would seem the height of presumption to assume even the unconsidered dignity of a " steward of science," may well find this conflict of apparently equal ecclesiastical authorities perplexing — suggestive, indeed, of the wisdom of postponing attention to either, until the question of precedence between them is settled. And this course will probably appear the more advisable, the more closely the fundamental position of the memorialists is examined.

" No opinion of the fact or form of Divine Revelation, founded on literary criticism [and I suppose I may add historical, or physical, criticism] of the Scriptures themselves, can be admitted to interfere with the traditionary testimony of the Church, when that has been once ascertained and verified by appeal to antiquity." [1]

Grant that it is " the traditionary testimony of the Church " which guarantees the canonicity of each and all of the books of the Old and New Testaments. Grant also that canonicity means infallibility; yet, according to the thirty-eight, this "traditionary

[1] *Declaration*, Article 10.

testimony" has to be "ascertained and verified by appeal to antiquity." But "ascertainment and verification" are purely intellectual processes, which must be conducted according to the strict rules of scientific investigation, or be self-convicted of worthlessness. Moreover, before we can set about the appeal to "antiquity," the exact sense of that usefully vague term must be defined by similar means. "Antiquity" may include any number of centuries, great or small; and whether "antiquity" is to comprise the Council of Trent, or to stop a little beyond that of Nicæa, or to come to an end in the time of Irenæus, or in that of Justin Martyr, are knotty questions which can be decided, if at all, only by those critical methods which the signataries treat so cavalierly. And yet the decision of these questions is fundamental, for as the limits of the canonical scriptures vary, so may the dogmas deduced from them require modification. Christianity is one thing, if the fourth Gospel, the Epistle to the Hebrews, the pastoral Epistles, and the Apocalypse are canonical and (by the hypothesis) infallibly true; and another thing, if they are not. As I have already said, whoso defines the canon defines the creed.

Now it is quite certain with respect to some of these books, such as the Apocalypse and the Epistle to the Hebrews, that the Eastern and the Western Church differed in opinion for centuries; and yet neither the one branch, nor the other, can have considered its judgment infallible, since they eventually agreed to a transaction, by which each gave up its

objection to the book patronised by the other.   Moreover, the "fathers" argue (in a more or less rational
manner) about the canonicity of this or that book,
and are by no means above producing evidence, internal and external, in favour of the opinions they
advocate.   In fact, imperfect as their conceptions of
scientific method may be, they not unfrequently used
it to the best of their ability.   Thus it would appear
that though science, like Nature, may be driven out
with a fork, ecclesiastical or other, yet she surely
comes back again.   The appeal to "antiquity" is, in
fact, an appeal to science, first to define what antiquity
is ; secondly, to determine what "antiquity," so defined, says about canonicity ; thirdly, to prove that
canonicity means infallibility.   And when science,
largely in the shape of the abhorred "criticism," has
done this, and has shown that "antiquity" used her
own methods, however clumsily and imperfectly, she
naturally turns round upon the appealers to "antiquity," and demands that they should show cause why,
in these days, science should not resume the work the
ancients did so imperfectly, and carry it out efficiently.

But no such cause can be shown.   If "antiquity"
permitted Eusebius, Origen, Tertullian, Irenæus, to
argue for the reception of this book into the canon
and the rejection of that, upon rational grounds,
"antiquity" admitted the whole principle of modern
criticism.   If Irenæus produces ridiculous reasons
for limiting the Gospels to four, it was open to any
one else to produce good reasons (if he had them) for
cutting them down to three, or increasing them to

five. If the Eastern branch of the Church had a right to reject the Apocalypse and accept the Epistle to the Hebrews, and the Western an equal right to accept the Apocalypse and reject the Epistle, down to the fourth century, any other branch would have an equal right, on cause shown, to reject both, or, as the Catholic Church afterwards actually did, to accept both.

Thus I cannot but think that the thirty-eight are hoist with their own petard. Their "appeal to antiquity" turns out to be nothing but a round-about way of appealing to the tribunal, the jurisdiction of which they affect to deny. Having rested the world of Christian supernaturalism on the elephant of biblical infallibility, and furnished the elephant with standing ground on the tortoise of " antiquity," they, like their famous Hindoo analogue, have been content to look no further ; and have thereby been spared the horror of discovering that the tortoise rests on a grievously fragile construction, to a great extent the work of that very intellectual operation which they anathematise and repudiate.

Moreover, there is another point to be considered. It is of course true that a Christian Church (whether the Christian Church, or not, depends on the connotation of the definite article) existed before the Christian scriptures ; and that the infallibility of these depends upon the infallibility of the judgment of the persons who selected the books, of which they are composed, out of the mass of literature current among the early Christians. The logical acumen of Augustine showed him that the authority of the Gospel he preached must rest on that of the Church to which he belonged.[1] But it is no less true that the Hebrew and the Septuagint versions of most, if not all, of the Old Testament books existed before the birth of Jesus of Nazareth ; and that their divine authority is presupposed by, and therefore can hardly depend upon, the religious body constituted by his disciples. As everybody knows, the very conception of a " Christ " is purely Jewish. The validity of the argument from the Messianic prophecies vanishes unless their infallible authority is granted ; and, as a matter of fact, whether we turn to the Gospels, the Epistles, or the writings of the early Apologists, the Jewish scriptures are recognised as the highest court of appeal of the Christian.

The proposal to cite Christian "antiquity" as a witness to the infallibility of the Old Testament, when its own claims to authority vanish, if certain propositions contained in the Old Testament are erroneous, hardly satisfies the requirements of lay logic. It is as if a claimant to be sole legatee, under another kind of testament, should offer his assertion as sufficient evidence of the validity of the will. And, even were not such a circular, or rather rotatory, argument, that the infallibility of the Bible is testified by the infallible Church, whose infallibility is testified by the infallible Bible, too absurd for serious consideration, it remains permissible to ask ; Where and when the Church, during the period of its infallibility, as

---

[1] Ego vero evangelio non crederem, nisi ecclesiæ Catholicæ me commoveret auctoritas.—*Contra Epistolam Manichæi*, cap. v.

limited by Anglican dogmatic necessities, has officially decreed the " actual historical truth of all records " in the Old Testament ? Was Augustine heretical when he denied the actual historical truth of the record of the Creation ? Father Suarez, standing on later Roman tradition, may have a right to declare that he was ; but it does not lie in the mouth of those who limit their appeal to that early " antiquity," in which Augustine played so great a part, to say so.

Among the watchers of the course of the world of thought, some view with delight and some with horror, the recrudescence of Supernaturalism which manifests itself among us, in shapes ranged along the whole flight of steps, which, in this case, separates the sublime from the ridiculous—from Neo-Catholicism and Inner-light mysticism, at the top, to unclean things, not worthy of mention in the same breath, at the bottom. In my poor opinion, the importance of these manifestations is often greatly over-estimated. The extant forms of Supernaturalism have deep roots in human nature, and will un-doubtedly die hard ; but, in these latter days, they have to cope with an enemy whose full strength is only just beginning to be put out, and whose forces, gathering strength year by year, are hemming them round on every side. This enemy is Science, in the acceptation of systematised natural knowledge, which, during the last two centuries, has extended those methods of investigation, the worth of which is confirmed by daily appeal to Nature, to every

region in which the Supernatural has hitherto
been recognised.

When scientific historical criticism reduced the
annals of heroic Greece and of regal Rome to the level of
fables; when the unity of authorship of the *Iliad* was
successfully assailed by scientific literary criticism;
when scientific physical criticism, after exploding the
geocentric theory of the universe, and reducing the
solar system itself to one of millions of groups of like
cosmic specks, circling, at unimaginable distances
from one another, through infinite space, showed
the supernaturalistic theories of the duration of the
earth and of life upon it, to be as inadequate as those
of its relative dimensions and importance had been;
it needed no prophetic gift to see that, sooner or later,
the Jewish and the early Christian records would be
treated in the same manner; that the authorship of
the Hexateuch and of the Gospels would be as
severely tested; and that the evidence in favour of
the veracity of many of the statements found in the
Scriptures would have to be strong indeed, if they
were to be opposed to the conclusions of physical
science. In point of fact, so far as I can discover,
no one competent to judge of the evidential strength
of these conclusions, ventures now to say that the
biblical accounts of the creation and of the deluge
are true in the natural sense of the words of the
narratives. The most the modern Reconciler ventures
upon is to affirm, that some quite different sense may
be put upon the words; and that this non-natural
sense may, with a little trouble, be manipulated

into some sort of non-contradiction of scientific truth.

My purpose, in the essay (XVI.) which treats of the narrative of the Deluge, was to prove, by physical criticism, that no such event as that described ever took place ; to exhibit the untrustworthy character of the narrative demonstrated by literary criticism ; and, finally, to account for its origin, by producing a form of those ancient legends of pagan Chaldæa, from which the biblical compilation is manifestly derived. I have yet to learn that the main propositions of this essay can be seriously challenged.

In the essays (II., III.) on the narrative of the Creation, I have endeavoured to controvert the assertion that modern science supports, either the interpretation put upon it by Mr. Gladstone, or any interpretation which is compatible with the general sense of the narrative, quite apart from particular details. The first chapter of Genesis teaches the supernatural creation of the present forms of life ; modern science teaches that they have come about by evolution. The first chapter of Genesis teaches the successive origin—firstly, of all the plants, secondly, of all the aquatic and aerial animals, thirdly, of all the terrestrial animals, which now exist — during distinct intervals of time ; modern science teaches that, throughout all the duration of an immensely long past, so far as we have any adequate knowledge of it (that is as far back as the Silurian epoch), plants, aquatic, aerial, and terrestrial animals have co - existed ; that the earliest known are unlike those which at present exist ; and that the modern species have come into existence as the last terms of a series, the members of which have appeared one after another.   Thus, far from confirming the account in Genesis, the results of modern science, so far as they go, are in principle, as in detail, hopelessly discordant with it.

Yet, if the pretensions to infallibility set up, not by the ancient Hebrew writings themselves, but by the ecclesiastical champions and friends from whom they may well pray to be delivered, thus shatter themselves against the rock of natural know-ledge, in respect of the two most important of all events, the origin of things and the palingenesis of terrestrial life, what historical credit dare any serious thinker attach to the narratives of the fabrication of Eve, of the Fall, of the commerce between the *Bene Elohim* and the daughters of men, which lie between the creational and the diluvial legends ?   And, if these are to lose all historical worth, what becomes of the infallibility of those who, according to the later scriptures, have accepted them, argued from them, and staked far-reaching dogmatic conclusions upon their historical accuracy ?

It is the merest ostrich policy for contemporary ecclesiasticism to try to hide its Hexateuchal head— in the hope that the inseparable connection of its body with pre-Abrahamic legends may be overlooked. The question will still be asked, if the first nine chapters of the Pentateuch are unhistorical, how is the historical accuracy of the remainder to be guaranteed ? What more intrinsic claim has the story of the Exodus than that of the Deluge, to belief ? If God did not walk in the Garden of Eden, how can we be assured that he spoke from Sinai ?

In some other of the following essays (IX., X., XI., XII., XIV., XV.) I have endeavoured to show that sober and well-founded physical and literary criticism plays no less havoc with the doctrine that the canonical scriptures of the New Testament " declare incontrovertibly the actual historical truth in all records." We are told that the Gospels contain a true revelation of the spiritual world — a proposition which, in one sense of the word " spiritual," I should not think it necessary to dispute. But, when it is taken to signify that everything we are told about the world of spirits in these books is infallibly true; that we are bound to accept the demonology which constitutes an inseparable part of their teaching; and to profess belief in a Supernaturalism as gross as that of any primitive people— it is at any rate permissible to ask why ? Science may be unable to define the limits of possibility, but it cannot escape from the moral obligation to weigh the evidence in favour of any alleged wonderful occurrence; and I have endeavoured to show that the evidence for the Gadarene miracle is altogether worthless. We have simply three, partially discrepant, versions of a story, about the primitive form, the origin, and the authority for which we know absolutely nothing. But the

D

evidence in favour of the Gadarene miracle is as good
as that for any other.

Elsewhere, I have pointed out that it is utterly
beside the mark to declaim against these conclusions on
the ground of their asserted tendency to deprive mankind of the consolations of the Christian faith, and
to destroy the foundations of morality; still less to
brand them with the question-begging vituperative
appellation of "infidelity." The point is not whether
they are wicked; but, whether, from the point of view
of scientific method, they are irrefragably true. If
they are, they will be accepted in time, whether they
are wicked, or not wicked. Nature, so far as we have
been able to attain to any insight into her ways,
recks little about consolation and makes for righteousness by very round-about paths. And, at any
rate, whatever may be possible for other people, it is
becoming less and less possible for the man who puts
his faith in scientific methods of ascertaining truth,
and is accustomed to have that faith justified by
daily experience, to be consciously false to his principle in any matter. But the number of such men,
driven into the use of scientific methods of inquiry
and taught to trust them, by their education, their
daily professional and business needs, is increasing
and will continually increase. The phraseology of
Supernaturalism may remain on men's lips, but in
practice they are Naturalists. The magistrate who
listens with devout attention to the precept "Thou
shalt not suffer a witch to live" on Sunday, on
Monday, dismisses, as intrinsically absurd, a charge

of bewitching a cow brought against some old woman; the superintendent of a lunatic asylum who substituted exorcism for rational modes of treatment would have but a short tenure of office; even parish clerks doubt the utility of prayers for rain, so long as the wind is in the east; and an outbreak of pestilence sends men, not to the churches, but to the drains. In spite of prayers for the success of our arms and *Te Deums* for victory, our real faith is in big battalions and keeping our powder dry; in knowledge of the science of warfare; in energy, courage, and discipline. In these, as in all other practical affairs, we act on the aphorism "*Laborare est orare*"; we admit that intelligent work is the only acceptable worship; and that, whether there be a Supernature or not, our business is with Nature.

It is important to note that the principle of the scientific Naturalism of the latter half of the nineteenth century, in which the intellectual movement of the Renascence has culminated, and which was first clearly formulated by Descartes, leads not to the denial of the existence of any Supernature;[1] but simply to the denial of the validity of the evidence adduced in favour of this, or of that, extant form of Supernaturalism.

[1] I employ the words "Supernature" and "Supernatural" in their popular senses. For myself, I am bound to say that the term "Nature" covers the totality of that which is. The world of psychical phenomena appears to me to be as much part of "Nature" as the world of physical phenomena; and I am unable to perceive any justification for cutting the Universe into two halves, one natural and one supernatural.

Looking at the matter from the most rigidly scientific point of view, the assumption that, amidst the myriads of worlds scattered through endless space, there can be no intelligence, as much greater than man's as his is greater than a blackbeetle's; no being endowed with powers of influencing the course of nature as much greater than his, as his is greater than a snail's, seems to me not merely baseless, but impertinent. Without stepping beyond the analogy of that which is known, it is easy to people the cosmos with entities, in ascending scale, until we reach something practically indistinguishable from omnipotence, omnipresence, and omniscience. If our intelligence can, in some matters, surely reproduce the past of thousands of years ago and anticipate the future, thousands of years hence, it is clearly within the limits of possibility that some greater intellect, even of the same order, may be able to mirror the whole past and the whole future; if the universe is penetrated by a medium of such a nature that a magnetic needle on the earth answers to a commotion in the sun, an omnipresent agent is also conceivable; if our insignificant knowledge gives us some influence over events, practical omniscience may confer indefinably greater power. Finally, if evidence that a thing may be, were equivalent to proof that it is, analogy might justify the construction of a naturalistic theology and demonology not less wonderful than the current supernatural; just as it might justify the peopling of Mars, or of Jupiter, with living forms to which terrestrial biology offers no

parallel.   Until human life is longer and the duties
of the present press less heavily, I do not think that
wise men will occupy themselves with Jovian, or
Martian, natural history; and they will probably
agree to a verdict of "not proven" in respect of
naturalistic theology; taking refuge in that agnostic
confession, which appears to me to be the only position for people who object to say that they know
what they are quite aware they do not know. As
to the interests of morality, I am disposed to think
that if mankind could be got to act up to this last
principle in every relation of life, a reformation
would be effected such as the world has not yet
seen; an approximation to the millennium, such as
no supernaturalistic religion has ever yet succeeded,
or seems likely ever to succeed, in effecting.

I have hitherto dwelt upon scientific Naturalism chiefly in its critical and destructive aspect.
But the present incarnation of the spirit of the
Renascence differs from its predecessor in the eighteenth century, in that it builds up, as well as
pulls down.

That of which it has laid the foundation, of which
it is already raising the superstructure, is the doctrine
of evolution.   But so many strange misconceptions
are current about this doctrine — it is attacked on
such false grounds by its enemies, and made to cover
so much that is disputable by some of its friends,
that I think it well to define as clearly as I can, what
I do not and what I do understand by the doctrine.

I have nothing to say to any "Philosophy of Evolution." Attempts to construct such a philosophy may be as useful, nay, even as admirable, as was the attempt of Descartes to get at a theory of the universe by the same *à priori* road; but, in my judgment, they are as premature. Nor, for this purpose, have I to do with any theory of the "Origin of Species," much as I value that which is known as the Darwinian theory. That the doctrine of natural selection presupposes evolution is quite true; but it is not true that evolution necessarily implies natural selection. In fact, evolution might conceivably have taken place, without the development of groups possessing the characters of species.

For me, the doctrine of evolution is no speculation, but a generalisation of certain facts, which may be observed by any one who will take the necessary trouble. These facts are those which are classed by biologists under the heads of Embryology and of Palæontology. Embryology proves that every higher form of individual life becomes what it is by a process of gradual differentiation from an extremely low form; palæontology proves, in some cases, and renders probable in all, that the oldest types of a group are the lowest; and that they have been followed by a gradual succession of more and more differentiated forms. It is simply a fact, that evolution of the individual animal and plant is taking place, as a natural process, in millions and millions of cases every day; it is a fact, that the species which have succeeded one another in the past, do, in many cases, present just those morphological relations, which they must possess, if they had proceeded, one from the other, by an analogous process of evolution.

The alternative presented, therefore, is: either the forms of one and the same type—say, *e.g.*, that of the Horse tribe[1]—arose successively, but independently of one another, at intervals, during myriads of years; or, the later forms are modified descendants of the earlier. And the latter supposition is so vastly more probable than the former, that rational men will adopt it, unless satisfactory evidence to the contrary can be produced. The objection sometimes put forward, that no one yet professes to have seen one species pass into another, comes oddly from those who believe that mankind are all descended from Adam. Has any one then yet seen the production of negroes from a white stock, or *vice versa*? Moreover, is it absolutely necessary to have watched every step of the progress of a planet, to be justified in concluding that it really does go round the sun? If so, astronomy is in a bad way.

I do not, for a moment, presume to suggest that some one, far better acquainted than I am with astronomy and physics; or that a master of the new chemistry, with its extraordinary revelations; or that a student of the development of human society, of language, and of religions, may not find a sufficient foundation for the doctrine of evolution in these several regions.

[1] The general reader will find an admirably clear and concise statement of the evidence in this case, in Professor Flower's recently published work *The Horse: a Study in Natural History.*

On the contrary, I rejoice to see that scientific investigation, in all directions, is tending to the same result. And it may well be, that it is only my long occupation with biological matters that leads me to feel safer among them than anywhere else. Be that as it may, I take my stand on the facts of embryology and of palæontology; and I hold that our present knowledge of these facts is sufficiently thorough and extensive to justify the assertion that all future philosophical and theological speculations will have to accommodate themselves to some such common body of established truths as the following:—

1. Plants and animals have existed on our planet for many hundred thousand, probably millions of years. During this time, their forms, or species, have undergone a succession of changes, which eventually gave rise to the species which constitute the present living population of the earth. There is no evidence, nor any reason to suspect, that this secular process of evolution is other than a part of the ordinary course of nature; there is no more ground for imagining the occurrence of supernatural intervention, at any moment in the development of species in the past, than there is for supposing such intervention to take place, at any moment in the development of an individual animal or plant, at the present day.

2. At present, every individual animal or plant commences its existence as an organism of extremely simple anatomical structure; and it acquires all the complexity it ultimately possesses by gradual differen-

tiation into parts of various structure and function. When a series of specific forms of the same type, extending over a long period of past time, is examined, the relation between the earlier and the later forms is analogous to that between earlier and later stages of individual development. Therefore, it is a probable conclusion that, if we could follow living beings back to their earliest states, we should find them to present forms similar to those of the individual germ, or, what comes to the same thing, of those lowest known organisms which stand upon the boundary line between plants and animals. At present, our knowledge of the ancient living world stops very far short of this point.

3. It is generally agreed, and there is certainly no evidence to the contrary, that all plants are devoid of consciousness; that they neither feel, desire, nor think. It is conceivable that the evolution of the primordial living substance should have taken place only along the plant line. In that case, the result might have been a wealth of vegetable life, as great, perhaps as varied, as at present, though certainly widely different from the present flora, in the evolution of which animals have played so great a part. But the living world thus constituted would be simply an admirable piece of unconscious machinery, the working out of which lay potentially in its primitive composition; pleasure and pain would have no place in it; it would be a veritable Garden of Eden without any tree of the knowledge of good and evil. The question of the moral government of such a world

could no more be asked, than we could reasonably seek for a moral purpose in a kaleidoscope.

4. How far down the scale of animal life the phenomena of consciousness are manifested, it is impossible to say. No one doubts their presence in his fellow-men; and, unless any strict Cartesians are left, no one doubts that mammals and birds are to be reckoned creatures that have feelings analogous to our smell, taste, sight, hearing, touch, pleasure, and pain. For my own part, I should be disposed to extend this analogical judgment a good deal further. On the other hand, if the lowest forms of plants are to be denied consciousness, I do not see on what ground it is to be ascribed to the lowest animals. I find it hard to believe that an infusory animalcule, a foraminifer, or a fresh-water polype is capable of feeling; and, in spite of Shakspere, I have doubts about the great sensitiveness of the "poor beetle that we tread upon." The question is equally perplexing when we turn to the stages of development of the individual. Granted a fowl feels; that the chick just hatched feels; that the chick when it chirps within the egg may possibly feel; what is to be said of it, on the fifth day, when the bird is there, but with all its tissues nascent? Still more, on the first day, when it is nothing but a flat cellular disk? I certainly cannot bring myself to believe that this disk feels. Yet if it does not, there must be some time in the three weeks, between the first day and the day of hatching, when, as a concomitant, or a consequence, of the attainment by the brain of

the chick of a certain stage of structural evolution, consciousness makes its appearance. I have frequently expressed my incapacity to understand the nature of the relation between consciousness and a certain anatomical tissue, which is thus established by observation. But the fact remains that, so far as observation and experiment go, they teach us that the psychical phenomena are dependent on the physical.

In like manner, if fishes, insects, scorpions, and such animals as the pearly nautilus, possess feeling, then undoubtedly consciousness was present in the world as far back as the Silurian epoch. But, if the earliest animals were similar to our rhizopods and monads, there must have been some time, between the much earlier epoch in which they constituted the whole animal population and the Silurian, in which feeling dawned, in consequence of the organism having reached the stage of evolution on which it depends.

5. Consciousness has various forms, which may be manifested independently of one another. The feelings of light and colour, of sound, of touch, though so often associated with those of pleasure and pain, are, by nature, as entirely independent of them as is thinking. An animal devoid of the feelings of pleasure and of pain, may nevertheless exhibit all the effects of sensation and purposive action. Therefore, it would be a justifiable hypothesis that, long after organic evolution had attained to consciousness, pleasure and pain were still absent. Such a world

would be without either happiness or misery ; no act could be punished and none could be rewarded ; and it could have no moral purpose.

6. Suppose, for argument's sake, that all mammals and birds are subjects of pleasure and pain. Then we may be certain that these forms of consciousness were in existence at the beginning of the Mesozoic epoch. From that time forth, pleasure has been distributed without reference to merit, and pain inflicted without reference to demerit, throughout all but a mere fraction of the higher animals. Moreover, the amount and the severity of the pain, no less than the variety and acuteness of the pleasure, have increased with every advance in the scale of evolution. As suffering came into the world, not in consequence of a fall, but of a rise, in the scale of being, so every further rise has brought more suffering. As the evidence stands, it would appear that the sort of brain which characterises the highest mammals and which, so far as we know, is the indispensable condition of the highest sensibility, did not come into existence before the Tertiary epoch. The primordial anthropoid was probably, in this respect, on much the same footing as his pithecoid kin. Like them he stood upon his "natural rights," gratified all his desires to the best of his ability, and was as incapable of either right or wrong doing as they. It would be as absurd as in their case, to regard his pleasures, any more than theirs, as moral rewards, and his pains, any more than theirs, as moral punishments.

7. From the remotest ages of which we have any cognizance, death has been the natural and, apparently. the necessary coneomitant of life. In our hypothetical world (3), inhabited by nothing but plants, death must have very early resulted from the struggle for existence : many of the crowd must have jostled one another out of the conditions on which life depends. The occurrence of death, as far back as we have any fossil record of life, however, needs not to be proved by such arguments ; for, if there had been no death there would have been no fossil remains, such as the great majority of those we meet with. Not only was there death in the world, as far as the record of life takes us ; but, ever since mammals and birds have been preyed upon by carnivorous animals, there has been painful death, inflicted by mechanisms specially adapted for inflicting it.

8. Those who are acquainted with the closeness of the structural relations between the human organisation and that of the mammals which come nearest to him, on the one hand ; and with the palæontological history of such animals as horses and dogs, on the other ; will not be disposed to question the origin of man from forms which stand in the same sort of relation to *Homo sapiens*, as *Hipparion* does to *Equus*. I think it a conclusion, fully justified by analogy, that, sooner or later, we shall discover the remains of our less specialised primatic ancestors in the strata which have yielded the less specialised equine and canine quadrupeds. At present, fossil remains of men do not take us back further than the later

part of the Quaternary epoch; and, as was to be expected, they do not differ more from existing men, than Quaternary horses differ from existing horses. Still earlier we find traces of man, in implements, such as are used by the ruder savages at the present day. Later, the remains of the palæolithic and neolithic conditions take us gradually from the savage state to the civilisations of Egypt and of Mycenæ; though the true chronological order of the remains actually discovered may be uncertain.

9. Much has yet to be learned, but, at present, natural knowledge affords no support to the notion that men have fallen from a higher to a lower state. On the contrary, everything points to a slow natural evolution ; which, favoured by the surrounding conditions in such localities as the valleys of the Yang-tse-kang, the Euphrates, and the Nile, reached a relatively high pitch, five or six thousand years ago ; while, in many other regions, the savage condition has persisted down to our day. In all this vast lapse of time there is not a trace of the occurrence of any general destruction of the human race ; not the smallest indication that man has been treated on any other principles than the rest of the animal world.

10. The results of the process of evolution in the case of man, and in that of his more nearly allied contemporaries, have been marvellously different. Yet it is easy to see that small primitive differences of a certain order, must, in the long run, bring about a wide divergence of the human stock from the others.

It is a reasonable supposition that, in the earliest human organisms, an improved brain, a voice more capable of modulation and articulation, limbs which lent themselves better to gesture, a more perfect hand, capable among other things of imitating form in plastic or other material, were combined with the curiosity, the mimetic tendency, the strong family affection of the next lower group ; and that they were accompanied by exceptional length of life and a prolonged minority. The last two peculiarities are obviously calculated to strengthen the family organisation, and to give great weight to its educative influences. The potentiality of language, as the vocal symbol of thought, lay in the faculty of modulating and articulating the voice. The potentiality of writing, as the visual symbol of thought, lay in the hand that could draw; and in the mimetic tendency, which, as we know, was gratified by drawing, as far back as the days of Quaternary man. With speech as the record, in tradition, of the experience of more than one generation ; with writing as the record of that of any number of generations ; the experience of the race, tested and corrected generation after generation, could be stored up and made the starting point for fresh progress. Having these perfectly natural factors of the evolutionary process in man before us, it seems unnecessary to go further a-field in search of others.

11. That the doctrine of evolution implies a former state of innocence of mankind is quite true ; but, as I have remarked, it is the innocence of the ape and of

the tiger, whose acts, however they may run counter
to the principles of morality, it would be absurd to
blame.   The lust of the one and the ferocity of the
other are as much provided for in their organisation,
are as clear evidences of design, as any other features
that can be named.

Observation and experiment upon the phenomena
of society soon taught men that, in order to obtain
the advantages of social existence, certain rules must
be observed.   Morality commenced with society.
Society is possible only upon the condition that the
members of it shall surrender more or less of their
individual freedom of action.   In primitive societies,
individual selfishness is a centrifugal force of such
intensity that it is constantly bringing the social
organisation to the verge of destruction.   Hence the
prominence of the positive rules of obedience to the
elders; of standing by the family or the tribe in all
emergencies; of fulfilling the religious rites, non-
observance of which is conceived to damage it with
the supernatural powers, belief in whose existence is
one of the earliest products of human thought; and of
the negative rules, which restrain each from meddling
with the life or property of another.

12. The highest conceivable form of human society
is that in which the desire to do what is best for the
whole, dominates and limits the action of every member of that society.   The more complex the social
organisation the greater the number of acts from
which each man must abstain, if he desires to do
that which is best for all.   Thus the progressive

evolution of society means increasing restriction of individual freedom in certain directions.

With the advance of civilisation, and the growth of cities and of nations by the coalescence of families and of tribes, the rules which constitute the common foundation of morality and of law became more numerous and complicated, and the temptations to break or evade many of them stronger. In the absence of a clear apprehension of the natural sanctions of these rules, a supernatural sanction was assumed; and imagination supplied the motives which reason was supposed to be incompetent to furnish. Religion, at first independent of morality, gradually took morality under its protection; and the supernaturalists have ever since tried to persuade mankind that the existence of ethics is bound up with that of supernaturalism.

I am not of that opinion. But, whether it is correct or otherwise, it is very clear to me that, as Beelzebub is not to be cast out by the aid of Beelzebub, so morality is not to be established by immorality. It is, we are told, the special peculiarity of the devil that he was a liar from the beginning. If we set out in life with pretending to know that which we do not know; with professing to accept for proof evidence which we are well aware is inadequate; with wilfully shutting our eyes and our ears to facts which militate against this or that comfortable hypothesis; we are assuredly doing our best to deserve the same character.

I have not the presumption to imagine that, in

E

spite of all my efforts, errors may not have crept
into these propositions. But I am tolerably confident that time will prove them to be substantially
correct. And if they are so, I confess I do not see how
any extant supernaturalistic system can also claim
exactness. That they are irreconcilable with the
biblical cosmogony, anthropology, and theodicy is
obvious; but they are no less inconsistent with the
sentimental Deism of the "Vicaire Savoyard" and
his numerous modern progeny. It is as impossible,
to my mind, to suppose that the evolutionary process
was set going with full foreknowledge of the result
and yet with what we should understand by a purely
benevolent intention, as it is to imagine that the intention was purely malevolent. And the prevalence of
dualistic theories from the earliest times to the
present day—whether in the shape of the doctrine of
the inherently evil nature of matter; of an Ahriman;
of a hard and cruel Demiurge; of a diabolical " prince
of this world," show how widely this difficulty has
been felt.

Many seem to think that, when it is admitted that
the ancient literature, contained in our Bibles, has no
more claim to infallibility than any other ancient
literature; when it is proved that the Israelites and
their Christian successors accepted a great many
supernaturalistic theories and legends which have
no better foundation than those of heathenism,
nothing remains to be done but to throw the Bible
aside as so much waste paper.

I have always opposed this opinion. It appears to

me that if there is anybody more objectionable than the orthodox Bibliolater it is the heterodox Philistine, who can discover in a literature which, in some respects, has no superior, nothing but a subject for scoffing and an occasion for the display of his conceited ignorance of the debt he owes to former generations.

Twenty-two years ago I pleaded for the use of the Bible as an instrument of popular education, and I venture to repeat what I then said :

" Consider the great historical fact that, for three centuries, this book has been woven into the life of all that is best and noblest in English history ; that it has become the national epic of Britain and is as familiar to gentle and simple, from John o' Groat's House to Land's End, as Dante and Tasso once were to the Italians ; that it is written in the noblest and purest English and abounds in exquisite beauties of mere literary form ; and, finally, that it forbids the veriest hind, who never left his village, to be ignorant of the existence of other countries and other civilisations and of a great past, stretching back to the furthest limits of the oldest nations in the world. By the study of what other book could children be so much humanised and made to feel that each figure in that vast historical procession fills, like themselves, but a momentary space in the interval between the Eternities ; and earns the blessings or the curses of all time, according to its effort to do good and hate evil, even as they also are earning their payment for their work ? " [1]

[1] " The School Boards : What they can do and what they may do," 1870. *Critiques and Addresses*, p. 51.

At the same time, I laid stress upon the necessity
of placing such instruction in lay hands; in the hope
and belief, that it would thus gradually accommodate
itself to the coming changes of opinion; that the
theology and the legend would drop more and
more out of sight, while the perennially interesting
historical, literary, and ethical contents would come
more and more into view.

I may add yet another claim of the Bible to the
respect and the attention of a democratic age.
Throughout the history of the western world, the
Scriptures, Jewish and Christian, have been the great
instigators of revolt against the worst forms of clerical
and political despotism. The Bible has been the
Magna Charta of the poor and of the oppressed;
down to modern times, no State has had a constitution
in which the interests of the people are so largely
taken into account, in which the duties, so much more
than the privileges, of rulers are insisted upon, as that
drawn up for Israel in Deuteronomy and in Leviticus;
nowhere is the fundamental truth that the welfare of
the State, in the long run, depends on the uprightness
of the citizen so strongly laid down. Assuredly, the
Bible talks no trash about the rights of man; but it
insists on the equality of duties, on the liberty to bring
about that righteousness which is somewhat different
from struggling for "rights;" on the fraternity of
taking thought for one's neighbour as for oneself.

So far as such equality, liberty, and fraternity
are included under the democratic principles which
assume the same names, the Bible is the most demo-

cratic book in the world. As such it began, through the heretical sects, to undermine the clerico-political despotism of the middle ages, almost as soon as it was formed, in the eleventh century; Pope and King had as much as they could do to put down the Albigenses and the Waldenses in the twelfth and thirteenth centuries; the Lollards and the Hussites gave them still more trouble in the fourteenth and fifteenth; from the sixteenth century onward, the Protestant sects have favoured political freedom in proportion to the degree in which they have refused to acknowledge any ultimate authority save that of the Bible.

But the enormous influence which has thus been exerted by the Jewish and Christian Scriptures has had no necessary connection with cosmogonies, demonologies, and miraculous interferences. Their strength lies in their appeals, not to the reason, but to the ethical sense. I do not say that even the highest biblical ideal is exclusive of others or needs no supplement. But I do believe that the human race is not yet, possibly may never be, in a position to dispense with it.

I

THE RISE AND PROGRESS OF PALÆONTOLOGY

THAT application of the sciences of biology and geology, which is commonly known as palæontology, took its origin in the mind of the first person who, finding something like a shell, or a bone, naturally embedded in gravel or rock, indulged in speculations upon the nature of this thing which he had dug out — this " fossil "—and upon the causes which had brought it into such a position.   In this rudimentary form, a high antiquity may safely be ascribed to palæontology, inasmuch as we know that, 500 years before the Christian era, the philosophic doctrines of Xenophanes were influenced by his observations upon the fossil remains exposed in the quarries of Syracuse.   From this time forth not only the philosophers, but the poets, the historians, the geographers of antiquity occasionally refer to fossils ; and, after the revival of learning, lively controversies arose respecting their real nature.   But hardly more than two centuries have elapsed since this fundamental problem was first exhaustively treated ; it was only in the last century that the archæological value of fossils — their importance, I mean, as records of the history of the earth—was fully

recognised ; the first adequate investigation of the fossil
remains of any large group of vertebrated animals is
to be found in Cuvier's *Recherches sur les Ossemens
Fossiles*, completed in 1822 ; and, so modern is strati-
graphical palæontology, that its founder, William
Smith, lived to receive the just recognition of his services by the award of the first Wollaston Medal in
1831.

But, although palæontology is a comparatively
youthful scientific speciality, the mass of materials
with which it has to deal is already prodigious.   In
the last fifty years the number of known fossil remains
of invertebrated animals has been trebled or quad-
rupled.   The work of interpretation of vertebrate
fossils, the foundations of which were so solidly laid
by Cuvier, was carried on, with wonderful vigour and
success, by Agassiz in Switzerland, by Von Meyer in
Germany, and last, but not least, by Owen in this
country, while, in later years, a multitude of workers
have laboured in the same field.   In many groups of
the animal kingdom the number of fossil forms already
known is as great as that of the existing species.   In
some cases it is much greater ; and there are entire
orders of animals of the existence of which we should
know nothing except for the evidence afforded by fossil
remains.   With all this it may be safely assumed that,
at the present moment, we are not acquainted with a
tithe of the fossils which will sooner or later be discovered.   If we may judge by the profusion yielded
within the last few years by the Tertiary formations
of North America, there seems to be no limit to the

multitude of Mammalian remains to be expected from that continent; and analogy leads us to expect similar riches in Eastern Asia, whenever the Tertiary formations of that region are as carefully explored. Again, we have as yet almost everything to learn respecting the terrestrial population of the Mesozoic epoch—and it seems as if the Western territories of the United States were about to prove as instructive in regard to this point as they have in respect of tertiary life. My friend Professor Marsh informs me that, within two years, remains of more than 160 distinct individuals of mammals, belonging to twenty species and nine genera, have been found in a space not larger than the floor of a good-sized room; while beds of the same age have yielded 300 reptiles, varying in size from a length of 60 feet or 80 feet to the dimensions of a rabbit.

The task which I have set myself to-night is to endeavour to lay before you, as briefly as possible, a sketch of the successive steps by which our present knowledge of the facts of palæontology and of those conclusions from them which are indisputable, has been attained; and I beg leave to remind you, at the outset, that in attempting to sketch the progress of a branch of knowledge to which innumerable labours have contributed, my business is rather with generalisations than with details. It is my object to mark the epochs of palæontology, not to recount all the events of its history.

That which I just now called the fundamental problem of palæontology, the question which has to be

settled before any other can be profitably discussed,
is this, What is the nature of fossils? Are they, as
the healthy common sense of the ancient Greeks
appears to have led them to assume without hesitation, the remains of animals and plants? Or are they,
as was so generally maintained in the fifteenth, sixteenth, and seventeenth centuries, mere figured stones,
portions of mineral matter which have assumed the
forms of leaves and shells and bones, just as those
portions of mineral matter which we call crystals take
on the form of regular geometrical solids? Or,
again, are they, as others thought, the products of the
germs of animals and of the seeds of plants which
have lost their way, as it were, in the bowels of the
earth, and have achieved only an imperfect and abor-
tive development? It is easy to sneer at our ancestors
for being disposed to reject the first in favour of one
or other of the last two hypotheses; but it is much
more profitable to try to discover why they, who
were really not one whit less sensible persons than
our excellent selves, should have been led to entertain
views which strike us as absurd. The belief in what
is erroneously called spontaneous generation, that is
to say, in the development of living matter out
of mineral matter, apart from the agency of pre-
existing living matter, as an ordinary occurrence at
the present day—which is still held by some of us,
was universally accepted as an obvious truth by them.
They could point to the arborescent forms assumed
by hoar-frost and by sundry metallic minerals as
evidence of the existence in nature of a " plastic

force" competent to enable inorganic matter to assume the form of organised bodies. Then, as every one who is familiar with fossils knows, they present innumerable gradations, from shells and bones which exactly resemble the recent objects, to masses of mere stone which, however accurately they repeat the outward form of the organic body, have nothing else in common with it; and, thence, to mere traces and faint impressions in the continuous substance of the rock. What we now know to be the results of the chemical changes which take place in the course of fossilisation, by which mineral is substituted for organic substance, might, in the absence of such knowledge, be fairly interpreted as the expression of a process of development in the opposite direction—from the mineral to the organic. Moreover, in an age when it would have seemed the most absurd of paradoxes to suggest that the general level of the sea is constant, while that of the solid land fluctuates up and down through thousands of feet in a secular ground swell, it may well have appeared far less hazardous to conceive that fossils are sports of nature than to accept the necessary alternative, that all the inland regions and highlands, in the rocks of which marine shells had been found, had once been covered by the ocean. It is not so surprising, therefore, as it may at first seem, that although such men as Leonardo da Vinci and Bernard Palissy took just views of the nature of fossils, the opinion of the majority of their contemporaries set strongly the other way; nor even that error maintained itself long after the scientific grounds of the true interpretation of fossils had been stated, in a manner that left nothing to be desired, in the latter half of the seventeenth century. The person who rendered this good service to palæontology was Nicolas Steno, professor of anatomy in Florence, though a Dane by birth. Collectors of fossils at that day were familiar with certain bodies termed " glossopetræ," and specu- lation was rife as to their nature. In the first half of the seventeenth century, Fabio Colonna had tried to convince his colleagues of the famous Accademia dei Lincei that the glossopetræ were merely fossil sharks' teeth, but his arguments made no impression. Fifty years later, Steno reopened the question, and, by dissecting the head of a shark and pointing out the very exact correspondence of its teeth with the glosso- petræ, left no rational doubt as to the origin of the latter. Thus far, the work of Steno went little further than that of Colonna, but it fortunately occurred to him to think out the whole subject of the interpreta- tion of fossils, and the result of his meditations was the publication, in 1669, of a little treatise with the very quaint title of *De Solido intra Solidum natur- aliter contento.* The general course of Steno's argu- ment may be stated in a few words. Fossils are solid bodies which, by some natural process, have come to be contained within other solid bodies, namely, the rocks in which they are embedded ; and the funda- mental problem of palæontology, stated generally, is this : " Given a body endowed with a certain shape and produced in accordance with natural laws, to find in that body itself the evidence of the place and manner of its production."[1]   The only way of solving this problem is by the application of the axiom that "like effects imply like causes," or as Steno puts it, in reference to this particular case, that "bodies which are altogether similar have been produced in the same way."[2]   Hence, since the glossopetræ are altogether similar to sharks' teeth, they must have been produced by sharklike fishes ; and since many fossil shells correspond, down to the minutest details of structure, with the shells of existing marine or freshwater animals, they must have been produced by similar animals ; and the like reasoning is applied by Steno to the fossil bones of vertebrated animals, whether aquatic or terrestrial.   To the obvious objection that many fossils are not altogether similar to their living analogues, differing in substance while agreeing in form, or being mere hollows or impressions, the surfaces of which are figured in the same way as those of animal or vegetable organisms, Steno replies by pointing out the changes which take place in organic remains embedded in the earth, and how their solid substance may be dissolved away entirely, or replaced by mineral matter, until nothing is left of the original but a cast, an impression, or a mere trace of its contours.   The principles of investigation thus excellently stated and illustrated by Steno in 1669, are those which have,

---

[1] *De Solido intra Solidum*, p. 5.—"Dato corpore certâ figurâ prædito et juxta leges naturæ producto, in ipso corpore argumenta invenire locum et modum productionis detegentia."

[2] "Corpora sibi invicem omnino similia simili etiam modo producta sunt."

consciously or unconsciously, guided the researches of palæontologists ever since. Even that feat of palæontology which has so powerfully impressed the popular imagination, the reconstruction of an extinct animal from a tooth or a bone, is based upon the simplest imaginable application of the logic of Steno. A moment's consideration will show, in fact, that Steno's conclusion that the glossopetræ are sharks' teeth implies the reconstruction of an animal from its tooth. It is equivalent to the assertion that the animal of which the glossopetræ are relics had the form and organisation of a shark; that it had a skull, a vertebral column, and limbs similar to those which are characteristic of this group of fishes; that its heart, gills, and intestines presented the peculiarities which those of all sharks exhibit; nay, even that any hard parts which its integument contained were of a totally different character from the scales of ordinary fishes. These conclusions are as certain as any based upon probable reasonings can be. And they are so, simply because a very large experience justifies us in believing that teeth of this particular form and structure are invariably associated with the peculiar organisation of sharks, and are never found in connection with other organisms. Why this should be we are not at present in a position even to imagine; we must take the fact as an empirical law of animal morphology, the reason of which may possibly be one day found in the history of the evolution of the shark tribe, but for which it is hopeless to seek for an explanation in ordinary physiological reasonings. Every one practically acquainted with palæontology is aware that it is not every tooth, nor every bone, which enables us to form a judgment of the character of the animal to which it belonged; and that it is possible to possess many teeth, and even a large portion of the skeleton of an extinct animal, and yet be unable to reconstruct its skull or its limbs. It is only when the tooth or bone presents peculiarities, which we know by previous experience to be characteristic of a certain group, that we can safely predict that the fossil belonged to an animal of the same group. Any one who finds a cow's grinder may be perfectly sure that it belonged to an animal which had two complete toes on each foot and ruminated; any one who finds a horse's grinder may be as sure that it had one complete toe on each foot and did not ruminate; but if ruminants and horses were extinct animals of which nothing but the grinders had ever been discovered, no amount of physiological reasoning could have enabled us to reconstruct either animal, still less to have divined the wide differences between the two. Cuvier, in the *Discours sur les Révolutions de la Surface du Globe*, strangely credits himself, and has ever since been credited by others, with the invention of a new method of palæontological research. But if you will turn to the *Recherches sur les Ossemens Fossiles* and watch Cuvier, not speculating, but working, you will find that his method is neither more nor less than that of Steno. If he was able to make his famous prophecy from the jaw which lay upon the surface of a block of stone to the pelvis of the same animal which lay hidden in it, it was not because either he, or any one else, knew, or knows, why a certain form of jaw is, as a rule, constantly accompanied by the presence of marsupial bones, but simply because experience has shown that these two structures are co-ordinated.

The settlement of the nature of fossils led at once to the next advance of palæontology, viz. its application to the deciphering of the history of the earth. When it was admitted that fossils are remains of animals and plants, it followed that, in so far as they resemble terrestrial, or freshwater, animals and plants, they are evidences of the existence of land, or fresh water; and, in so far as they resemble marine organisms, they are evidences of the existence of the sea at the time at which they were parts of actually living animals and plants. Moreover, in the absence of evidence to the contrary, it must be admitted that the terrestrial or the marine organisms implied the existence of land or sea at the place in which they were found while they were yet living. In fact, such conclusions were immediately drawn by everybody, from the time of Xenophanes downwards, who believed that fossils were really organic remains. Steno discusses their value as evidence of repeated alteration of marine and terrestrial conditions upon the soil of Tuscany in a manner worthy of a modern geologist. The speculations of De Maillet in the beginning of the eighteenth century turn upon fossils; and Buffon follows him very closely in those two remarkable works, the *Théorie de la Terre* and the *Époques de la Nature*, with which he commenced and ended his career as a naturalist.

The opening sentences of the *Époques de la Nature* show us how fully Buffon recognised the analogy of geological with archæological inquiries. " As in civil history we consult deeds, seek for coins, or decipher antique inscriptions in order to determine the epochs of human revolutions and fix the date of moral events ; so, in natural history, we must search the archives of the world, recover old monuments from the bowels of the earth, collect their fragmentary remains, and gather into one body of evidence all the signs of physical change which may enable us to look back upon the different ages of nature. It is our only means of fixing some points in the immensity of space, and of setting a certain number of waymarks along the eternal path of time."

Buffon enumerates five classes of these monuments of the past history of the earth, and they are all facts of palæontology. In the first place, he says, shells and other marine productions are found all over the surface and in the interior of the dry land ; and all calcareous rocks are made up of their remains. Secondly, a great many of these shells which are found in Europe are not now to be met with in the adjacent seas ; and, in the slates and other deep-seated deposits, there are remains of fishes and of plants of which no species now exist in our latitudes, and which are either extinct, or exist only in more northern climates. Thirdly, in Siberia and in other northern regions of Europe and of Asia, bones and teeth of elephants, rhinoceroses, and hippopotamuses occur in such numbers that these animals must once have lived and multiplied in those regions, although at the present day they are confined to southern climates. The deposits in which these remains are found are superficial, while those which contain shells and other marine remains lie much deeper. Fourthly, tusks and bones of elephants and hippopotamuses are found not only in the northern regions of the old world, but also in those of the new world, although, at present, neither elephants nor hippopotamuses occur in America. Fifthly, in the middle of the continents, in regions most remote from the sea, we find an infinite number of shells, of which the most part belong to animals of those kinds which still exist in southern seas, but of which many others have no living analogues; so that these species appear to be lost, destroyed by some unknown cause. It is needless to inquire how far these statements are strictly accurate; they are sufficiently so to justify Buffon's conclusions that the dry land was once beneath the sea; that the formation of the fossiliferous rocks must have occupied a vastly greater lapse of time than that traditionally ascribed to the age of the earth; that fossil remains indicate different climatal conditions to have obtained in former times, and especially that the polar regions were once warmer; that many species of animals and plants have become extinct; and that geological change has had something to do with geographical distribution.

But these propositions almost constitute the framework of palæontology. In order to complete it but

one addition was needed, and that was made, in the
last years of the eighteenth century, by William
Smith, whose work comes so near our own times that
many living men may have been personally acquainted
with him. This modest land-surveyor, whose business
took him into many parts of England, profited by the
peculiarly favourable conditions offered by the arrangement of our secondary strata to make a careful examination and comparison of their fossil contents at
different points of the large area over which they
extend. The result of his accurate and widely-
extended observations was to establish the important
truth that each stratum contains certain fossils which
are peculiar to it; and that the order in which
the strata, characterised by these fossils, are super-
imposed one upon the other is always the same.
This most important generalisation was rapidly verified and extended to all parts of the world accessible
to geologists; and now it rests upon such an immense
mass of observations as to be one of the best established truths of natural science. To the geologist
the discovery was of infinite importance, as it enabled
him to identify rocks of the same relative age, however their continuity might be interrupted or their
composition altered. But to the biologist it had a still
deeper meaning, for it demonstrated that, throughout
the prodigious duration of time registered by the
fossiliferous rocks, the living population of the earth
had undergone continual changes, not merely by the
extinction of a certain number of the species which
had at first existed, but by the continual generation

of new species, and the no less constant extinction of
old ones.

Thus the broad outlines of palæontology, in so far
as it is the common property of both the geologist and
the biologist, were marked out at the close of the
last century. In tracing its subsequent progress I
must confine myself to the province of biology, and,
indeed, to the influence of palæontology upon zoological morphology. And I accept this limitation the
more willingly as the no less important topic of the
bearing of geology and of palæontology upon distribution has been luminously treated in the address of the
President of the Geographical Section.[1]

The succession of the species of animals and plants
in time being established, the first question which the
zoologist or the botanist had to ask himself was, What
is the relation of these successive species one to
another? And it is a curious circumstance that the
most important event in the history of palæontology
which immediately succeeded William Smith's generalisation was a discovery which, could it have been
rightly appreciated at the time, would have gone far
towards suggesting the answer, which was in fact
delayed for more than half a century. I refer to
Cuvier's investigation of the Mammalian fossils yielded
by the quarries in the older tertiary rocks of Mont-
martre, among the chief results of which was the
bringing to light of two genera of extinct hoofed
quadrupeds, the *Anoplotherium* and the *Palæotherium*.
The rich materials at Cuvier's disposition enabled him

---

[1] [Sir J. D. Hooker.]

to obtain a full knowledge of the osteology and of the
dentition of these two forms, and consequently to
compare their structure critically with that of existing
hoofed animals.   The effect of this comparison was to
prove that the *Anoplotherium*, though it presented
many points of resemblance with the pigs on the one
hand and with the ruminants on the other, differed
from both to such an extent that it could find a place
in neither group.   In fact, it held, in some respects,
an intermediate position, tending to bridge over the
interval between these two groups, which in the
existing fauna are so distinct.   In the same way, the
Palæotherium tended to connect forms so different as
the tapir, the rhinoceros, and the horse.   Subsequent
investigations have brought to light a variety of facts
of the same order, the most curious and striking of
which are those which prove the existence, in the
mesozoic epoch, of a series of forms intermediate
between birds and reptiles—two classes of vertebrate
animals which at present appear to be more widely
separated than any others.   Yet the interval between
them is completely filled, in the mesozoic fauna, by
birds which have reptilian characters on the one side,
and reptiles which have ornithic characters on the
other.   So again, while the group of fishes termed
ganoids is at the present time so distinct from that of
the dipnoi, or mudfishes, that they have been reckoned
as distinct orders, the Devonian strata present us
with forms of which it is impossible to say with certainty whether they are dipnoi or whether they are
ganoids.

Agassiz's long and elaborate researches upon fossil fishes, published between 1833 and 1842, led him to suggest the existence of another kind of relation between ancient and modern forms of life. He observed that the oldest fishes present many characters which recall the embryonic conditions of existing fishes ; and that, not only among fishes, but in several groups of the invertebrata which have a long palæontological history, the latest forms are more modified, more specialised, than the earlier. The fact that the dentition of the older tertiary ungulate and carnivorous mammals is always complete, noticed by Professor Owen, illustrated the same generalisation.

Another no less suggestive observation was made by Mr. Darwin, whose personal investigations during the voyage of the *Beagle* led him to remark upon the singular fact, that the fauna, which immediately precedes that at present existing in any geographical province of distribution, presents the same peculiarities as its successor. Thus, in South America and in Australia, the later tertiary or quaternary fossils show that the fauna which immediately preceded that of the present day was, in the one case, as much characterised by edentates and, in the other, by marsupials as it is now, although the species of the older are largely different from those of the newer fauna.

However clearly these indications might point in one direction, the question of the exact relation of the successive forms of animal and vegetable life could be satisfactorily settled only in one way ; namely, by comparing, stage by stage, the series of

forms presented by one and the same type throughout
a long space of time.   Within the last few years this
has been done fully in the case of the horse, less completely in the case of the other principal types of the
ungulata and of the carnivora; and all these investigations tend to one general result, namely, that, in any
given series, the successive members of that series
present a gradually increasing specialisation of structure.   That is to say, if any such mammal at present
existing has specially modified and reduced limbs or
dentition and complicated brain, its predecessors in
time show less and less modification and reduction in
limbs and teeth and a less highly developed brain.
The labours of Gaudry, Marsh, and Cope furnish
abundant illustrations of this law from the marvellous
fossil wealth of Pikermi and the vast uninterrupted
series of tertiary rocks in the territories of North
America.

I will now sum up the results of this sketch of the
rise and progress of palæontology.   The whole fabric
of palæontology is based upon two propositions : the
first is, that fossils are the remains of animals and
plants ; and the second is, that the stratified rocks in
which they are found are sedimentary deposits ; and
each of these propositions is founded upon the same
axiom, that like effects imply like causes.   If there is
any cause competent to produce a fossil stem, or shell,
or bone, except a living being, then palæontology has
no foundation ; if the stratification of the rocks is not
the effect of such causes as at present produce stratifi-

cation, we have no means of judging of the duration
of past time, or of the order in which the forms of
life have succeeded one another. But if these two
propositions are granted, there is no escape, as it
appears to me, from three very important conclusions.
The first is that living matter has existed upon the
earth for a vast length of time, certainly for millions
of years. The second is that, during this lapse of
time, the forms of living matter have undergone
repeated changes, the effect of which has been that
the animal and vegetable population, at any period of
the earth's history, contains some species which did
not exist at some antecedent period, and others which
ceased to exist at some subsequent period. The third
is that, in the case of many groups of mammals and
some of reptiles, in which one type can be followed
through a considerable extent of geological time, the
series of different forms by which the type is represented, at successive intervals of this time, is exactly
such as it would be, if they had been produced by the
gradual modification of the earliest forms of the series.
These are facts of the history of the earth guaranteed
by as good evidence as any facts in civil history.

Hitherto I have kept carefully clear of all the
hypotheses to which men have at various times endeavoured to fit the facts of palæontology, or by
which they have endeavoured to connect as many of
these facts as they happened to be acquainted with.
I do not think it would be a profitable employment of
our time to discuss conceptions which doubtless have
had their justification and even their use, but which

are now obviously incompatible with the well-ascertained truths of palæontology. At present these truths leave room for only two hypotheses. The first is that, in the course of the history of the earth, innumerable species of animals and plants have come into existence, independently of one another, innumerable times. This, of course, implies either that spontaneous generation on the most astounding scale, and of animals such as horses and elephants, has been going on, as a natural process, through all the time recorded by the fossiliferous rocks; or it necessitates the belief in innumerable acts of creation repeated innumerable times. The other hypothesis is, that the successive species of animals and plants have arisen, the later by the gradual modification of the earlier. This is the hypothesis of evolution; and the palæontological discoveries of the last decade are so completely in accordance with the requirements of this hypothesis that, if it had not existed, the palæontologist would have had to invent it.

I have always had a certain horror of presuming to set a limit upon the possibilities of things. Therefore I will not venture to say that it is impossible that the multitudinous species of animals and plants may have been produced, one separately from the other, by spontaneous generation; nor that it is impossible that they should have been independently originated by an endless succession of miraculous creative acts. But I must confess that both these hypotheses strike me as so astoundingly improbable, so devoid of a shred of either scientific or traditional support, that even if there were no other evidence than that of palæontology in its favour, I should feel compelled to adopt the hypothesis of evolution. Happily, the future of palæontology is independent of all hypothetical considerations. Fifty years hence, whoever undertakes to record the progress of palæontology will note the present time as the epoch in which the law of succession of the forms of the higher animals was determined by the observation of palæontological facts. He will point out that, just as Steno and as Cuvier were enabled from their knowledge of the empirical laws of coexistence of the parts of animals to conclude from a part to the whole, so the knowledge of the law of succession of forms empowered their successors to conclude, from one or two terms of such a succession, to the whole series ; and thus to divine the existence of forms of life, of which, perhaps, no trace remains, at epochs of inconceivable remoteness in the past.

II

THE INTERPRETERS OF GENESIS AND THE INTERPRETERS OF NATURE

OUR fabulist warns "those who in quarrels interpose" of the fate which is probably in store for them; and, in venturing to place myself between so powerful a controversialist as Mr. Gladstone and the eminent divine whom he assaults with such vigour in the last number of this Review,[1] I am fully aware that I run great danger of verifying Gay's prediction. Moreover, it is quite possible that my zeal in offering aid to a combatant so extremely well able to take care of himself as M. Réville may be thought to savour of indiscretion.

Two considerations, however, have led me to face the double risk. The one is that though, in my judgment, M. Réville is wholly in the right in that part of the controversy to which I propose to restrict my observations, nevertheless he, as a foreigner, has very little chance of making the truth prevail with Englishmen against the authority and the dialectic skill of the greatest master of persuasive rhetoric among English-speaking men of our time.

[1] *The Nineteenth Century.*

As the Queen's proctor intervenes, in certain cases, between two litigants in the interests of justice, so it may be permitted me to interpose as a sort of uncommissioned science proctor. My second excuse for my meddlesomeness is, that important questions of natural science—respecting which neither of the combatants professes to speak as an expert—are involved in the controversy ; and I think it is desirable that the public should know what it is ,that natural science really has to say on these topics, to the best belief of one who has been a diligent student of natural science for the last forty years.

The original *Prolégomènes de l'histoire des Religions* has not come in my way ; but I have read the translation of M. Réville's work, published in England under the auspices of Professor Max Müller, with very great interest. It puts more fairly and clearly than any book previously known to me, the view which a man of strong religious feelings, but at the same time possessing the information and the reasoning power which enable him to estimate the strength of scientific methods of inquiry and the weight of scientific truth, may be expected to take of the relation between science and religion.

In the chapter on " The Primitive Revelation " the scientific worth of the account of the Creation given in the book of Genesis is estimated in terms which are as unquestionably respectful as, in my judgment, they are just ; and, at the end of the chapter on " Primitive Tradition," M. Réville appraises the value of pentateuchal anthropology in a way which I should have thought sure of enlisting the assent of all competent judges, even if it were extended to the whole of the cosmogony and biology of Genesis :—

As, however, the original traditions of nations sprang up in an epoch less remote than our own from the primitive life, it is indispensable to consult them, to compare them, and to associate them with other sources of information which are available. From this point of view, the traditions recorded in Genesis possess, in addition to their own peculiar charm, a value of the highest order ; but we cannot ultimately see in them more than a venerable fragment, well deserving attention, of the great genesis of mankind.

Mr. Gladstone is of a different mind. He dissents from M. Réville's views respecting the proper estimation of the pentateuchal traditions, no less than he does from his interpretation of those Homeric myths which have been the object of his own special study. In the latter case, Mr. Gladstone tells M. Réville that he is wrong on his own authority, to which, in such a matter, all will pay due respect : in the former, he affirms himself to be " wholly destitute of that kind of knowledge which carries authority," and his rebuke is administered in the name and by the authority of natural science.

An air of magisterial gravity hangs about the following passage :—

But the question is not here of a lofty poem, or a skilfully constructed narrative : it is whether natural science, in the patient exercise of its high calling to examine facts, finds that the works of God cry out against what we have fondly believed to be His word and tell another tale ; or whether, in this nineteenth century of Christian progress, it substantially echoes back the majestic sound, which, before it existed as a pursuit, went forth into all lands.

First, looking largely at the latter portion of the narrative, which describes the creation of living organisms, and waiving details, on some of which (as in v. 24) the Septuagint seems to vary from the Hebrew, there is a grand fourfold division, set forth in an orderly succession of times as follows : on the fifth day

1. The water-population ;
2. The air-population :
and, on the sixth day,
3. The land-population of animals ;
4. The land-population consummated in man.

Now this same fourfold order is understood to have been so affirmed in our time by natural science, that it may be taken as a demonstrated conclusion and established fact (p. 696).

"Understood?" By whom? I cannnot bring myself to imagine that Mr. Gladstone has made so solemn and authoritative a statement on a matter of this importance without due inquiry—without being able to found himself upon recognised scientific authority. But I wish he had thought fit to name the source from whence he has derived his information, as, in that case, I could have dealt with his authority, and I should have thereby escaped the appearance of making an attack on Mr. Gladstone himself, which is in every way distasteful to me.

For I can meet the statement in the last paragraph of the above citation with nothing but a direct negative. If I know anything at all about the results attained by the natural science of our time, it is "a demonstrated conclusion and established fact" that the "fourfold order" given by Mr. Gladstone is not that in which the evidence at our disposal tends to show that the water, air, and land-populations of the globe have made their appearance.

Perhaps I may be told that Mr. Gladstone does give his authority—that he cites Cuvier, Sir John Herschel, and Dr. Whewell in support of his case. If that has been Mr. Gladstone's intention in mentioning these eminent names, I may remark that, on this particular question, the only relevant authority is that of Cuvier. But great as Cuvier was, it is to be remembered that, as Mr. Gladstone incidentally remarks, he cannot now be called a recent authority. In fact, he has been dead more than half a century; and the palæontology of our day is related to that of his, very much as the geography of the sixteenth century is related to that of the fourteenth. Since 1832, when Cuvier died, not only a new world, but new worlds, of ancient life have been discovered; and those who have most faithfully carried on the work of the chief founder of palæontology have done most to invalidate the essentially negative grounds of his speculative adherence to tradition.

If Mr. Gladstone's latest information on these matters is derived from the famous discourse prefixed to the *Ossemens Fossiles*, I can understand the position he has taken up; if he has ever opened a respectable modern manual of palæontology, or geology, I cannot. For the facts which demolish his whole argument are of the commonest notoriety. But before proceeding to consider the evidence for this assertion we must be clear about the meaning of the phraseology employed.

I apprehend that when Mr. Gladstone uses the term "water-population" he means those animals which in Genesis i. 21 (Revised Version) are spoken of as "the great sea monsters and every living creature that moveth, which the waters brought forth abundantly, after their kind." And I presume that it will be agreed that whales and porpoises, sea fishes, and the innumerable hosts of marine invertebrated animals, are meant thereby. So "air-population" must be the equivalent of "fowl" in verse 20, and "every winged fowl after its kind," verse 21. I suppose I may take it for granted that by "fowl" we have here to understand birds—at any rate primarily. Secondarily, it may be that the bats and the extinct pterodactyles, which were flying reptiles, come under the same head. But whether all insects are "creeping things" of the land-population, or whether flying insects are to be included under the denomination of "winged fowl," is a point for the decision of Hebrew exegetes. Lastly, I suppose I may assume that "land-population" signifies "the cattle" and "the beast of the earth," and "every creeping thing that creepeth upon the earth," in verses 25 and 26; presumably, it comprehends all kinds of terrestrial animals, vertebrate and invertebrate, except such as may be comprised under the head of the "air-population."

Now what I want to make clear is this: that if the terms "water-population," "air-population," and "land-population" are understood in the senses here defined, natural science has nothing to say in favour of the proposition that they succeeded one another in the order given by Mr. Gladstone ; but that, on the contrary, all the evidence we possess goes to prove that they did not.   Whence it will follow that, if Mr. Gladstone has interpreted Genesis rightly (on which point I am most anxious to be understood to offer no opinion), that interpretation is wholly irreconcilable with the conclusions at present accepted by the interpreters of nature—with everything that can be called "a demonstrated conclusion and established fact" of natural science.   And be it observed that I am not here dealing with a question of speculation, but with a question of fact.

Either the geological record is sufficiently complete to afford us a means of determining the order in which animals have made their appearance on the globe or it is not.   If it is, the determination of that order is little more than a mere matter of observation ; if it is not, then natural science neither affirms nor refutes the "fourfold order," but is simply silent.

The series of the fossiliferous deposits, which contain the remains of the animals which have lived on the earth in past ages of its history, and which can alone afford the evidence required by natural science of the order of appearance of their different species, may be grouped in the manner shown in the left-hand column of the following table, the oldest being at the bottom :—

Formations	First known appearance of
Quaternary.	
Pliocene.	
Miocene.	
Eocene . . .	Vertebrate *air*-population (Bats).

Formations	First known appearance of
Cretaceous.	
Jurassic . .	Vertebrate *air*-population (Birds and Pterodactyles).
Triassic.	
Upper **Palæozoic.**	
Middle Palæozoic .	Vertebrate *land* - population (Amphibia, Reptilia [?]).
Lower Palæozoic.	
Silurian . .	Vertebrate *water*-population (Fishes). Invertebrate *air* and *land*-population (Flying Insects and Scorpions).
Cambrian .	Invertebrate *water* - population (much earlier, if *Eozoon* is animal).

In the right-hand column I have noted the group of strata in which, according to our present information, the *land, air,* and *water*-populations respectively appear for the first time; and in consequence of the ambiguity about the meaning of "fowl," I have separately indicated the first appearance of bats, birds, flying reptiles, and flying insects. It will be observed that, if "fowl" means only "bird," or at most flying vertebrate, then the first certain evidence of the latter, in the Jurassic epoch, is posterior to the first appearance of truly terrestrial *Amphibia,* and possibly of true reptiles, in the Carboniferous epoch (Middle Palæozoic) by a prodigious interval of time.

The water-population of vertebrated animals first appears in the Upper Silurian.[1] Therefore, if we found ourselves on vertebrated animals and take "fowl" to mean birds only, or, at most, flying vertebrates, natural science says that the order of succession was water, land, and air-population, and not—

[1 Earlier, if more recent announcements are correct.]

as Mr. Gladstone, founding himself on Genesis, says —water, air, land-population. If a chronicler of Greece affirmed that the age of Alexander preceded that of Pericles and immediately succeeded that of the Trojan war, Mr. Gladstone would hardly say that this order is "understood to have been so affirmed by historical science that it may be taken as a demonstrated conclusion and established fact." Yet natural science "affirms" his "fourfold order" to exactly the same extent—neither more nor less.

Suppose, however, that "fowl" is to be taken to include flying insects. In that case, the first appearance of an air-population must be shifted back for long ages, recent discovery having shown that they occur in rocks of Silurian age. Hence there might still have been hope for the fourfold order, were it not that the fates unkindly determined that scorpions —"creeping things that creep on the earth" *par excellence*—turned up in Silurian strata nearly at the same time. So that, if the word in the original Hebrew translated "fowl" should really after all mean "cockroach"—and I have great faith in the elasticity of that tongue in the hands of Biblical exegetes—the order primarily suggested by the existing evidence—

      2. Land and air-population ;

      1. Water-population ;

and Mr. Gladstone's order—

      3. Land-population ;

      2. Air-population ;

      1. Water-population ;

can by no means be made to coincide. As a matter
of fact, then, the statement so confidently put forward turns out to be devoid of foundation and in
direct contradiction of the evidence at present at our
disposal.[1]

If, stepping beyond that which may be learned
from the facts of the successive appearance of the
forms of animal life upon the surface of the globe, in
so far as they are yet made known to us by natural
science, we apply our reasoning faculties to the task
of finding out what those observed facts mean, the
present conclusions of the interpreters of nature
appear to be no less directly in conflict with those of
the latest interpreter of Genesis.

Mr. Gladstone appears to admit that there is some
truth in the doctrine of evolution, and indeed places
it under very high patronage.

I contend that evolution in its highest form has not been a

[1] It may be objected that I have not put the case fairly, inasmuch
as the solitary insect's wing which was discovered twelve months ago
in Silurian rocks, and which is, at present, the sole evidence of insects
older than the Devonian epoch, came from strata of Middle Silurian
age, and is therefore older than the scorpions which, within the last
two years, have been found in Upper Silurian strata in Sweden,
Britain, and the United States. But no one who comprehends the
nature of the evidence afforded by fossil remains would venture to
say that the non-discovery of scorpions in the Middle Silurian strata,
up to this time, affords any more ground for supposing that they did
not exist, than the non-discovery of flying insects in the Upper
Silurian strata, up to this time, throws any doubt on the certainty
that they existed, which is derived from the occurrence of the wing in
the Middle Silurian. In fact, I have stretched a point in admitting
that these fossils afford a colourable pretext for the assumption that
the land and air-population were of contemporaneous origin.

thing heretofore unknown to history, to philosophy, or to theology.   I contend that it was before the mind of Saint Paul when he taught that in the fulness of time God sent forth His Son, and of Eusebius when he wrote the *Preparation for the Gospel*, and of Augustine when he composed the *City of God* (p. 706).

Has any one ever disputed the contention, thus solemnly enunciated, that the doctrine of evolution was not invented the day before yesterday? Has any one ever dreamed of claiming it as a modern innovation?   Is there any one so ignorant of the history of philosophy as to be unaware that it is one of the forms in which speculation embodied itself long before the time either of the Bishop of Hippo or of the Apostle to the Gentiles?   Is Mr. Gladstone, of all people in the world, disposed to ignore the founders of Greek philosophy, to say nothing of Indian sages to whom evolution was a familiar notion ages before Paul of Tarsus was born?   But it is ungrateful to cavil at even the most oblique admission of the possible value of one of those affirmations of natural science which really may be said to be " a demonstrated conclusion and established fact."   I note it with pleasure, if only for the purpose of introducing the observation that, if there is any truth whatever in the doctrine of evolution as applied to animals, Mr. Gladstone's gloss on Genesis in the following passage is hardly happy :—

God created
(*a*) The water-population ;
(*b*) The air-population.
And they receive His benediction (v. 20-23).

6. Pursuing this regular progression from the lower to the higher, from the simple to the complex, the text now gives us the work of the sixth " day," which supplies the land-population, air and water having been already supplied (pp. 695, 696).

The gloss to which I refer is the assumption that the " air-population " forms a term in the order of progression from lower to higher, from simple to complex—the place of which lies between the water-population below and the land-population above—and I speak of it as a "gloss," because the pentateuchal writer is nowise responsible for it.

But it is not true that the air-population, as a whole, is "lower" or less "complex" than the land-population. On the contrary, every beginner in the study of animal morphology is aware that the organisation of a bat, of a bird, or of a pterodactyle presupposes that of a terrestrial quadruped; and that it is intelligible only as an extreme modification of the organisation of a terrestrial mammal or reptile. In the same way winged insects (if they are to be counted among the "air-population") presuppose insects which were wingless, and, therefore, as "creeping things," were part of the land-population. Thus theory is as much opposed as observation to the admission that natural science endorses the succession of animal life which Mr. Gladstone finds in Genesis. On the contrary, a good many representatives of natural science would be prepared to say, on theoretical grounds alone, that it is incredible that the " air-population " should have appeared before the " land-population "—and that, if this assertion is to be found in Genesis, it merely demonstrates the scientific worthlessness of the story of which it forms a part.

Indeed, we may go further. It is not even admissible to say that the water-population, as a whole, appeared before the air and the land-populations. According to the Authorised Version, Genesis especially mentions, among the animals created on the fifth day, " great whales," in place of which the Revised Version reads " great sea monsters." Far be it from me to give an opinion which rendering is right, or whether either is right. All I desire to remark is, that if whales and porpoises, dugongs and manatees, are to be regarded as members of the water-population (and if they are not, what animals can claim the designation ?), then that much of the water-population has, as certainly, originated later than the land-population as bats and birds have. For I am not aware that any competent judge would hesitate to admit that the organisation of these animals shows the most obvious signs of their descent from terrestrial quadrupeds.

A similar criticism applies to Mr. Gladstone's assumption that, as the fourth act of that " orderly succession of times" enunciated in Genesis, " the land-population consummated in man."

If this means simply that man is the final term in the evolutional series of which he forms a part, I do not suppose that any objection will be raised to that statement on the part of students of natural science. But if the pentateuchal author goes further than this, and intends to say that which is ascribed to him by

Mr. Gladstone, I think natural science will have to enter a *caveat*. It is by not any means certain that man—I mean the species *Homo sapiens* of zoological terminology—has " consummated " the land-population in the sense of appearing at a later period of time than any other. Let me make my meaning clear by an example. From a morphological point of view, our beautiful and useful contemporary—I might almost call him colleague—the horse (*Equus caballus*), is the last term of the evolutional series to which he belongs, just as *Homo sapiens* is the last term of the series of which he is a member. If I want to know whether the species *Equus caballus* made its appearance on the surface of the globe before or after *Homo sapiens*, deduction from known laws does not help me. There is no reason, that I know of, why one should have appeared sooner or later than the other. If I turn to observation, I find abundant remains of *Equus caballus* in Quaternary strata, perhaps a little earlier. The existence of *Homo sapiens* in the Quaternary epoch is also certain. Evidence has been adduced in favour of man's existence in the Pliocene, or even in the Miocene epoch. It does not satisfy me ; but I have no reason to doubt that the fact may be so, nevertheless. Indeed, I think it is quite possible that further research will show that *Homo sapiens* existed, not only before *Equus caballus*, but before many other of the existing forms of animal life ; so that, if all the species of animals have been separately created, man, in this case, would by no means be the " consummation " of the land-population.

I am raising no objection to the position of the fourth term in Mr. Gladstone's "order"—on the facts, as they stand, it is quite open to any one to hold, as a pious opinion, that the fabrication of man was the acme and final achievement of the process of peopling the globe. But it must not be said that natural science counts this opinion among her "demonstrated conclusions and established facts," for there would be just as much, or as little, reason for ranging the contrary opinion among them.

It may seem superfluous to add to the evidence that Mr. Gladstone has been utterly misled in supposing that his interpretation of Genesis receives any support from natural science. But it is as well to do one's work thoroughly while one is about it; and I think it may be advisable to point out that the facts, as they are at present known, not only refute Mr. Gladstone's interpretation of Genesis in detail, but are opposed to the central idea on which it appears to be based.

There must be some position from which the reconcilers of science and Genesis will not retreat, some central idea the maintenance of which is vital and its refutation fatal. Even if they now allow that the words "the evening and the morning" have not the least reference to a natural day, but mean a period of any number of millions of years that may be necessary; even if they are driven to admit that the word "creation," which so many millions of pious Jews and Christians have held, and still hold, to mean a sudden act of the Deity, signifies a process of gradual evolu-

tion of one species from another, extending through
immeasurable time ; even if they are willing to grant
that the asserted coincidence of the order of Nature
with the "fourfold order" ascribed to Genesis is an
obvious error instead of an established truth ; they are
surely prepared to make a last stand upon the conception which underlies the whole, and which constitutes the essence of Mr. Gladstone's "fourfold division,
set forth in an orderly succession of times." It is,
that the animal species which compose the water-
population, the air-population, and the land-population
respectively, originated during three distinct and successive periods of time, and only during those periods
of time.

This statement appears to me to be the interpretation of Genesis which Mr. Gladstone supports, reduced
to its simplest expression. "Period of time" is substituted for "day"; "originated" is substituted for
"created"; and "any order required" for that adopted
by Mr. Gladstone. It is necessary to make this pro-
viso, for if "day" may mean a few million years, and
"creation" may mean evolution, then it is obvious
that the order (1) water-population, (2) air-population, (3) land-population, may also mean (1) water-
population, (2) land-population, (3) air-population ;
and it would be unkind to bind down the reconcilers
to this detail when one has parted with so many
others to oblige them.

But even this sublimated essence of the pentateuchal doctrine (if it be such) remains as discordant
with natural science as ever.

It is not true that the species composing any one of the three populations originated during any one of three successive periods of time, and not at any other of these.

Undoubtedly, it is in the highest degree probable that animal life appeared first under aquatic conditions; that terrestrial forms appeared later, and flying animals only after land animals; but it is, at the same time, testified by all the evidence we possess, that the great majority, if not the whole, of the primordial species of each division have long since died out and have been replaced by a vast succession of new forms. Hundreds of thousands of animal species, as distinct as those which now compose our water, land, and air-populations, have come into existence and died out again, throughout the æons of geological time which separate us from the lower Palæozoic epoch, when, as I have pointed out, our present evidence of the existence of such distinct populations commences. If the species of animals have all been separately created, then it follows that hundreds of thousands of acts of creative energy have occurred, at intervals, throughout the whole time recorded by the fossiliferous rocks; and, during the greater part of that time, the " creation" of the members of the water, land, and air-populations must have gone on contemporaneously.

If we represent the water, land, and air-populations by a, b, and c respectively, and take vertical succession on the page to indicate order in time, then the following schemes will roughly shadow forth the contrast I have been endeavouring to explain :—

Genesis (as interpreted by Mr. Gladstone).	Nature (as interpreted by natural science).
$b\ b\ b$	$c^1\ a^3\ b^2$
$c\ c\ c$	$c\ a^2\ b^1$
$a\ a\ a$	$b\ a^1\ b$
	$a\ a\ a$

So far as I can see, there is only one resource left for those modern representatives of Sisyphus, the reconcilers of Genesis with science; and it has the advantage of being founded on a perfectly legitimate appeal to our ignorance. It has been seen that, on any interpretation of the terms water-population and land-population, it must be admitted that invertebrate representatives of these populations existed during the lower Palæozoic epoch. No evolutionist can hesitate to admit that other land animals (and possibly vertebrates among them) may have existed during that time, of the history of which we know so little; and, further, that scorpions are animals of such high organisation that it is highly probable their existence indicates that of a long antecedent land-population of a similar character.

Then, since the land-population is said not to have been created until the sixth day, it necessarily follows that the evidence of the order in which animals appeared must be sought in the record of those older Palæozoic times in which only traces of the water-population have as yet been discovered.

Therefore, if any one chooses to say that the creative work took place in the Cambrian or Laurentian epoch, in exactly that manner which Mr. Gladstone does, and natural science does not, affirm, natural science is not in a position to disprove the accuracy of the statement. Only one cannot have one's cake and eat it too, and such safety from the contradiction of science means the forfeiture of her support.

Whether the account of the work of the first, second, and third days in Genesis would be confirmed by the demonstration of the truth of the nebular hypothesis; whether it is corroborated by what is known of the nature and probable relative antiquity of the heavenly bodies; whether, if the Hebrew word translated "firmament" in the Authorised Version really means "expanse," the assertion that the waters are partly under this "expanse" and partly above it would be any more confirmed by the ascertained facts of physical geography and meteorology than it was before; whether the creation of the whole vegetable world, and especially of "grass, herb yielding seed after its kind, and tree bearing fruit," before any kind of animal, is "affirmed" by the apparently plain teaching of botanical palæontology, that grasses and fruit-trees originated long subsequently to animals— all these are questions which, if I mistake not, would be answered decisively in the negative by those who are specially conversant with the sciences involved. And it must be recollected that the issue raised by Mr. Gladstone is not whether, by some effort of ingenuity, the pentateuchal story can be shown to be not disprovable by scientific knowledge, but whether it is supported thereby.

There is nothing, then, in the criticisms of Dr. Réville but what rather tends to confirm than to impair the old-fashioned belief that there is a revelation in the book of Genesis (p. 694).

The form into which Mr. Gladstone has thought fit to throw this opinion leaves me in doubt as to its substance. I do not understand how a hostile criticism can, under any circumstances, tend to confirm that which it attacks. If, however, Mr. Gladstone merely means to express his personal impression, " as one wholly destitute of that kind of knowledge which carries authority," that he has destroyed the value of these criticisms, I have neither the wish nor the right to attempt to disturb his faith. On the other hand, I may be permitted to state my own conviction that, so far as natural science is involved, M. Réville's observations retain the exact value they possessed before Mr. Gladstone attacked them.

Trusting that I have now said enough to secure the author of a wise and moderate disquisition upon a topic which seems fated to stir unwisdom and fanaticism to their depths, a fuller measure of justice than has hitherto been accorded to him, I retire from my self-appointed championship, with the hope that I shall not hereafter be called upon by M. Réville to apologise for damage done to his strong case by imperfect or impulsive advocacy. But, perhaps, I may be permitted to add a word or two, on my own account, in reference to the great question of the relations between science and religion; since it is one about which I have thought a good deal ever since I have been able to think at all; and about which I have ventured to express my views publicly, more than once, in the course of the last thirty years.

The antagonism between science and religion, about which we hear so much, appears to me to be purely factitious—fabricated, on the one hand, by short-sighted religious people who confound a certain branch of science, theology, with religion; and, on the other, by equally short-sighted scientific people who forget that science takes for its province only that which is susceptible of clear intellectual comprehension; and that, outside the boundaries of that province, they must be content with imagination, with hope, and with ignorance.

It seems to me that the moral and intellectual life of the civilised nations of Europe is the product of that interaction, sometimes in the way of antagonism, sometimes in that of profitable interchange, of the Semitic and the Aryan races, which commenced with the dawn of history, when Greek and Phœnician came in contact, and has been continued by Carthaginian and Roman, by Jew and Gentile, down to the present day. Our art (except, perhaps, music) and our science are the contributions of the Aryan; but the essence of our religion is derived from the Semite. In the eighth century B.C., in the heart of a world of idolatrous polytheists, the Hebrew prophets put forth a conception of religion which appears to me to be as wonderful an inspiration of genius as the art of Pheidias or the science of Aristotle.

" And what doth the Lord require of thee, but to

do justly, and to love mercy, and to walk humbly
with thy God?"

If any so-called religion takes away from this great
saying of Micah, I think it wantonly mutilates, while,
if it adds thereto, I think it obscures, the perfect ideal
of religion.

But what extent of knowledge, what acuteness of
scientific criticism, can touch this, if any one possessed
of knowledge, or acuteness, could be absurd enough to
make the attempt? Will the progress of research
prove that justice is worthless and mercy hateful;
will it ever soften the bitter contrast between our
actions and our aspirations; or show us the bounds of
the universe, and bid us say, Go to, now we comprehend the infinite? A faculty of wrath lay in those
ancient Israelites, and surely the prophet's staff would
have made swift acquaintance with the head of the
scholar who had asked Micah whether, peradventure,
the Lord further required of him an implicit belief
in the accuracy of the cosmogony of Genesis!

What we are usually pleased to call religion nowadays is, for the most part, Hellenised Judaism; and,
not unfrequently, the Hellenic element carries with it
a mighty remnant of old-world paganism and a great
infusion of the worst and weakest products of Greek
scientific speculation; while fragments of Persian and
Babylonian, or rather Accadian, mythology burden
the Judaic contribution to the common stock.

The antagonism of science is not to religion, but to
the heathen survivals and the bad philosophy under
which religion herself is often wellnigh crushed.

And, for my part, I trust that this antagonism will never cease ; but that, to the end of time, true science will continue to fulfil one of her most beneficent functions, that of relieving men from the burden of false science which is imposed upon them in the name of religion.

This is the work that M. Réville and men such as he are doing for us ; this is the work which his opponents are endeavouring, consciously or unconsciously, to hinder.

III

MR. GLADSTONE AND GENESIS

In controversy, as in courtship, the good old rule to
be off with the old before one is on with the new,
greatly commends itself to my sense of expediency.
And, therefore, it appears to me desirable that I
should preface such observations as I may have to
offer upon the cloud of arguments (the relevancy of
which to the issue which I had ventured to raise is
not always obvious) put forth by Mr. Gladstone in the
January number of this Review,[1] by an endeavour to
make clear to such of our readers as have not had the
advantage of a forensic education the present net
result of the discussion.

I am quite aware that, in undertaking this task, I
run all the risks to which the man who presumes to
deal judicially with his own cause is liable. But it is
exactly because I do not shun that risk, but, rather,
earnestly desire to be judged by him who cometh
after me, provided that he has the knowledge and
impartiality appropriate to a judge, that I adopt my
present course.

[1 *The Nineteenth Century*, 1886.]

In the article on "The Dawn of Creation and Worship," it will be remembered that Mr. Gladstone unreservedly commits himself to three propositions. The first is that, according to the writer of the Pentateuch, the "water-population," the "air-population," and the "land-population" of the globe were created successively, in the order named. In the second place, Mr. Gladstone authoritatively asserts that this (as part of his "fourfold order") has been "so affirmed in our time by natural science, that it may be taken as a demonstrated conclusion and established fact." In the third place, Mr. Gladstone argues that the fact of this coincidence of the pentateuchal story with the results of modern investigation makes it "impossible to avoid the conclusion, first, that either this writer was gifted with faculties passing all human experience, or else his knowledge was divine." And having settled to his own satisfaction that the first "branch of the alternative is truly nominal and unreal," Mr. Gladstone continues, "So stands the plea for a revelation of truth from God, a plea only to be met by questioning its possibility" (p. 697).

I am a simple-minded person, wholly devoid of subtlety of intellect, so that I willingly admit that there may be depths of alternative meaning in these propositions out of all soundings attainable by my poor plummet. Still there are a good many people who suffer under a like intellectual limitation; and, for once in my life, I feel that I have the chance of attaining that position of a representative of average opinion which appears to be the modern ideal of a leader of men, when I make free confession that, after turning the matter over in my mind, with all the aid derived from a careful consideration of Mr. Gladstone's reply, I cannot get away from my original conviction that, if Mr. Gladstone's second proposition can be shown to be not merely inaccurate, but directly contradictory of facts known to every one who is acquainted with the elements of natural science, the third proposition collapses of itself.

And it was this conviction which led me to enter upon the present discussion. I fancied that if my respected clients, the people of average opinion and capacity, could once be got distinctly to conceive that Mr. Gladstone's views as to the proper method of dealing with grave and difficult scientific and religious problems had permitted him to base a solemn " plea for a revelation of truth from God " upon an error as to a matter of fact, from which the intelligent perusal of a manual of palæontology would have saved him, I need not trouble myself to occupy their time and attention with further comments upon his contribution to apologetic literature. It is for others to judge whether I have efficiently carried out my project or not. It certainly does not count for much that I should be unable to find any flaw in my own case, but I think it counts for a good deal that Mr. Gladstone appears to have been equally unable to do so. He does, indeed, make a great parade of authorities, and I have the greatest respect for those authorities whom Mr. Gladstone mentions. If he will get them to sign a joint memorial to the effect that our present palæontological evidence proves that birds appeared before the "land-population" of terrestrial reptiles, I shall think it my duty to reconsider my position—but not till then.

It will be observed that I have cautiously used the word "appears" in referring to what seems to me to be absence of any real answer to my criticisms in Mr. Gladstone's reply. For I must honestly confess that, notwithstanding long and painful strivings after clear insight, I am still uncertain whether Mr. Gladstone's "Defence" means that the great "plea for a revelation from God" is to be left to perish in the dialectic desert; or whether it is to be withdrawn under the protection of such skirmishers as are available for covering retreat.

In particular, the remarkable disquisition which covers pages 11 to 14 of Mr. Gladstone's last contribution has greatly exercised my mind. Socrates is reported to have said of the works of Heraclitus that he who attempted to comprehend them should be a "Delian swimmer," but that, for his part, what he could understand was so good that he was disposed to believe in the excellence of that which he found unintelligible. In endeavouring to make myself master of Mr. Gladstone's meaning in these pages, I have often been overcome by a feeling analogous to that of Socrates, but not quite the same. That which I do understand, in fact, has appeared to me so very much the reverse of good, that I have sometimes permitted myself to doubt the value of that which I do not understand.

In this part of Mr. Gladstone's reply, in fact, I find nothing of which the bearing upon my arguments is clear to me, except that which relates to the question whether reptiles, so far as they are represented by tortoises and the great majority of lizards and snakes, which are land animals, are creeping things in the sense of the pentateuchal writer or not.

I have every respect for the singer of the Song of the Three Children (whoever he may have been); I desire to cast no shadow of doubt upon, but, on the contrary, marvel at, the exactness of Mr. Gladstone's information as to the considerations which " affected the method of the Mosaic writer"; nor do I venture to doubt that the inconvenient intrusion of these contemptible reptiles—" a family fallen from greatness " (p. 14), a miserable decayed aristocracy reduced to mere " skulkers about the earth " (*ibid.*)—in consequence, apparently, of difficulties about the occupation of land arising out of the earth-hunger of their former serfs, the mammals—into an apologetic argument, which otherwise would run quite smoothly, is in every way to be deprecated.   Still, the wretched creatures stand there, importunately demanding notice; and, however different may be the practice in that contentious atmosphere with which Mr. Gladstone expresses and laments his familiarity, in the atmosphere of science it really is of no avail whatever to shut one's eyes to facts, or to try to bury them out of sight under a tumulus of rhetoric.   That is my experience of " the Elysian regions of Science," wherein it is a pleasure to me to think that a man of Mr. Gladstone's intimate knowledge of English life, during the last quarter of a century, believes my philosophic existence to have been rounded off in unbroken equanimity.

However reprehensible, and indeed contemptible, terrestrial reptiles may be, the only question which appears to me to be relevant to my argument is whether these creatures are or are not comprised under the denomination of "everything that creepeth upon the ground."

Mr. Gladstone speaks of the author of the first chapter of Genesis as " the Mosaic writer"; I suppose, therefore, that he will admit that it is equally proper to speak of the author of Leviticus as the "Mosaic writer." Whether such a phrase would be used by any one who had an adequate conception of the assured results of modern Biblical criticism is another matter; but, at any rate, it cannot be denied that Leviticus has as much claim to Mosaic authorship as Genesis. Therefore, if one wants to know the sense of a phrase used in Genesis, it will be well to see what Leviticus has to say on the matter. Hence, I commend the following extract from the eleventh chapter of Leviticus to Mr. Gladstone's serious attention :—

And these are they which are unclean unto you among the creeping things that creep upon the earth : the weasel, and the mouse, and the great lizard after its kind, and the gecko, and the land-crocodile, and the sand-lizard, and the chameleon. These are they which are unclean to you among all that creep (v. 29-31).

The merest Sunday-school exegesis therefore suffices

to prove that when the "Mosaic writer" in Genesis i. 24 speaks of "creeping things," he means to include lizards among them.

This being so, it is agreed, on all hands, that terrestrial lizards, and other reptiles allied to lizards, occur in the Permian strata. It is further agreed that the Triassic strata were deposited after these. Moreover, it is well known that, even if certain footprints are to be taken as unquestionable evidence of the existence of birds, they are not known to occur in rocks earlier than the Trias, while indubitable remains of birds are to be met with only much later. Hence it follows that natural science does not "affirm" the statement that birds were made on the fifth day, and "everything that creepeth on the ground" on the sixth, on which Mr. Gladstone rests his order; for, as is shown by Leviticus, the "Mosaic writer" includes lizards among his "creeping things."

Perhaps I have given myself superfluous trouble in the preceding argument, for I find that Mr. Gladstone is willing to assume (he does not say to admit) that the statement in the text of Genesis as to reptiles cannot "in all points be sustained" (p. 16). But my position is that it cannot be sustained in any point, so that, after all, it has perhaps been as well to go over the evidence again. And then Mr. Gladstone proceeds as if nothing had happened to tell us that—

There remain great unshaken facts to be weighed. First, the fact that such a record should have been made at all.

As most peoples have their cosmogonies, this "fact" does not strike me as having much value.

Secondly, the fact that, instead of dwelling in generalities, it has placed itself under the severe conditions of a chronological order reaching from the first *nisus* of chaotic matter to the consummated production of a fair and goodly, a furnished and a peopled world.

This " fact " can be regarded as of value only by ignoring the fact demonstrated in my previous paper, that natural science does not confirm the order asserted so far as living things are concerned; and by upsetting a fact to be brought to light presently, to wit, that, in regard to the rest of the pentateuchal cosmogony, prudent science has very little to say one way or the other.

Thirdly, the fact that its cosmogony seems, in the light of the nineteenth century, to draw more and more of countenance from the best natural philosophy.

I have already questioned the accuracy of this statement, and I do not observe that mere repetition adds to its value.

And, fourthly, that it has described the successive origins of the five great categories of present life with which human experience was and is conversant, in that order which geological authority confirms.

By comparison with a sentence on page 14, in which a fivefold order is substituted for the " fourfold order," on which the " plea for revelation " was originally founded, it appears that these five categories are " plants, fishes, birds, mammals, and man," which, Mr. Gladstone affirms, " are given to us in Genesis in the order of succession in which they are also given by the latest geological authorities."

I must venture to demur to this statement. I showed, in my previous paper, that there is no reason to doubt that the term "great sea monster" (used in Gen. i. 21) includes the most conspicuous of great sea animals—namely, whales, dolphins, porpoises, manatees, and dugongs;[1] and, as these are indubitable mammals, it is impossible to affirm that mammals come after birds, which are said to have been created on the same day. Moreover, I pointed out that as these Cetacea and Sirenia are certainly modified land animals, their existence implies the antecedent existence of land mammals.

Furthermore, I have to remark that the term "fishes," as used, technically, in zoology, by no means covers all the moving creatures that have life, which are bidden to "fill the waters in the seas" (Gen. i. 20-22). Marine mollusks and crustacea, echinoderms, corals, and foraminifera are not technically fishes. But they are abundant in the palæozoic rocks, ages upon ages older than those in which the first evidences of true fishes appear. And if, in a geological book, Mr. Gladstone finds the quite true statement that plants appeared before fishes, it is only by a complete misunderstanding that he can be led to imagine it serves his purpose. As a matter of fact, at the present moment, it is a question whether, on the bare evidence afforded by fossils, the marine creeping thing or the marine plant has the seniority.

---

[1] Both dolphins and dugongs occur in the Red Sea, porpoises and dolphins in the Mediterranean ; so that the "Mosaic writer" may well have been acquainted with them.

No cautious palæontologist would express a decided opinion on the matter. But, if we are to read the pentateuchal statement as a scientific document (and, in spite of all protests to the contrary, those who bring it into comparison with science do seek to make a scientific document of it), then, as it is quite clear that only terrestrial plants of high organisation are spoken of in verses 11 and 12, no palæontologist would hesitate to say that, at present, the records of sea animal life are vastly older than those of any land plant describable as " grass, herb yielding seed, or fruit-tree."

Thus, although, in Mr. Gladstone's "Defence," the "old order passeth into new," his case is not improved. The fivefold order is no more "affirmed in our time by natural science" to be "a demonstrated conclusion and established fact" than the fourfold order was. Natural science appears to me to decline to have anything to do with either; they are as wrong in detail as they are mistaken in principle.

There is another change of position, the value of which is not so apparent to me, as it may well seem to be to those who are unfamiliar with the subject under discussion. Mr. Gladstone discards his three groups of "water-population," "air-population," and "land-population," and substitutes for them (1) fishes, (2) birds, (3) mammals, (4) man. Moreover, it is assumed, in a note, that "the higher or ordinary mammals" alone were known to the "Mosaic writer" (p. 6). No doubt it looks, at first, as if something were gained by this alteration; for, as I have just pointed out, the word "fishes" can be used in two senses, one of which has a deceptive appearance of adjustability to the "Mosaic" account. Then the inconvenient reptiles are banished out of sight; and, finally, the question of the exact meaning of "higher" and "ordinary" in the case of mammals opens up the prospect of a hopeful logomachy. But what is the good of it all in the face of Leviticus on the one hand and of palæontology on the other?

As, in my apprehension, there is not a shadow of justification for the suggestion that when the penta-teuchal writer says "fowl" he excludes bats (which, as we shall see directly, are expressly included under "fowl" in Leviticus), and as I have already shown that he demonstrably includes reptiles, as well as mammals, among the creeping things of the land, I may be permitted to spare my readers further discussion of the "fivefold order." On the whole, it is seen to be rather more inconsistent with Genesis than its fourfold predecessor.

But I have yet a fresh order to face. Mr. Gladstone (p. 11) understands "the main statements of Genesis in successive order of time, but without any measurement of its divisions, to be as follows:—

1. A period of land, anterior to all life (v. 9, 10).
2. A period of vegetable life, anterior to animal life (v. 11, 12).
3. A period of animal life, in the order of fishes (v. 20).
4. Another stage of animal life, in the order of birds.
5. Another, in the order of beasts (v. 24, 25).
6. Last of all, man (v. 26, 27)."

Mr. Gladstone then tries to find the proof of the occurrence of a similar succession in sundry excellent works on geology.

I am really grieved to be obliged to say that this third (or is it fourth?) modification of the foundation of the " plea for revelation " originally set forth, satisfies me as little as any of its predecessors.

For, in the first place, I cannot accept the assertion that this order is to be found in Genesis. With respect to No. 5, for example, I hold, as I have already said, that " great sea monsters " includes the Cetacea, in which case mammals (which is what, I suppose, Mr. Gladstone means by " beasts ") come in under head No. 3, and not under No. 5. Again, " fowl " are said in Genesis to be created on the same day as fishes; therefore I cannot accept an order which makes birds succeed fishes. Once more, as it is quite certain that the term " fowl " includes the bats,—for in Leviticus xi. 13-19 we read, " And these shall ye have in abomination among the fowls . . . the heron after its kind, and the hoopoe, and the bat,"—it is obvious that bats are also said to have been created at stage No. 3. And as bats are mammals, and their existence obviously presupposes that of terrestrial " beasts," it is quite clear that the latter could not have first appeared as No. 5. I need not repeat my reasons for doubting whether man came " last of all."

As the latter half of Mr. Gladstone's sixfold order thus shows itself to be wholly unauthorised by, and inconsistent with, the plain language of the Pentateuch,

I might decline to discuss the admissibility of its former half.

But I will add one or two remarks on this point also. Does Mr. Gladstone mean to say that in any of the works he has cited, or indeed anywhere else, he can find scientific warranty for the assertion that there was a period of land—by which I suppose he means dry land (for submerged land must needs be as old as the separate existence of the sea)—" anterior to all life " ?

It may be so, or it may not be so ; but where is the evidence which would justify any one in making a positive assertion on the subject? What competent palæontologist will affirm, at this present moment, that he knows anything about the period at which life originated, or will assert more than the extreme probability that such origin was a long way antecedent to any traces of life at present known? What physical geologist will affirm that he knows when dry land began to exist, or will say more than that it was probably very much earlier than any extant direct evidence of terrestrial conditions indicates?

I think I know pretty well the answers which the authorities quoted by Mr. Gladstone would give to these questions ; but I leave it to them to give them if they think fit.

If I ventured to speculate on the matter at all, I should say it is by no means certain that sea is older than dry land, inasmuch as a solid terrestrial surface may very well have existed before the earth was cool enough to allow of the existence of fluid water. And, in this case, dry land may have existed before the sea. As to the first appearance of life, the whole argument of analogy, whatever it may be worth in such a case, is in favour of the absence of living beings until long after the hot water seas had constituted themselves; and of the subsequent appearance of aquatic before terrestrial forms of life. But whether these "proto-plasts" would, if we could examine them, be reckoned among the lowest microscopic algæ, or fungi; or among those doubtful organisms which lie in the debatable land between animals and plants, is, in my judgment, a question on which a prudent biologist will reserve his opinion.

I think that I have now disposed of those parts of Mr Gladstone's defence in which I seem to discover a design to rescue his solemn "plea for revelation." But a great deal of the "Proem to Genesis" remains which I would gladly pass over in silence, were such a course consistent with the respect due to so dis-tinguished a champion of the "reconcilers."

I hope that my clients—the people of average opinions—have by this time some confidence in me; for when I tell them that, after all, Mr. Gladstone is of opinion that the "Mosaic record" was meant to give moral, and not scientific, instruction to those for whom it was written, they may be disposed to think that I must be misleading them. But let them listen further to what Mr. Gladstone says in a compendious but not exactly correct statement respecting my opinions :—

He holds the writer responsible for scientific precision : I look for nothing of the kind, but assign to him a statement general, which admits exceptions : popular, which aims mainly at producing moral impression ; summary, which cannot but be open to more or less of criticism of detail. He thinks it is a lecture. I think it is a sermon (p. 5).

I note, incidentally, that Mr. Gladstone appears to consider that the *differentia* between a lecture and a sermon is, that the former, so far as it deals with matters of fact, may be taken seriously, as meaning exactly what it says, while a sermon may not. I have quite enough on my hands without taking up the cudgels for the clergy, who will probably find Mr. Gladstone's definition unflattering.

But I am diverging from my proper business, which is to say that I have given no ground for the ascription of these opinions ; and that, as a matter of fact, I do not hold them and never have held them. It is Mr. Gladstone, and not I, who will have it that the pentateuchal cosmogony is to be taken as science.

My belief, on the contrary, is, and long has been, that the pentateuchal story of the creation is simply a myth. I suppose it to be an hypothesis respecting the origin of the universe which some ancient thinker found himself able to reconcile with his knowledge, or what he thought was knowledge, of the nature of things, and therefore assumed to be true. As such, I hold it to be not merely an interesting, but a venerable, monument of a stage in the mental progress of mankind ; and I find it difficult to suppose that any one who is acquainted with the cosmogonies of other

nations—and especially with those of the Egyptians
and the Babylonians, with whom the Israelites were
in such frequent and intimate communication—
should consider it to possess either more, or less,
scientific importance than may be allotted to these.

Mr. Gladstone's definition of a sermon permits me
to suspect that he may not see much difference between that form of discourse and what I call a
myth; and I hope it may be something more than the
slowness of apprehension, to which I have confessed,
which leads me to imagine that a statement
which is " general " but " admits exceptions," which is
" popular " and " aims mainly at producing moral
impression," " summary " and therefore open to " criticism of detail," amounts to a myth, or perhaps less
than a myth.   Put algebraically, it comes to this,
$x = a + b + c$; always remembering that there is nothing to show the exact value of either a, or b, or c.
It is true that a is commonly supposed to equal 10,
but there are exceptions, and these may reduce it to
8, or 3, or 0 ; b also popularly means 10, but being
chiefly used by the algebraist as a " moral " value,
you cannot do much with it in the addition or sub-
traction of mathematical values; c also is quite
" summary," and if you go into the details of which
it is made up, many of them may be wrong, and their
sum total equal to 0, or even to a minus quantity.

Mr. Gladstone appears to wish that I should (1)
enter upon a sort of essay competition with the
author of the pentateuchal cosmogony ; (2) that I
should make a further statement about some ele-

mentary facts in the history of Indian and Greek philosophy ; and (3) that I should show cause for my hesitation in accepting the assertion that Genesis is supported, at any rate to the extent of the first two verses, by the nebular hypothesis.

A certain sense of humour prevents me from accepting the first invitation. I would as soon attempt to put Hamlet's soliloquy into a more scientific shape. But if I supposed the " Mosaic writer" to be inspired, as Mr. Gladstone does, it would not be consistent with my notions of respect for the Supreme Being to imagine Him unable to frame a form of words which should accurately, or, at least, not inaccurately, express His own meaning. It is sometimes said that, had the statements contained in the first chapter of Genesis been scientifically true, they would have been unintelligible to ignorant people ; but how is the matter mended if, being scientifically untrue, they must needs be rejected by instructed people ?

With respect to the second suggestion, it would be presumptuous in me to pretend to instruct Mr. Gladstone in matters which lie as much within the province of Literature and History as in that of Science ; but if any one desirous of further knowledge will be so good as to turn to that most excellent and by no means recondite source of information, the *Encyclopædia Britannica*, he will find, under the letter E, the word " Evolution," and a long article on that subject. Now, I do not recommend him to read the first half of the article ; but the second half, by my friend Mr. Sully, is really very good.   He will there find it said that in some of the philosophies of ancient India, the idea of evolution is clearly expressed: "Brahma is conceived as the eternal self-existent being, which, on its material side, unfolds itself to the world by gradually condensing itself to material objects through the gradations of ether, fire, water, earth, and other elements."   And again: "In the later system of emanation of Sankhya there is a more marked approach to a materialistic doctrine of evolution."   What little knowledge I have of the matter—chiefly derived from that very instructive book, *Die Religion des Buddha*, by C. F. Koeppen, supplemented by Hardy's interesting works—leads me to think that Mr. Sully might have spoken much more strongly as to the evolutionary character of Indian philosophy, and especially of that of the Buddhists. But the question is too large to be dealt with incidentally.

And, with respect to early Greek philosophy,[1] the seeker after additional enlightenment need go no further than the same excellent storehouse of information :—

The early Ionian physicists, including Thales, Anaximander, and Anaximenes, seek to explain the world as generated out of a primordial matter which is at the same time the universal support of things.   This substance is endowed with a generative or transmutative force by virtue of which it passes into a

---

[1] I said nothing about "the greater number of schools of Greek philosophy," as Mr. Gladstone implies that I did, but expressly spoke of the "founders of Greek philosophy."

succession of forms.   They thus resemble modern evolutionists, since they regard the world, with its infinite variety of forms, as issuing from a simple mode of matter.

Further on, Mr. Sully remarks that "Heraclitus deserves a prominent place in the history of the idea of evolution," and he states, with perfect justice, that Heraclitus has foreshadowed some of the special peculiarities of Mr. Darwin's views. It is indeed a very strange circumstance that the philosophy of the great Ephesian more than adumbrates the two doctrines which have played leading parts, the one in the development of Christian dogma, the other in that of natural science. The former is the conception of the Word (λόγος) which took its Jewish shape in Alexandria, and its Christian form [1] in that Gospel which is usually referred to an Ephesian source of some five centuries later date; and the latter is that of the struggle for existence.   The saying that "strife is father and king of all" (πόλεμος πάντων μὲν πατήρ ἐστι, πάντων δὲ βασιλεύς), ascribed to Heraclitus, would be a not inappropriate motto for the "Origin of Species."

I have referred only to Mr. Sully's article, because his authority is quite sufficient for my purpose.   But the consultation of any of the more elaborate histories of Greek philosophy, such as the great work of Zeller, for example, will only bring out the same fact into still more striking prominence.   I have professed no "minute acquaintance" with either Indian or Greek philosophy, but I have taken a great deal of pains to

---

[1] See Heinze, *Die Lehre vom Logos*, p. 9 *et seq.*

secure that such knowledge as I do possess shall be accurate and trustworthy.

In the third place, Mr. Gladstone appears to wish that I should discuss with him the question whether the nebular hypothesis is, or is not, confirmatory of the pentateuchal account of the origin of things. Mr. Gladstone appears to be prepared to enter upon this campaign with a light heart. I confess I am not, and my reason for this backwardness will doubtless surprise Mr. Gladstone. It is that, rather more than a quarter of a century ago (namely, in February 1859), when it was my duty, as President of the Geological Society, to deliver the Anniversary Address,[1] I chose a topic which involved a very careful study of the remarkable cosmogonical speculation, originally promulgated by Immanuel Kant and, subsequently, by Laplace, which is now known as the nebular hypothesis. With the help of such little acquaintance with the principles of physics and astronomy as I had gained, I endeavoured to obtain a clear understanding of this speculation in all its bearings. I am not sure that I succeeded; but of this I am certain, that the problems involved are very difficult, even for those who possess the intellectual discipline requisite for dealing with them. And it was this conviction that led me to express my desire to leave the discussion of the question of the asserted harmony between Genesis and the nebular hypothesis to experts in the appropriate branches of knowledge. And I think my course was a

---

[1] Reprinted in *Lay Sermons, Addresses, and Reviews*, 1870.

wise one ; but as Mr. Gladstone evidently does
not understand how there can be any hesitation
on my part, unless it arises from a conviction that
he is in the right, I may go so far as to set out my
difficulties.

They are of two kinds—exegetical and scientific.
It appears to me that it is vain to discuss a
supposed coincidence between Genesis and science
unless we have first settled, on the one hand, what
Genesis says, and, on the other hand, what science
says.

In the first place, I cannot find any consensus
among Biblical scholars as to the meaning of the
words, " In the beginning God created the heaven and
the earth."   Some say that the Hebrew word *bara*,
which is translated " create," means " made out of
nothing."   I venture to object to that rendering,
not on the ground of scholarship, but of common
sense.   Omnipotence itself can surely no more make
something " out of " nothing than it can make a
triangular circle.   What is intended by " made out
of nothing " appears to be " caused to come into
existence," with the implication that nothing of the
same kind previously existed.   It is further usually
assumed that " the heaven and the earth " means the
material substance of the universe.   Hence the
" Mosaic writer " is taken to imply that where
nothing of a material nature previously existed, this
substance appeared.   That is perfectly conceivable,
and therefore no one can deny that it may have
happened.   But there are other very authoritative

critics who say that the ancient Israelite[1] who wrote the passage was not likely to have been capable of such abstract thinking; and that, as a matter of philology, *bara* is commonly used to signify the "fashioning," or "forming," of that which already exists. Now it appears to me that the scientific investigator is wholly incompetent to say anything at all about the first origin of the material universe. The whole power of his organon vanishes when he has to step beyond the chain of natural causes and effects. No form of the nebular hypothesis, that I know of, is necessarily connected with any view of the origination of the nebular substance. Kant's form of it expressly supposes that the nebular material from which one stellar system starts may be nothing but the disintegrated substance of a stellar and planetary system which has just come to an end. Therefore, so far as I can see, one who believes that matter has existed from all eternity has just as much right to hold the nebular hypothesis as one who believes that matter came into existence at a specified epoch. In other words, the nebular hypothesis and the creation hypothesis, up to this point, neither confirm nor oppose one another.

Next, we read in the revisers' version, in which I suppose the ultimate results of critical scholarship to be embodied: "And the earth was waste ['without form,' in the Authorised Version] and void." Most

---

[1] "Ancient," doubtless, but his antiquity must not be exaggerated. For example, there is no proof that the "Mosaic" cosmogony was known to the Israelites of Solomon's time.

people seem to think that this phraseology intends to imply that the matter out of which the world was to be formed was a veritable " chaos," devoid of law and order.   If this interpretation is correct, the nebular hypothesis can have nothing to say to it.   The scientific thinker cannot admit the absence of law and order, anywhere or anywhen, in nature.   Sometimes law and order are patent and visible to our limited vision; sometimes they are hidden.   But every particle of the matter of the most fantastic-looking nebula in the heavens is a realm of law and order in itself; and, that it is so, is the essential condition of the possibility of solar and planetary evolution from the apparent chaos.[1]

"Waste" is too vague a term to be worth consideration.   "Without form," intelligible enough as a metaphor, if taken literally, is absurd; for a material thing existing in space must have a superficies, and if it has a superficies it has a form.   The wildest streaks of marestail clouds in the sky, or the most irregular heavenly nebulæ, have surely just as much form as a geometrical tetrahedron; and as for "void," how can that be void which is full of matter?   As poetry, these lines are vivid and admirable; as a scientific statement, which they must be taken to be if any one is justified in comparing them with another scientific statement, they fail to convey any intelligible conception to my mind.

[1] When Jeremiah (iv. 23) says, "I beheld the earth, and, lo, it was waste and void," he certainly does not mean to imply that the form of the earth was less definite, or its substance less solid, than before.

The account proceeds : " And darkness was upon the face of the deep." So be it ; but where, then, is the likeness to the celestial nebulæ, of the existence of which we should know nothing unless they shone with a light of their own ? " And the spirit of God moved upon the face of the waters." I have met with no form of the nebular hypothesis which involves anything analogous to this process.

I have said enough to explain some of the difficulties which arise in my mind, when I try to ascertain whether there is any foundation for the contention that the statements contained in the first two verses of Genesis are supported by the nebular hypothesis. The result does not appear to me to be exactly favourable to that contention. The nebular hypothesis assumes the existence of matter, having definite properties, as its foundation. Whether such matter was created a few thousand years ago, or whether it has existed through an eternal series of metamorphoses of which our present universe is only the last stage, are alternatives, neither of which is scientifically untenable, and neither scientifically demonstrable. But science knows nothing of any stage in which the universe could be said, in other than a metaphorical and popular sense, to be formless or empty; or in any respect less the seat of law and order than it is now. One might as well talk of a fresh-laid hen's egg being " without form and void," because the chick therein is potential and not actual, as apply such terms to the nebulous mass which contains a potential solar system.

Until some further enlightenment comes to me, then, I confess myself wholly unable to understand the way in which the nebular hypothesis is to be converted into an ally of the " Mosaic writer." [1]

But Mr. Gladstone informs us that Professor Dana and Professor Guyot are prepared to prove that the " first or cosmogonical portion of the Proem not only accords with, but teaches, the nebular hypothesis." There is no one to whose authority on geological questions I am more readily disposed to bow than that of my eminent friend Professor Dana. But I am familiar with what he has previously said on this topic in his well-known and standard work, into which, strangely enough, it does not seem to have occurred to Mr. Gladstone to look before he set out upon his present undertaking ; and unless Professor Dana's latest contribution (which I have not yet met with) takes up altogether new ground, I am afraid I shall

---

[1] In looking through the delightful volume recently published by the Astronomer Royal for Ireland, a day or two ago, I find the following remarks on the nebular hypothesis, which I should have been glad to quote in my text if I had known them sooner :—

" Nor can it be ever more than a speculation ; it cannot be established by observation, nor can it be proved by calculation. It is merely a conjecture, more or less plausible, but perhaps, in some degree, necessarily true, if our present laws of heat, as we understand them, admit of the extreme application here required, and if the present order of things has reigned for sufficient time without the intervention of any influence at present known to us " (*The Story of the Heavens*, p. 506).

Would any prudent advocate base a plea, either for or against revelation, upon the coincidence, or want of coincidence, of the declarations of the latter with the requirements of an hypothesis thus guardedly dealt with by an astronomical expert ?

not be able to extricate myself, by its help, from my present difficulties.

It is a very long time since I began to think about the relations between modern scientifically ascertained truths and the cosmogonical speculations of the writer of Genesis; and, as I think that Mr. Gladstone might have been able to put his case with a good deal more force if he had thought it worth while to consult the last chapter of Professor Dana's admirable *Manual of Geology*, so I think he might have been made aware that he was undertaking an enterprise of which he had not counted the cost, if he had chanced upon a discussion of the subject which I published in 1877.[1]

Finally, I should like to draw the attention of those who take interest in these topics to the weighty words of one of the most learned and moderate of Biblical critics :—

A propos de cette première page de la Bible, on a coutume de nos jours de disserter, à perte de vue, sur l'accord du récit mosaïque avec les sciences naturelles ; et comme celles-ci, tout éloignées qu'elles sont encore de la perfection absolue, ont rendu populaires et en quelque sorte irréfragables un certain nombre de faits généraux ou de thèses fondamentales de la cosmologie et de la géologie, c'est le texte sacré qu'on s'évertue à torturer pour le faire concorder avec ces données.[2]

In my paper on the " Interpreters of Nature and the Interpreters of Genesis," while freely availing myself of the rights of a scientific critic, I endeavoured to keep the expression of my views well within those

[1] Lectures on Evolution delivered in New York (American Addresses).

[2] Reuss, *L'Histoire Sainte et la Loi*, vol. i. p. 275.

bounds of courtesy which are set by self-respect and consideration for others. I am therefore glad to be favoured with Mr. Gladstone's acknowledgment of the success of my efforts. I only wish that I could accept all the products of Mr. Gladstone's gracious appreciation, but there is one about which, as a matter of honesty, I hesitate. In fact, if I had expressed my meaning better than I seem to have done, I doubt if this particular proffer of Mr. Gladstone's thanks would have been made.

To my mind, whatever doctrine professes to be the result of the application of the accepted rules of inductive and deductive logic to its subject-matter; and accepts, within the limits which it sets to itself, the supremacy of reason, is Science. Whether the subject-matter consists of realities or unrealities, truths or falsehoods, is quite another question. I conceive that ordinary geometry is science, by reason of its method, and I also believe that its axioms, definitions, and conclusions are all true. However, there is a geometry of four dimensions, which I also believe to be science, because its method professes to be strictly scientific. It is true that I cannot conceive four dimensions in space, and therefore, for me, the whole affair is unreal. But I have known men of great intellectual powers who seemed to have no difficulty either in conceiving them, or, at any rate, in imagining how they could conceive them; and, therefore, four-dimensioned geometry comes under my notion of science. So I think astrology is a science, in so far as it professes to reason logically from principles established by just inductive methods. To prevent misunderstanding, perhaps I had better add that I do not believe one whit in astrology ; but no more do I believe in Ptolemaic astronomy, or in the catastrophic geology of my youth, although these, in their day, claimed—and, to my mind, rightly claimed—the name of science. If nothing is to be called science but that which is exactly true from beginning to end, I am afraid there is very little science in the world outside mathematics. Among the physical sciences, I do not know that any could claim more than that it is true within certain limits, so narrow that, for the present at any rate, they may be neglected. If such is the case, I do not see where the line is to be drawn between exactly true, partially true, and mainly untrue forms of science. And what I have said about the current theology at the end of my paper [p. 95] leaves, I think, no doubt as to the category in which I rank it. For all that, I think it would be not only unjust, but almost impertinent, to refuse the name of science to the *Summa* of St. Thomas or to the *Institutes* of Calvin.

In conclusion, I confess that my supposed "un-jaded appetite" for the sort of controversy in which it needed not Mr. Gladstone's express declaration to tell us he is far better practised than I am (though probably, without another express declaration, no one would have suspected that his controversial fires are burning low) is already satiated.

In "Elysium" we conduct scientific discussions in

a different medium, and we are liable to threatenings
of asphyxia in that " atmosphere of contention " in
which Mr. Gladstone has been able to live, alert and
vigorous beyond the common race of men, as if it
were purest mountain air.   I trust that he may long
continue to seek truth, under the difficult conditions
he has chosen for the search, with unabated energy—
I had almost said fire—

> May age not wither him, nor custom stale
> His infinite variety.

But Elysium suits my less robust constitution better,
and I beg leave to retire thither, not sorry for my
experience of the other region—no one should regret
experience—but determined not to repeat it, at any
rate in reference to the " plea for revelation."

NOTE ON THE PROPER SENSE OF THE " MOSAIC " NARRATIVE
OF THE CREATION.

It has been objected to my argument from Leviticus (p. 103),
that the Hebrew words translated by " creeping things " in
Genesis i. 24 and Leviticus xi. 29, are different; namely,
" reh-mes " in the former, " sheh-retz " in the latter.   The obvious
reply to this objection is that the question is not one of words
but of the meaning of words.   To borrow an illustration from
our own language, if " crawling things " had been used by the
translators in Genesis and " creeping things " in Leviticus, it
would not have been necessarily implied that they intended to
denote different groups of animals.   " Sheh-retz " is employed in
a wider sense than " reh-mes."   There are " sheh-retz " of the
waters, of the earth, of the air, and of the land.   Leviticus speaks
of land reptiles, among other animals, as " sheh-retz " ; Genesis
speaks of all creeping land animals, among which land reptiles are
necessarily included, as " reh-mes."   Our translators, therefore,

have given the true sense when they render both "sheh-retz" and " reh-mes " by "creeping things."

Having taken a good deal of trouble to show what Genesis i.-ii. 4 does not mean, in the preceding pages, perhaps it may be well that I should briefly give my opinion as to what it does mean. I conceive that the unknown author of this part of the Hexateuchal compilation believed, and meant his readers to believe, that his words, as they understood them—that is to say, in their ordinary natural sense—conveyed the "actual historical truth." When he says that such and such things happened, I believe him to mean that they actually occurred and not that he imagined or dreamed them ; when he says "day," I believe he uses the word in the popular sense ; when he says "made " or "created," I believe he means that they came into being by a process analogous to that which the people whom he addressed called "making" or "creating"; and I think that, unless we forget our present knowledge of nature, and, putting ourselves back into the position of a Phœnician or a Chaldæan philosopher, start from his conception of the world, we shall fail to grasp the meaning of the Hebrew writer. We must conceive the earth to be an immovable, more or less flattened, body, with the vault of heaven above, the watery abyss below and around. We must imagine sun, moon, and stars to be "set " in a "firmament " with, or in, which they move ; and above which is yet another watery mass. We must consider "light" and "darkness" to be things, the alternation of which constitutes day and night, independently of the existence of sun, moon, and stars. We must further suppose that, as in the case of the story of the deluge, the Hebrew writer was acquainted with a Gentile (probably Chaldæan or Accadian) account of the origin of things, in which he substantially believed, but which he stripped of all its idolatrous associations by substituting "Elohim " for Ea, Anu, Bel, and the like.

From this point of view the first verse strikes the keynote of the whole. In the beginning "Elohim [1] created the heaven and the earth." Heaven and earth were not primitive existences from which the gods proceeded, as the Gentiles taught ; on the contrary, the "Powers " preceded and created heaven and earth.

---

[1] For the sense of the term "Elohim," see p. 141.

Whether by "creation" is meant "causing to be where nothing was before" or "shaping of something which pre-existed," seems to me to be an insoluble question.

As I have pointed out, the second verse has an interesting parallel in Jeremiah iv. 23 : "I beheld the earth, and, lo, it was waste and void; and the heavens, and they had no light." I conceive that there is no more allusion to chaos in the one than in the other. The earth-disk lay in its watery envelope, like the yolk of an egg in the *glaire*, and the spirit, or breath, of Elohim stirred the mass. Light was created as a thing by itself ; and its antithesis "darkness" as another thing. It was supposed to be the nature of these two to alternate, and a pair of alternations constituted a "day" in the sense of an unit of time.

The next step was, necessarily, the formation of that "firmament," or dome over the earth-disk, which was supposed to support the celestial waters ; and in which sun, moon, and stars were conceived to be set, as in a sort of orrery. The earth was still surrounded and covered by the lower waters, but the upper were separated from it by the "firmament," beneath which what we call the air lay. A second alternation of darkness and light marks the lapse of time.

After this, the waters which covered the earth-disk, under the firmament, were drawn away into certain regions, which became seas, while the part laid bare became dry land. In accordance with the notion, universally accepted in antiquity, that moist earth possesses the potentiality of giving rise to living beings, the land, at the command of Elohim, "put forth" all sorts of plants. They are made to appear thus early, not, I apprehend, from any notion that plants are lower in the scale of being than animals (which would seem to be inconsistent with the prevalence of tree worship among ancient people), but rather because animals obviously depend on plants ; and because, without crops and harvests, there seemed to be no particular need of heavenly signs for the seasons.

These were provided by the fourth day's work. Light existed already ; but now vehicles for the distribution of light, in a special manner and with varying degrees of intensity, were provided. I conceive that the previous alternations of light and darkness were supposed to go on ; but that the "light" was strengthened during the daytime by the sun, which, as a source of heat as well as of light, glided up the firmament from the east, and slid down in the west, each day. Very probably each day's sun was supposed to be a new one. And, as the light of the day was strengthened by the sun, so the darkness of the night was weakened by the moon, which regularly waxed and waned every month. The stars are, as it were, thrown in. And nothing can more sharply mark the doctrinal purpose of the author, than the manner in which he deals with the heavenly bodies, which the Gentiles identified so closely with their gods, as if they were mere accessories to the almanac.

Animals come next in order of creation, and the general notion of the writer seems to be that they were produced by the medium in which they live; that is to say, the aquatic animals by the waters and the terrestrial animals by the land. But there was a difficulty about flying things, such as bats, birds, and insects. The cosmogonist seems to have had no conception of " air " as an elemental body. His " elements " are earth and water, and he ignores air as much as he does fire. Birds " fly above the earth in the open firmament " or " on the face of the expanse " of heaven. They are not said to fly through the air. The choice of a generative medium for flying things, there-fore, seemed to lie between water and earth ; and, if we take into account the conspicuousness of the great flocks of water-birds and the swarms of winged insects, which appear to arise from water, I think the preference of water becomes intelligible. However, I do not put this forward as more than a probable hypothesis. As to the creation of aquatic animals on the fifth, that of land animals on the sixth day, and that of man last of all, I presume the order was determined by the fact that man could hardly receive dominion over the living world before it existed ; and that the " cattle " were not wanted until he was about to make his appearance. The other terrestrial animals would naturally be associated with the cattle.

The absurdity of imagining that any conception, analogous to that of a zoological classification, was in the mind of the writer will be apparent, when we consider that the fifth day's work must include the zoologist's *Cetacea*, *Sirenia*, and seals,[1] all of which are

---

[1] Perhaps even hippopotamuses and otters !

Mammalia; all birds, turtles, sea-snakes and, presumably, the fresh
water *Reptilia* and *Amphibia*; with the great majority of *Invertebrata*.

The creation of man is announced as a separate act, resulting
from a particular resolution of Elohim to "make man in our
image, after our likeness." To learn what this remarkable
phrase means we must turn to the fifth chapter of Genesis, the
work of the same writer. "In the day that Elohim created
man, in the likeness of Elohim made he him; male and female
created he them; and blessed them and called their name Adam
in the day when they were created. And Adam lived an
hundred and thirty years and begat *a son* in his own likeness,
after his image; and called his name Seth." I find it impossible
to read this passage without being convinced that, when the
writer says Adam was made in the likeness of Elohim, he means
the same sort of likeness as when he says that Seth was be-
gotten in the likeness of Adam. Whence it follows that his
conception of Elohim was completely anthropomorphic.

In all this narrative I can discover nothing which differen-
tiates it, in principle, from other ancient cosmogonies, except
the rejection of all gods, save the vague, yet anthropomorphic,
Elohim, and the assigning to them anteriority and superiority to
the world. It is as utterly irreconcilable with the assured truths of
modern science, as it is with the account of the origin of man,
plants, and animals given by the writer of the second chief
constituent of the Hexateuch in the second chapter of Genesis.
This extraordinary story starts with the assumption of the
existence of a rainless earth, devoid of plants and herbs of the
field. The creation of living beings begins with that of a solitary
man; the next thing that happens is the laying out of the
Garden of Eden, and the causing the growth from its soil of
every tree "that is pleasant to the sight and good for food";
the third act is the formation out of the ground of "every beast
of the field, and every fowl of the air"; the fourth and last, the
manufacture of the first woman from a rib, extracted from Adam,
while in a state of anæsthesia.

Yet there are people who not only profess to take this monstrous legend seriously; but who declare it to be reconcilable
with the Elohistic account of the creation!

THE EVOLUTION OF THEOLOGY: AN ANTHRO-POLOGICAL STUDY

I CONCEIVE that the origin, the growth, the decline, and the fall of those speculations respecting the existence, the powers, and the dispositions of beings analogous to men, but more or less devoid of corporeal qualities, which may be broadly included under the head of theology, are phenomena the study of which legitimately falls within the province of the anthropologist. And it is purely as a question of anthropology (a department of biology to which I have at various times given a good deal of attention) that I propose to treat of the evolution of theology in the following pages.

With theology as a code of dogmas which are to be believed, or at any rate repeated, under penalty of present or future punishment, or as a storehouse of anæsthetics for those who find the pains of life too hard to bear, I have nothing to do; and, so far as it may be possible, I shall avoid the expression of any opinion as to the objective truth or falsehood of the systems of theological speculation of which I may find occasion to speak. From my present point of view, theology is regarded as a natural product of the operations of the human mind, under the conditions of its existence, just as any other branch of science, or the arts of architecture, or music, or painting are such products. Like them, theology has a history. Like them also, it is to be met with in certain simple and rudimentary forms; and these can be connected by a multitude of gradations, which exist or have existed, among people of various ages and races, with the most highly developed theologies of past and present times. It is not my object to interfere, even in the slightest degree, with beliefs which anybody holds sacred; or to alter the conviction of any one who is of opinion that, in dealing with theology, we ought to be guided by considerations different from those which would be thought appropriate if the problem lay in the province of chemistry or of mineralogy. And if people of these ways of thinking choose to read beyond the present paragraph, the responsibility for meeting with anything they may dislike rests with them and not with me.

We are all likely to be more familiar with the theological history of the Israelites than with that of any other nation. We may therefore fitly make it the first object of our studies; and it will be convenient to commence with that period which lies between the invasion of Canaan and the early days of the monarchy, and answers to the eleventh and twelfth centuries B.C. or thereabouts. The evidence on which any conclusion as to the nature of Israelitic theology in those days must be based is wholly contained in the Hebrew Scriptures — an agglomeration of documents which certainly belong to very different ages, but of the exact dates and authorship of any one of which (except perhaps one or two of the prophetical writings) there is no evidence, either internal or external, so far as I can discover, of such a nature as to justify more than a confession of ignorance, or, at most, an approximate conclusion. In this venerable record of ancient life, miscalled a book, when it is really a library comparable to a selection of works from English literature between the times of Beda and those of Milton, we have the stratified deposits (often confused and even with their natural order inverted) left by the stream of the intellectual and moral life of Israel during many centuries. And, embedded in these strata, there are numerous remains of forms of thought which once lived, and which, though often unfortunately mere fragments, are of priceless value to the anthropologist. Our task is to rescue these from their relatively unimportant surroundings, and by careful comparison with existing forms of theology to make the dead world which they record live again. In other words, our problem is palæontological, and the method pursued must be the same as that employed in dealing with other fossil remains.

Among the richest of the fossiliferous strata to which I have alluded are the books of Judges and

Samuel.[1]   It has often been observed that these
writings stand out, in marked relief from those which
precede and follow them, in virtue of a certain archaic
freshness and of a greater freedom from traces of
late interpolation and editorial trimming.   Jephthah,
Gideon, and Samson are men of old heroic stamp, who
would look as much in place in a Norse Saga as where
they are ; and if the varnish-brush of later respect-
ability has passed over these memoirs of the mighty
men of a wild age, here and there, it has not succeeded in effacing, or even in seriously obscuring, the
essential characteristics of the theology traditionally
ascribed to their epoch.

There is nothing that I have met with in the
results of Biblical criticism inconsistent with the
conviction that these books give us a fairly trustworthy account of Israelitic life and thought in the
times which they cover ; and, as such, apart from the
great literary merit of many of their episodes, they
possess the interest of being, perhaps, the oldest
genuine history, as apart from mere chronicles on the
one hand and mere legends on the other, at present
accessible to us.

But it is often said with exultation by writers of

[1] Even the most sturdy believers in the popular theory that the
proper or titular names attached to the books of the Bible are those
of their authors will hardly be prepared to maintain that Jephthah,
Gideon, and their colleagues wrote the book of Judges.   Nor is it
easily admissible that Samuel wrote the two books which pass under
his name, one of which deals entirely with events which took place
after his death.   In fact, no one knows who wrote either Judges or
Samuel, nor when, within the range of 100 years, their present form
was given to these books.

one party, and often admitted, more or less unwillingly, by their opponents, that these books are untrustworthy, by reason of being full of obviously unhistoric tales. And, as a notable example, the narrative of Saul's visit to the so-called " witch of Endor " is often cited. As I have already intimated, I have nothing to do with theological partisanship, either heterodox or orthodox, nor, for my present purpose, does it matter very much whether the story is historically true, or whether it merely shows what the writer believed ; but, looking at the matter solely from the point of view of an anthropologist, I beg leave to express the opinion that the account of Saul's necromantic expedition is quite consistent with probability. That is to say, I see no reason whatever to doubt, firstly, that Saul made such a visit ; and, secondly, that he and all who were present, including the wise woman of Endor herself, would have given, with entire sincerity, very much the same account of the business as that which we now read in the twenty-eighth chapter of the first book of Samuel ; and I am further of opinion that this story is one of the most important of those fossils, to which I have referred, in the material which it offers for the reconstruction of the theology of the time. Let us therefore study it attentively—not merely as a narrative which, in the dramatic force of its gruesome simplicity, is not surpassed, if it is equalled, by the witch scenes in Macbeth—but as a piece of evidence bearing on an important anthropological problem.

We are told (1 Sam. xxviii.) that Saul, encamped

at Gilboa, became alarmed by the strength of the Philistine army gathered at Shunem. He therefore "inquired of Jahveh," but "Jahveh answered him not, neither by dreams, nor by Urim, nor by prophets." [1] Thus deserted by Jahveh, Saul, in his extremity, bethought him of "those that had familiar spirits, and the wizards," whom he is said, at some previous time, to have "put out of the land"; but who seem, nevertheless, to have been very imperfectly banished, since Saul's servants, in answer to his command to seek him a woman "that hath a familiar spirit," reply without a sign of hesitation or of fear, "Behold, there is a woman that hath a familiar spirit at Endor"; just as, in some parts of England, a countryman might tell any one who did not look like a magistrate or a policeman, where a "wise woman" was to be met with. Saul goes to this woman, who, after being assured of immunity, asks, "Whom shall I bring up to thee?" whereupon Saul says, "Bring me up Samuel." The woman immediately sees an apparition. But to Saul nothing is visible, for he asks, "What seest thou?" And the woman replies, "I see Elohim coming up out of the earth." Still the spectre remains invisible to Saul, for he asks, "What form is he of?" And she replies, "An old man cometh up, and he is covered with a robe." So far, therefore, the wise woman unquestionably plays the part of a "medium," and Saul is dependent upon her version of what happens.

---

[1] My citations are taken from the Revised Version, but for LORD and GOD I have substituted Jahveh and Elohim.

THE EVOLUTION OF THEOLOGY

The account continues :—

And Saul perceived that it was Samuel, and he bowed with his face to the ground and did obeisance. And Samuel said to Saul, Why hast thou disquieted me to bring me up? And Saul answered, I am sore distressed: for the Philistines make war against me, and Elohim is departed from me and answereth me no more, neither by prophets nor by dreams; therefore I have called thee that thou mayest make known unto me what I shall do. And Samuel said, Wherefore then dost thou ask of me, seeing that Jahveh is departed from thee and is become thine adversary? And Jahveh hath wrought for himself, as he spake by me, and Jahveh hath rent the kingdom out of thine hand and given it to thy neighbour, even to David. Because thou obeyedst not the voice of Jahveh and didst not execute his fierce wrath upon Amalek, therefore hath Jahveh done this thing unto thee this day. Moreover, Jahveh will deliver Israel also with thee into the hand of the Philistines; and to-morrow shalt thou and thy sons be with me: Jahveh shall deliver the host of Israel also into the hand of the Philistines. Then Saul fell straightway his full length upon the earth and was sore afraid because of the words of Samuel . . . (v. 14-20).

The statement that Saul "perceived" that it was Samuel is not to be taken to imply that, even now, Saul actually saw the shade of the prophet, but only that the woman's allusion to the prophetic mantle and to the aged appearance of the spectre convinced him that it was Samuel. Reuss [1] in fact translates the passage "Alors Saul reconnut que c'était Samuel."

[1] I need hardly say that I depend upon authoritative Biblical critics, whenever a question of interpretation of the text arises. As Reuss appears to me to be one of the most learned, acute, and fair-minded of those whose works I have studied, I have made most use of the commentary and dissertations in his splendid French edition of the Bible. But I have also had recourse to the works of Dillman, Kalisch, Kuenen, Thenius, Tuch, and others, in cases in which another opinion seemed desirable.

Nor does the dialogue between Saul and Samuel necessarily, or probably, signify that Samuel spoke otherwise than by the voice of the wise woman. The Septuagint does not hesitate to call her ἐγγαστρίμυθος, that is to say, a ventriloquist, implying that it was she who spoke—and this view of the matter is in harmony with the fact that the exact sense of the Hebrew words which are translated as " a woman that hath a familiar spirit " is " a woman mistress of *Ob.*" *Ob* means primitively a leather bottle, such as a wine skin, and is applied alike to the necromancer and to the spirit evoked. Its use, in these senses, appears to have been suggested by the likeness of the hollow sound emitted by a half-empty skin when struck, to the sepulchral tones in which the oracles of the evoked spirits were uttered by the medium. It is most probable that, in accordance with the general theory of spiritual influences which obtained among the old Israelites, the spirit of Samuel was conceived to pass into the body of the wise woman, and to use her vocal organs to speak in his own name—for I cannot discover that they drew any clear distinction between possession and inspiration.[1]

If the story of Saul's consultation of the occult powers is to be regarded as an authentic narrative, or, at any rate, as a statement which is perfectly veracious so far as the intention of the narrator goes—and, as I have said, I see no reason for refusing it this character—it will be found, on further consideration,

[1] See " Divination," by Hazoral, *Journal of Anthropology*, Bombay, vol. i. No. 1.

to throw a flood of light, both directly and indirectly, on the theology of Saul's countrymen—that is to say, upon their beliefs respecting the nature and ways of spiritual beings.

Even without the confirmation of other abundant evidences to the same effect, it leaves no doubt as to the existence, among them, of the fundamental doctrine that man consists of a body and of a spirit, which last, after the death of the body, continues to exist as a ghost. At the time of Saul's visit to Endor, Samuel was dead and buried; but that his spirit would be believed to continue to exist in Sheol may be concluded from the well-known passage in the song attributed to Hannah, his mother :—

> Jahveh killeth and maketh alive,
> He bringeth down to Sheol and bringeth up
> (1 Sam. ii. 6).

And it is obvious that this Sheol was thought to be a place underground in which Samuel's spirit had been disturbed by the necromancer's summons, and in which, after his return thither, he would be joined by the spirits of Saul and his sons when they had met with their bodily death on the hill of Gilboa. It is further to be observed that the spirit, or ghost, of the dead man presents itself as the image of the man himself—it is the man not merely in his ordinary corporeal presentment (even down to the prophet's mantle) but in his moral and intellectual characteristics. Samuel, who had begun as Saul's friend and ended as his bitter enemy, gives it to be understood that he is annoyed at Saul's presumption in disturbing him ; and that, in Sheol, he is as much the devoted servant of Jahveh and as much empowered to speak in Jahveh's name as he was during his sojourn in the upper air.

It appears now to be universally admitted that, before the exile, the Israelites had no belief in rewards and punishments after death, nor in anything similar to the Christian heaven and hell ; but our story proves that it would be an error to suppose that they did not believe in the continuance of individual existence after death by a ghostly simulacrum of life. Nay, I think it would be very hard to produce conclusive evidence that they disbelieved in immortality ; for I am not aware that there is anything to show that they thought the existence of the souls of the dead in Sheol ever came to an end. But they do not seem to have conceived that the condition of the souls in Sheol was in any way affected by their conduct in life. If there was immortality, there was no state of retribution in their theology. Samuel expects Saul and his sons to come to him in Sheol.

The next circumstance to be remarked is that the name of *Elohim* is applied to the spirit which the woman sees " coming up out of the earth," that is to say, from Sheol. The Authorised Version translates this in its literal sense " gods." The Revised Version gives " god " with " gods " in the margin. Reuss renders the word by " spectre," remarking in a note that it is not quite exact ; but that the word Elohim expresses " something divine, that is to say, superhuman, commanding respect and terror " (*Histoire*

des Israelites, p. 321). Tuch, in his commentary on
Genesis, and Thenius, in his commentary on Samuel,
express substantially the same opinion. Dr. Alexander (in Kitto's *Cyclopædia* s. v. " God ") has the
following instructive remarks :—

> [*Elohim* is] sometimes used vaguely to describe unseen powers
> or superhuman beings that are not properly thought of as divine.
> Thus the witch of Endor saw " Elohim ascending out of the
> earth" (1 Sam. xxviii. 13), meaning thereby some beings of an
> unearthly, superhuman character. So also in Zechariah xii. 8, it is
> said "the house of David shall be as Elohim, as the angel of
> the Lord," where, as the transition from Elohim to the angel of
> the Lord is a minori ad majus, we must regard the former as a
> vague designation of supernatural powers.

Dr. Alexander speaks here of " beings "; but there
is no reason to suppose that the wise woman of
Endor referred to anything but a solitary spectre;
and it is quite clear that Saul understood her in this
sense, for he asks, " What form is HE of? "

This fact, that the name of Elohim is applied to a
ghost, or disembodied soul, conceived as the image of
the body in which it once dwelt, is of no little
importance. For it is well known that the same
term was employed to denote the gods of the heathen,
who were thought to have definite quasi - corporeal
forms and to be as much real entities as any other
Elohim.[1] The difference which was supposed to exist

[1] See, for example, the message of Jephthah to the King of the
Ammonites : " So now Jahveh, the Elohim of Israel, hath dis-
possessed the Amorites from before his people Israel, and shouldest
thou possess them ? Wilt not thou possess that which Chemosh, thy
Elohim, giveth thee to possess ? " (Jud. xi. 23, 24). For Jephthah,
Chemosh is obviously as real a personage as Jahveh.

between the different Elohim was one of degree, not one of kind. Elohim was, in logical terminology, the genus of which ghosts, Chemosh, Dagon, Baal, and Jahveh were species. The Israelite believed Jahveh to be immeasurably superior to all other kinds of Elohim. The inscription on the Moabite stone shows that King Mesa held Chemosh to be, as unquestionably, the superior of Jahveh. But if Jahveh was thus supposed to differ only in degree from the undoubtedly zoomorphic or anthropomorphic " gods of the nations," why is it to be assumed that he also was not thought of as having a human shape? It is possible for those who forget that the time of the great prophetic writers is at least as remote from that of Saul as our day is from that of Queen Elizabeth, to insist upon interpreting the gross notions current in the earlier age and among the mass of the people by the refined conceptions promulgated by a few select spirits centuries later. But if we take the language constantly used concerning the Deity in the books of Genesis, Exodus, Joshua, Judges, Samuel, or Kings, in its natural sense (and I am aware of no valid reason which can be given for taking it in any other sense), there cannot, to my mind, be a doubt that Jahveh was conceived by those from whom the substance of these books is mainly derived, to possess the appearance and the intellectual and moral attributes of a man; and, indeed, of a man of just that type with which the Israelites were familiar in their stronger and intellectually abler rulers and leaders. In a well-known passage of Genesis (i. 27) Elohim is said to have "created man in his own image, in the image of Elohim created he him." It is "man" who is here said to be the image of Elohim—not man's soul alone, still less his "reason," but the whole man. It is obvious that for those who called a manlike ghost Elohim, there could be no difficulty in conceiving any other Elohim under the same aspect. And if there could be any doubt on this subject, surely it cannot stand in the face of what we find in the fifth chapter, where, immediately after a repetition of the statement that " Elohim created man, in the likeness of Elohim made he him," it is said that Adam begat Seth "in his own likeness, after his image." Does this mean that Seth resembled Adam only in a spiritual and figurative sense? And if that interpretation of the third verse of the fifth chapter of Genesis is absurd, why does it become reasonable in the first verse of the same chapter?

But let us go further. Is not the Jahveh who "walks in the garden in the cool of the day"; from whom one may hope to " hide oneself among the trees"; of whom it is expressly said that " Moses and Aaron, Nadab and Abihu, and seventy of the elders of Israel," saw the Elohim of Israel (Exod. xxiv. 9-11); and that, although the seeing Jahveh was understood to be a high crime and misdemeanour, worthy of death, under ordinary circumstances, yet, for this once, he " laid not his hand on the nobles of Israel"; " that they beheld Elohim and did eat and drink"; and that afterwards Moses saw his back (Exod. xxxiii. 23) —is not this Deity conceived as manlike in form?

Again, is not the Jahveh who eats with Abraham
under the oaks at Mamre, who is pleased with the
" sweet savour " of Noah's sacrifice, to whom sacrifices
are said to be " food "[1]—is not this Deity depicted
as possessed of human appetites? If this were not
the current Israelitish idea of Jahveh even in the
eighth century B.C., where is the point of Isaiah's
scathing admonitions to his countrymen : " To what
purpose is the multitude of your sacrifices unto me?
saith Jahveh : I am full of the burnt-offerings of rams
and the fat of fed beasts; and I delight not in the
blood of bullocks, or of lambs, or of he-goats "
(Isa. i. 11). Or of Micah's inquiry, " Will Jahveh
be pleased with thousands of rams or with ten thousands of rivers of oil ? " (vi. 7) And in the innumerable passages in which Jahveh is said to be jealous
of other gods, to be angry, to be appeased, and to
repent; in which he is represented as casting off
Saul because the king does not quite literally execute
a command of the most ruthless severity ; or as smit-
ing Uzzah to death because the unfortunate man
thoughtlessly, but naturally enough, put out his hand
to stay the ark from falling—can any one deny that
the old Israelites conceived Jahveh not only in the
image of a man, but in that of a changeable, irritable,
and, occasionally, violent man ? There appears to
me, then, to be no reason to doubt that the notion of
likeness to man, which was indubitably held of the

---

[1] For example : " My oblation, my food for my offerings made by
fire, of a sweet savour to me, shall ye observe to offer unto me in their
due season " (Num. xxviii. 2).

ghost Elohim, was carried out consistently throughout the whole series of Elohim, and that Jahveh-Elohim was thought of as a being of the same substantially human nature as the rest, only immeasurably
more powerful for good and for evil.

The absence of any real distinction between the
Elohim of different ranks is further clearly illustrated
by the corresponding absence of any sharp delimitation between the various kinds of people who serve
as the media of communication between them and
men. The agents through whom the lower Elohim
are consulted are called necromancers, wizards, and
diviners, and are looked down upon by the prophets
and priests of the higher Elohim ; but the " seer "
connects the two, and they are all alike in their essential characters of media. The wise woman of Endor was
believéd by others, and, I have little doubt, believed
herself, to be able to " bring up " whom she would
from Sheol, and to be inspired, whether in virtue of
actual possession by the evoked Elohim, or otherwise,
with a knowledge of hidden things. I am unable to
see that Saul's servant took any really different view
of Samuel's powers, though he may have believed
that he obtained them by the grace of the higher
Elohim. For when Saul fails to find his father's
asses, his servant says to him—

Behold, there is in this city a man of Elohim, and he is a
man that is held in honour ; all that he saith cometh surely to
pass : now let us go thither ; peradventure he can tell us concerning our journey whereon we go. Then said Saul to his
servant, But behold if we go, what shall we bring the man ? for
the bread is spent in our vessels and there is not a present to

bring to the man of Elohim. What have we ? And the servant answered Saul again and said, Behold I have in my hand the fourth part of a shekel of silver: that will I give to the man of Elohim to tell us our way. (Beforetime in Israel when a man went to inquire of Elohim, then he said, Come and let us go to the Seer: for he that is now called a Prophet was beforetime called a Seer [1]) (1 Sam. ix. 6-10).

In fact, when, shortly afterwards, Saul accidentally meets Samuel, he says, " Tell me, I pray thee, where the Seer's house is." Samuel answers, " I am the Seer." Immediately afterwards Samuel informs Saul that the asses are found, though how he obtained his knowledge of the fact is not stated. It will be observed that Samuel is not spoken of here as, in any special sense, a seer or prophet of Jahveh, but as a " man of Elohim "—that is to say, a seer having access to the " spiritual powers," just as the wise woman of Endor might have been said to be a " woman of Elohim "—and the narrator's or editor's explanatory note seems to indicate that " Prophet " is merely a name, introduced later than the time of Samuel, for a superior kind of " Seer," or " man of Elohim."[2]

Another very instructive passage shows that Samuel was not only considered to be diviner, seer, and prophet in one, but that he was also, to all intents and purposes, priest of Jahveh — though,

[1] In 2 Samuel xv. 27 David says to Zadok the priest, " Art thou not a seer ? " and Gad is called David's seer.

[2] This would at first appear to be inconsistent with the use of the word " prophetess " for Deborah. But it does not follow because the writer of Judges applies the name to Deborah that it was used in her day.

according to his biographer, he was not a member of the tribe of Levi.   At the outset of their acquaintance, Samuel says to Saul, "Go up before me into the high place," where, as the young maidens of the city had just before told Saul, the Seer was going, "for the people will not eat till he come, because he doth bless the sacrifice" (1 Sam. x. 12).   The use of the word "bless" here—as if Samuel were not going to sacrifice, but only to offer a blessing or thanksgiving—is curious.   But that Samuel really acted as priest seems plain from what follows.   For he not only asks Saul to share in the customary sacrificial feast, but he disposes in Saul's favour of that portion of the victim which the Levitical legislation, doubtless embodying old customs, recognises as the priest's special property.[1]

Although particular persons adopted the profession of media between men and Elohim, there was no limitation of the power, in the view of ancient Israel, to any special class of the population.   Saul inquires of Jahveh and builds him altars on his own account ; and in the very remarkable story told in the four-

---

[1] Samuel tells the cook, "Bring the portion which I gave thee, of which I said to thee, Set it by thee."   It was therefore Samuel's to give.   "And the cook took up the thigh (or shoulder) and that which was upon it and set it before Saul."   But, in the Levitical regulations, it is the thigh (or shoulder) which becomes the priest's own property. "And the right thigh (or shoulder) shall ye give unto the priest for an heave-offering," which is given along with the wave breast "unto Aaron the priest and unto his sons as a due for ever from the children of Israel" (Lev. vii. 31-34).   Reuss writes on this passage : "La cuisse n'est point agitée, mais simplement *prelevée* sur ce que les convives mangeront."

teenth chapter of the first book of Samuel (v. 37-46),
Saul appears to conduct the whole process of divination, although he has a priest at his elbow. David
seems to do the same.

Moreover, Elohim constantly appear in dreams
—which in old Israel did not mean that, as we
should say, the subject of the appearance "dreamed
he saw the spirit"; but that he veritably saw the
Elohim which, as a soul, visited his soul while his
body was asleep. And, in the course of the history
of Israel, Jahveh himself thus appears to all sorts of
persons, non-Israelites as well as Israelites. Again,
the Elohim possess, or inspire, people against their
will, as in the case of Saul and Saul's messengers, and
then these people prophesy—that is to say, "rave"—
and exhibit the ungoverned gestures attributed by a
later age to possession by malignant spirits. Apart
from other evidence to be adduced by and by, the
history of ancient demonology and of modern re-
vivalism does not permit me to doubt that the
accounts of these phenomena given in the history of
Saul may be perfectly historical.

In the ritual practices, of which evidence is to be
found in the books of Judges and Samuel, the chief
part is played by sacrifices, usually burnt offerings.
Whenever the aid of the Elohim of Israel is sought,
or thanks are considered due to him, an altar is built,
and oxen, sheep, and goats are slaughtered and
offered up. Sometimes the entire victim is burnt
as a holocaust; more frequently only certain parts,
notably the fat about the kidneys, are burnt on the

altar. The rest is properly cooked; and, after the reservation of a part for the priest, is made the foundation of a joyous banquet, in which the sacrificer, his family, and such guests as he thinks fit to invite, participate.[1] Elohim was supposed to share in the feast, and it has been already shown that that which was set apart on the altar, or consumed by fire, was spoken of as the food of Elohim, who was thought to be influenced by the costliness, or by the pleasant smell, of the sacrifice in favour of the sacrificer.

All this bears out the view that, in the mind of the old Israelite, there was no difference, save one of degree, between one Elohim and another. It is true that there is but little direct evidence to show that the old Israelites shared the widespread belief of their own, and indeed of all times, that the spirits of the dead not only continue to exist, but are capable of a ghostly kind of feeding and are grateful for such aliment as can be assimilated by their attenuated substance, and even for clothes, ornaments, and weapons.[2] That they were familiar with this doctrine in the time of the captivity is suggested by the well-known reference of Ezekiel (xxxii. 27) to the "mighty that are fallen of the uncircumcised, which

[1] See, for example, Elkanah's sacrifice, 1 Sam. i. 3-9.

[2] The ghost was not supposed to be capable of devouring the gross material substance of the offering; but his vaporous body appropriated the smoke of the burnt sacrifice, the visible and odorous exhalations of other offerings. The blood of the victim was particularly useful because it was thought to be the special seat of its soul or life. A West African negro replied to an European sceptic: "Of course, the spirit cannot eat corporeal food, but he extracts its spiritual part, and, as we see, leaves the material part behind" (Lippert, *Seelencult*, p. 16).

are gone down to [Sheol] hell with their weapons of war, and have laid their swords under their heads." Perhaps there is a still earlier allusion in the " giving of food for the dead" spoken of in Deuteronomy (xxvi. 14).[1]

It must be remembered that the literature of the old Israelites, as it lies before us, has been subjected to the revisal of strictly monotheistic editors, violently opposed to all kinds of idolatry, who are not likely to have selected from the materials at their disposal any obvious evidence, either of the practice under discussion, or of that ancestor-worship which is so closely related to it, for preservation in the permanent records of their people.

The mysterious objects known as *Teraphim*, which are occasionally mentioned in Judges, Samuel, and elsewhere, however, can hardly be interpreted otherwise than as indications of the existence both of ancestor-worship and of image-worship in old Israel.

[1] It is further well worth consideration whether indications of former ancestor-worship are not to be found in the singular weight attached to the veneration of parents in the fourth commandment. It is the only positive commandment, in addition to those respecting the Deity and that concerning the Sabbath, and the penalties for infringing it were of the same character. In China, a corresponding reverence for parents is part and parcel of ancestor-worship ; so in ancient Rome and in Greece (where parents were even called δεύτεροι καὶ ἐπίγεοι θεοί). The fifth commandment, as it stands, would be an excellent compromise between ancestor-worship and monotheism. The larger hereditary share allotted by Israelitic law to the eldest son reminds one of the privileges attached to primogeniture in ancient Rome, which were closely connected with ancestor-worship. There is a good deal to be said in favour of the speculation that the ark of the covenant may have been a relic of ancestor-worship ; but that topic is too large to be dealt with incidentally in this place.

The teraphim were certainly images of family gods, and, as such, in all probability represented deceased ancestors. Laban indignantly demands of his son-in-law, "Wherefore hast thou stolen my Elohim?" which Rachel, who must be assumed to have worshipped Jacob's God, Jahveh, had carried off, obviously because she, like her father, believed in their divinity. It is not suggested that Jacob was in any way scandalised by the idolatrous practices of his favourite wife, whatever he may have thought of her honesty when the truth came to light; for the teraphim seem to have remained in his camp, at least until he "hid" his strange gods "under the oak that was by Shechem" (Gen. xxxv. 4). And indeed it is open to question if he got rid of them then, for the subsequent history of Israel renders it more than doubtful whether the teraphim were regarded as "strange gods" even as late as the eighth century B.C.

The writer of the books of Samuel takes it quite as a matter of course that Michal, daughter of one royal Jahveh worshipper and wife of the servant of Jahveh *par excellence*, the pious David, should have her teraphim handy, in her and David's chamber, when she dresses them up in their bed into a simulation of her husband, for the purpose of deceiving her father's messengers. Even one of the early prophets, Hosea, when he threatens that the children of Israel shall abide many days without "ephod or teraphim" (iii. 4), appears to regard both as equally proper appurtenances of the suspended worship of Jahveh, and equally certain to be restored when that is resumed.

When we further take into consideration that only in
the reign of Hezekiah was the brazen serpent, preserved
in the temple and believed to be the work of Moses,
destroyed, and the practice of offering incense to it,
that is, worshipping it, abolished—that Jeroboam
could set up "calves of gold" for Israel to worship,
with apparently none but a political object, and
certainly with no notion of creating a schism among
the worshippers of Jahveh, or of repelling the men of
Judah from his standard—it seems obvious, either
that the Israelites of the tenth and eleventh centuries
B.C. knew not the second commandment, or that they
construed it merely as part of the prohibition to
worship any supreme god other than Jahveh, which
precedes it.

In seeking for information about the teraphim, I
lighted upon the following passage in the valuable
article on that subject by Archdeacon Farrar, in
Kitto's *Cyclopædia of Biblical Literature*, which is
so much to the purpose of my argument, that I
venture to quote it in full :—

The main and certain results of this review are that the
teraphim were rude human images; that the use of them was
an antique Aramaic custom; that there is reason to suppose
them to have been images of deceased ancestors; that they were
consulted oracularly; that they were not confined to Jews; that
their use continued down to the latest period of Jewish history;
and lastly, that although the enlightened prophets and strictest
later kings regarded them as idolatrous, the priests were much
less averse to such images, and their cult was not considered in
any way repugnant to the pious worship of Elohim, nay, even to
the worship of him :"under the awful title of Jehovah." In fact,
they involved *a monotheistic idolatry very different indeed from poly-*

theism ; and the tolerance of them by priests, as compared with the denunciation of them by the prophets, offers a close analogy to the views of the Roman Catholics respecting pictures and images as compared with the views of Protestants. It was against this use of idolatrous symbols and emblems in a mono-theistic worship that the *second* commandment was directed, whereas the first is aimed against the graver sin of direct poly-theism. But the whole history of Israel shows how utterly and how early the law must have fallen into desuetude. The worship of the golden calf and of the calves at Dan and Bethel, against which, so far as we know, neither Elijah nor Elisha said a single word ; the tolerance of high places, teraphim and betylia ; the offering of incense for centuries to the brazen serpent destroyed by Hezekiah ; the occasional glimpses of the most startling irregularities sanctioned apparently even in the temple worship itself, prove most decisively that a pure monotheism and an independence of symbols was the result of a slow and painful course of God's disciplinal dealings among the noblest thinkers of a single nation, and not, as is so constantly and erroneously urged, the instinct of the whole Semitic race ; in other words, one single branch of the Semites was under God's providence *educated* into pure monotheism only by centuries of misfortune and series of inspired men (vol. iii. p. 986).

It appears to me that the researches of the anthro-pologist lead him to conclusions identical in sub-stance, if not in terms, with those here enunciated as the result of a careful study of the same subject from a totally different point of view.

There is abundant evidence in the books of Samuel and elsewhere that an article of dress termed an *ephod* was supposed to have a peculiar efficacy in enabling the wearer to exercise divination by means of Jahveh-Elohim. Great and long continued have been the disputes as to the exact nature of the ephod —whether it always means something to wear, or whether it sometimes means an image. But the probabilities are that it usually signifies a kind of waistcoat or broad zone, with shoulder-straps, which the person who "inquired of Jahveh" put on. In 1 Samuel xxiii. 2 David appears to have inquired without an ephod, for Abiathar the priest is said to have "come down with an ephod in his hand" only subsequently. And then David asks for it before inquiring of Jahveh whether the men of Keilah would betray him or not. David's action is obviously divination pure and simple; and it is curious that he seems to have worn the ephod himself and not to have employed Abiathar as a medium. How the answer was given is not clear, though the probability is that it was obtained by casting lots. The *Urim* and *Thummim* seem to have been two such lots of a peculiarly sacred character, which were carried in the pocket of the high priest's "breastplate." This last was worn along with the ephod.

With the exception of one passage (1 Sam. xiv. 18) the ark is ignored in the history of Saul. But in this place the Septuagint reads "ephod" for ark, while in 1 Chronicles xiii. 3 David says that "we sought not unto it [the ark] in the days of Saul." Nor does Samuel seem to have paid any regard to the ark after its return from Philistia; though, in his childhood, he is said to have slept in "the temple of Jahveh, where the ark of Elohim was" (1 Sam. iii. 3), at Shiloh, and there to have been the seer of the earliest apparitions vouchsafed to him by Jahveh. The space between the cherubim or winged images on the canopy or cover (*Kapporeth*) of this holy chest was held to be the special seat of Jahveh—the place selected for a temporary residence of the Supreme Elohim who had, after Aaron and Phineas, Eli and his sons for priests and seers. And, when the ark was carried to the camp at Eben-ezer, there can be no doubt that the Israelites, no less than the Philistines, held that " Elohim is come into the camp" (iv. 7), and that the one, as much as the other, conceived that the Israelites had summoned to their aid a powerful ally in " these (or this) mighty Elohim"—elsewhere called Jahve-Sabaoth, the Jahveh of Hosts. If the " temple" at Shiloh was the pentateuchal tabernacle, as is suggested by the name of " tent of meeting" given to it in 1 Samuel ii. 22, it was essentially a large tent, though constituted of very expensive and ornate materials; if, on the other hand, it was a different edifice, there can be little doubt that this " house of Jahveh" was built on the model of an ordinary house of the time. But there is not the slightest evidence that, during the reign of Saul, any greater importance attached to this seat of the cult of Jahveh than to others. Sanctuaries, and " high places" for sacrifice, were scattered all over the country from Dan to Beer-sheba. And, as Samuel is said to have gone up to one of these high places to bless the sacrifice, it may be taken for tolerably certain that he knew nothing of the Levitical laws which severely condemn the high places and those who sacrifice away from the sanctuary hallowed by the presence of the ark.

There is no evidence that, during the time of the

Judges and of Samuel, any one occupied the position
of the high priest of later days. And persons who
were neither priests nor Levites sacrificed and divined
or " inquired of Jahveh," when they pleased and where
they pleased, without the least indication that they,
or any one else in Israel at that time, knew they were
doing wrong. There is no allusion to any special
observance of the Sabbath ; and the references to
circumcision are indirect.

Such are the chief articles of the theological creed
of the old Israelites, which are made known to us by
the direct evidence of the ancient record to which we
have had recourse, and they are as remarkable for
that which they contain as for that which is absent
from them. They reveal a firm conviction that, when
death takes place, a something termed a soul or spirit
leaves the body and continues to exist in Sheol for a
period of indefinite duration, even though there is no
proof of any belief in absolute immortality ; that such
spirits can return to earth to possess and inspire the
living ; that they are, in appearance and in disposition,
likenesses of the men to whom they belonged, but
that, as spirits, they have larger powers and are freer
from physical limitations ; that they thus form a group
among a number of kinds of spiritual existences known
as Elohim, of whom Jahveh, the national God of Israel,
is one ; that, consistently with this view, Jahveh was
conceived as a sort of spirit, human in aspect and in
senses, and with many human passions, but with
immensely greater intelligence and power than any

other Elohim, whether human or divine. Further, the evidence proves that this belief was the basis of the Jahveh-worship to which Samuel and his followers were devoted; that there is strong reason for believing, and none for doubting, that idolatry, in the shape of the worship of the family gods or teraphim, was practised by sincere and devout Jahveh-worshippers; that the ark, with its protective tent or tabernacle, was regarded as a specially, but by no means exclusively, favoured sanctuary of Jahveh; that the ephod appears to have had a particular value for those who desired to divine by the help of Jahveh; and that divination by lots was practised before Jahveh. On the other hand, there is not the slightest evidence of any belief in retribution after death, but the contrary; ritual obligations have at least as strong sanction as moral; there are clear indications that some of the most stringent of the Levitical laws were unknown even to Samuel; priests often appear to be superseded by laymen, even in the performance of sacrifices and divination; and no line of demarcation can be drawn between necromancer, wizard, seer, prophet, and priest, each of whom is regarded, like all the rest, as a medium of communication between the world of Elohim and that of living men.

The theological system thus defined offers to the anthropologist no feature which is devoid of a parallel in the known theologies of other races of mankind, even of those who inhabit parts of the world most remote from Palestine. And the foundation of the whole, the ghost theory, is exactly that theological speculation which is the most widely spread of all, and the most deeply rooted among uncivilised men. I am able to base this statement, to some extent, on facts within my own knowledge. In December 1848, H.M.S. *Rattlesnake*, the ship to which I then belonged, was anchored off Mount Ernest, an island in Torres Straits. The people were few and well disposed; and, when a friend of mine (whom I will call B.) and I went ashore, we made acquaintance with an old native, Paouda by name. In course of time we became quite intimate with the old gentleman, partly by the rendering of mutual good offices, but chiefly because Paouda believed he had discovered that B. was his father-in-law. And his grounds for this singular conviction were very remarkable. We had made a long stay at Cape York hard by; and, in accordance with a theory which is widely spread among the Australians, that white men are the reincarnated spirits of black men, B. was held to be the ghost, or *narki*, of a certain Mount Ernest native, one Antarki, who had lately died, on the ground of some real or fancied resemblance to the latter. Now Paouda had taken to wife a daughter of Antarki's, named Domani, and as soon as B. informed him that he was the ghost of Antarki, Paouda at once admitted the relationship and acted upon it. For, as all the women on the island had hidden away in fear of the ship, and we were anxious to see what they were like, B. pleaded pathetically with Paouda that it would be very unkind not to let him see his daughter and grandchildren.

After a good deal of hesitation and the exaction of pledges of deep secrecy, Paouda consented to take B., and myself as B.'s friend, to see Domani and the three daughters, by whom B. was received quite as one of the family, while I was courteously welcomed on his account.

This scene made an impression upon me which is not yet effaced. It left no question on my mind of the sincerity of the strange ghost theory of these savages, and of the influence which their belief has on their practical life. I had it in my mind, as well as many a like result of subsequent anthropological studies, when, in 1869,[1] I wrote as follows :—

> There are savages without God in any proper sense of the word, but none without ghosts. And the Fetishism, Ancestor-worship, Hero-worship, and Demonology of primitive savages are all, I believe, different manners of expression of their belief in ghosts, and of the anthropomorphic interpretation of out-of-the-way events which is its concomitant. Witchcraft and sorcery are the practical expressions of these beliefs ; and they stand in the same relation to religious worship as the simple anthropomorphism of children or savages does to theology.

I do not quote myself with any intention of making a claim to originality in putting forth this view ; for I have since discovered that the same conception is virtually contained in the great *Discours sur l'Histoire Universelle* of Bossuet, now more than two centuries old :—

> Le culte des hommes morts faisoit presque tout le fond de l'idolâtrie : presque tous les hommes sacrifioient aux mânes, c'est-

---

[1] " The Scientific Aspects of Positivism," *Fortnightly Review*, 1869, republished in *Lay Sermons.*

à-dire aux âmes des morts. De si anciennes erreurs nous font
voir à la vérité combien étoit ancienne la croyance de l'immor-
talité de l'âme, et nous montrent qu'elle doit être rangée parmi
les premières traditions du genre humain. Mais l'homme, qui
gâtoit tout, en avoit étrangement abusé, puisqu'elle le portoit à
sacrifier aux morts. On alloit même jusqu'à cet excès, de leur
sacrifier des hommes vivans : on tuoit leurs esclaves, et même
leurs femmes, pour les aller servir dans l'autre monde.[1]

Among more modern writers J. G. Müller, in his
excellent *Geschichte der amerikanischen Urreligionen*
(1855), clearly recognises " gespensterhafter Geister-
glaube " as the foundation of all savage and semi-civilised theology, and I need do no more than mention
the important developments of the same view which
are to be found in Mr. Tylor's *Primitive Culture*, and
in the writings of Mr. Herbert Spencer, especially his
recently-published *Ecclesiastical Institutions*.[2]
    It is a matter of fact that, whether we direct our
attention to the older conditions of civilised societies,
in Japan, in China, in Hindostan, in Greece, or in
Rome,[3] we find underlying all other theological
notions the belief in ghosts, with its inevitable concomitant sorcery; and a primitive cult in the shape
of a worship of ancestors, which is essentially an
attempt to please, or appease, their ghosts.   The

---

[1] *Œuvres de Bossuet*, ed. 1808, t. xxxv. p. 282.

[2] I should like further to add the expression of my indebtedness
to two works by Herr Julius Lippert, *Der Seelencult in seinen Bezie-
hungen zur alt-hebraischen Religion*, and *Die Religionen der europäischen
Culturvölker*, both published in 1881. I have found them full of
valuable suggestions.

[3] See among others the remarkable work of Fustel de Coulanges,
La cité antique, in which the social importance of the old Roman
ancestor-worship is brought out with great clearness.

same thing is true of old Mexico and Peru, and of every semi-civilised or savage people who have developed a definite cult; and in those who, like the natives of Australia, have not even a cult, the belief in, and fear of, ghosts is as strong as anywhere else. The most clearly demonstrable article of the theology of the Israelites in the eleventh and twelfth centuries B.C. is therefore simply the article which is to be found in all primitive theologies, namely, the belief that a man has a soul which continues to exist after death for a longer or shorter time, and may return, as a ghost, with a divine, or at least demonic, character, to influence for good or evil (and usually for evil) the affairs of the living. But the correspondence between the old Israelitic and other archaic forms of theology extends to details. If, in order to avoid all chance of direct communication, we direct our attention to the theology of semi-civilised people, such as the Polynesian Islanders, separated by the greatest possible distance, and by every conceivable barrier, from the inhabitants of Palestine, we shall find not merely that all the features of old-Israelitic theology, which are revealed in the records cited, are found among them; but that extant information as to the inner mind of these people tends to remove many of the difficulties which those who have not studied anthropology find in the Hebrew narrative.

One of the best sources, if not the best source, of information on these topics is Mariner's *Tonga Islands*, which tells us of the condition of Cook's "Friendly Islanders" eighty years ago, before European influence was sensibly felt among them. Mariner, a youth of fair education and of no inconsiderable natural ability (as the work which was drawn up from the materials he furnished shows), was about fifteen years of age when his ship was attacked and plundered by the Tongans : he remained four years in the islands, familiarised himself with the language, lived the life of the people, became intimate with many of them, and had every opportunity of acquainting himself with their opinions, as well as with their habits and customs. He seems to have been devoid of prejudices, theological or other, and the impression of strict accuracy which his statements convey has been justified by all the knowledge of Polynesian life which has been subsequently acquired.

It is desirable, therefore, to pay close attention to that which Mariner tells us about the theological views of these people :—

The human soul,[1] after its separation from the body, is termed a *hotooa* (a god or spirit), and is believed to exist in the shape of the body ; to have the same propensities as during life, but to be corrected by a more enlightened understanding, by which it readily distinguishes good from evil, truth from falsehood, right from wrong ; having the same attributes as the original gods, but in a minor degree, and having its dwelling for ever in the happy regions of Bolotoo, holding the same rank in regard to other souls as during this life ; it has, however, the power of returning to Tonga to inspire priests, relations, or others, or to appear in dreams to those it wishes to admonish ;

---

[1] Supposed to be "the finer or more aeriform part of the body," standing in "the same relation to the body as the perfume and the more essential qualities of a flower do to the more solid substances" (Mariner, vol. ii. p. 127).

and sometimes to the external eye in the form of a ghost or apparition ; but this power of reappearance at Tonga particularly belongs to the souls of chiefs rather than of matabooles (vol. ii. p. 130).

The word " hotooa " is the same as that which is usually spelt " atua " by Polynesian philologues, and it will be convenient to adopt this spelling. Now under this head of " *Atuas* or supernatural intelligent beings " the Tongans include :—

1. The original gods.   2. The souls of nobles that have all attributes in common with the first but inferior in degree.   3. The souls of matabooles [1] that are still inferior, and have not the power as the two first have of coming back to Tonga to inspire the priests, though they are supposed to have the power of appearing to their relatives.   4. The original attendants or servants, as it were, of the gods, who, although they had their origin and have ever since existed in Bolotoo, are still inferior to the third class.   5. The *Atua pow* or mischievous gods. 6. *Mooi*, or the god that supports the earth and does not belong to Bolotoo (vol. ii. pp. 103, 104).

From this it appears that the " Atuas " of the Polynesian are exactly equivalent to the " Elohim " of the old Israelite.[2]   They comprise everything spiritual, from a ghost to a god, and from " the merely tutelar gods to particular private families " (vol. ii. p. 104), to Tá-li-y-Tooboó, who was the national god of Tonga.   The Tongans had no doubt that these Atuas daily and hourly influenced their destinies and could, conversely, be influenced by them.

[1] A kind of " clients " in the Roman sense.
[2] It is worthy of remark that δαίμων among the Greeks, and *Deus* among the Romans, had the same wide signification.   The *dii manes* were ghosts of ancestors = Atuas of the family.

Hence their "piety," the incessant acts of sacrificial worship which occupied their lives, and their belief in omens and charms. Moreover, the Atuas were believed to visit particular persons,—their own priests in the case of the higher gods, but apparently anybody in that of the lower,—and to inspire them by a process which was conceived to involve the actual residence of the god, for the time being, in the person inspired, who was thus rendered capable of prophesying (vol. ii. p. 100). For the Tongan, therefore, inspiration indubitably was possession.

When one of the higher gods was invoked, through his priest, by a chief who wished to consult the oracle, or, in old Israelitic phraseology, to "inquire of," the god, a hog was killed and cooked over night, and, together with plantains, yams, and the materials for making the peculiar drink *kava* (of which the Tongans were very fond), was carried next day to the priest. A circle, as for an ordinary kava-drinking entertainment, was then formed; but the priest, as the representative of the god, took the highest place, while the chiefs sat outside the circle, as an expression of humility calculated to please the god.

As soon as they are all seated the priest is considered as inspired, the god being supposed to exist within him from that moment. He remains for a considerable time in silence with his hands clasped before him, his eyes are cast down and he rests perfectly still. During the time the victuals are being shared out and the kava preparing, the matabooles sometimes begin to consult him; sometimes he answers, and at other times not; in either case he remains with his eyes cast down. Frequently he will not utter a word till the repast is finished and the kava too. When he speaks he generally begins in a low

and very altered tone of voice, which gradually rises to nearly its
natural pitch, though sometimes a little above it. All that he
says is supposed to be the declaration of the god, and he accordingly speaks in the first person, as if he were the god. All this
is done generally without any apparent inward emotion or outward agitation; but, on some occasions, his countenance becomes
fierce, and as it were inflamed, and his whole frame agitated
with inward feeling; he is seized with an universal trembling,
the perspiration breaks out on his forehead, and his lips turning
black are convulsed; at length tears start in floods from his
eyes, his breast heaves with great emotion, and his utterance is
choked. These symptoms gradually subside. Before this par-
oxysm comes on, and after it is over, he often eats as much as
four hungry men under other circumstances could devour. The
fit being now gone off, he remains for some time calm and then
takes up a club that is placed by him for the purpose, turns it
over and regards it attentively; he then looks up earnestly, now
to the right, now to the left, and now again at the club; afterwards he looks up again and about him in like manner, and
then again fixes his eyes on the club, and so on for several times.
At length he suddenly raises the club, and, after a moment's
pause, strikes the ground or the adjacent part of the house with
considerable force; immediately the god leaves him, and he
rises up and retires to the back of the ring among the people
(vol. i. pp. 100, 101).

The phenomena thus described, in language which,
to any one who is familiar with the manifestations
of abnormal mental states among ourselves, bears
the stamp of fidelity, furnish a most instructive commentary upon the story of the wise woman of Endor.
As in the latter, we have the possession by the spirit
or soul (Atua, Elohim), the strange voice, the speaking
in the first person. Unfortunately nothing (beyond
the loud cry) is mentioned as to the state of the wise
woman of Endor. But what we learn from other
sources (e.g. 1 Sam. x. 20-24) respecting the physical

concomitants of inspiration among the old Israelites
has its exact equivalent in this and other accounts
of Polynesian prophetism.   An excellent authority,
Moerenhout, who lived among the people of the
Society Islands many years and knew them well,
says that, in Tahiti, the *rôle* of the prophet had very
generally passed out of the hands of the priests into
that of private persons who professed to represent
the god, often assumed his name, and in this capacity
prophesied.   I will not run the risk of weakening the
force of Moerenhout's description of the prophetic
state by translating it :—

Un individu, dans cet état, avait le bras gauche enveloppé
d'un morceau d'étoffe, signe de la présence de la Divinité.   Il ne
parlait que d'un ton impérieux et véhément.   Ses attaques,
quand il allait prophétiser, étaient aussi effroyables qu'impo-
santes.   Il tremblait d'abord de tous ses membres, la figure
enflée, les yeux hagards, rouges et étincelants d'une expression
sauvage.   Il gesticulait, articulait des mots vides de sens,
poussait des cris horribles qui faisaient tressaillir tous les
assistans, et s'exaltait parfois au point qu'on n'osait pas
l'approcher.   Autour de lui, le silence de la terreur et du respect. . . .   C'est alors qu'il répondait aux questions, annonçait
l'avenir, le destin des batailles, la volonté des dieux ; et, chose
étonnante ! au sein de ce délire, de cet enthousiasme religieux,
son langage était grave, imposant, son éloquence noble et persuasive.[1]

Just so Saul strips off his clothes, " prophesies " before
Samuel, and lies down " naked all that day and
night."

Both Mariner and Moerenhout refuse to have
recourse to the hypothesis of imposture in order to

---

[1] *Voyages aux îles du Grand Ocean,* t. i. p. 482.

account for the inspired state of the Polynesian prophets.   On the contrary, they fully believe in their
sincerity.   Mariner tells the story of a young chief,
an acquaintance of his, who thought himself possessed
by the Atua of a dead woman who had fallen in love
with him, and who wished him to die that he might
be near her in Bolotoo.   And he died accordingly.
But the most valuable evidence on this head is contained in what the same authority says about King
Finow's son.   The previous king, Toogoo Ahoo, had
been assassinated by Finow, and his soul, become an
Atua of divine rank in Bolotoo, had been pleased to
visit and inspire Finow's son—with what particular
object does not appear.

When this young chief returned to Hapai, Mr. Mariner, who
was upon a footing of great friendship with him, one day asked
him how he felt himself when the spirit of Toogoo Ahoo visited
him ; he replied that he could not well describe his feelings, but
the best he could say of it was, that he felt himself all over in a
glow of heat and quite restless and uncomfortable, and did not
feel his own personal identity, as it were, but seemed to have a
mind different from his own natural mind, his thoughts wandering upon strange and unusual subjects, although perfectly sensible of surrounding objects.   He next asked him how he knew
it was the spirit of Toogoo Ahoo ?   His answer was, "There's
a fool!   How can I tell you *how* I knew it ?   I felt and knew
it was so by a kind of consciousness ; my *mind* told me that it
was Toogoo Ahoo" (vol. i. pp. 104, 105).

Finow's son was evidently made for a theological
disputant, and fell back at once on the inexpugnable
stronghold of faith when other evidence was lacking.
"There's a fool!   I know it is true, because I
know it," is the exemplar and epitome of the sceptic-

crushing process in other places than the Tonga
Islands.

The island of Bolotoo, to which all the souls (of
the upper classes at any rate) repair after the death
of the body, and from which they return at will to
interfere, for good or evil, with the lives of those
whom they have left behind, obviously answers to
Sheol. In Tongan tradition this place of souls is a
sort of elysium above ground, and pleasant enough
to live in. But, in other parts of Polynesia, the corresponding locality, which is called Po, has to be
reached by descending into the earth, and is represented dark and gloomy like Sheol. But it was
not looked upon as a place of rewards and punishments in any sense. Whether in Bolotoo or in Po,
the soul took the rank it had in the flesh ; and, a
shadow, lived among the shadows of the friends
and houses and food of its previous life.

The Tongan theologians recognised several hundred
gods ; but there was one, already mentioned as their
national god, whom they regarded as far greater
than any of the others, " as a great chief from the
top of the sky down to the bottom of the earth"
(Mariner, vol. ii. p. 106). He was also god of war,
and the tutelar deity of the royal family, whoever happened to be the incumbent of the royal office for the
time being. He had no priest except the king himself, and his visits, even to royalty, were few and far
between. The name of this supreme deity was Tá-li-
y-Tooboó, the literal meaning of which is said to be
" Wait there, Tooboó," from which it would appear

that the peculiar characteristic of Tá-li-y-Tooboó, in the eyes of his worshippers, was persistence of duration. And it is curious to notice, in relation to this circumstance, that many Hebrew philologers have thought the meaning of Jahveh to be best expressed by the word "Eternal." It would probably be difficult to express the notion of an eternal being, in a dialect so little fitted to convey abstract conceptions as Tongan, better than by that of one who always "waits there."

The characteristics of the gods in Tongan theology are exactly those of men whose shape they are supposed to possess, only they have more intelligence and greater power. The Tongan belief that, after death, the human Atua more readily distinguishes good from evil, runs parallel with the old Israelitic conception of Elohim expressed in Genesis, "Ye shall be as Elohim, knowing good from evil." They further agreed with the old Israelites, that "all rewards for virtue and punishments for vice happen to men in this world only, and come immediately from the gods" (vol. ii. p. 100). Moreover, they were of opinion that though the gods approve of some kinds of virtue and are displeased with some kinds of vice, and, to a certain extent, protect or forsake their worshippers according to their moral conduct, yet neglect to pay due respect to the deities, and forgetfulness to keep them in good humour, might be visited with even worse consequences than moral delinquency. And those who will carefully study the so - called "Mosaic code" contained in the

books of Exodus, Leviticus, and Numbers, will see
that, though Jahveh's prohibitions of certain forms
of immorality are strict and sweeping, his wrath is
quite as strongly kindled against infractions of ritual
ordinances. Accidental homicide may go unpunished,
and reparation may be made for wilful theft. On
the other hand, Nadab and Abihu, who " offered
strange fire before Jahveh, which he had not commanded them," were swiftly devoured by Jahveh's
fire ; he who sacrificed anywhere except at the allotted
place was to be "cut off from his people"; so was he
who eat blood ; and the details of the upholstery of the
Tabernacle, of the millinery of the priests' vestments,
and of the cabinet work of the ark, can plead direct
authority from Jahveh, no less than moral commands.

Amongst the Tongans, the sacrifices were regarded
as gifts of food and drink offered to the divine Atuas,
just as the articles deposited by the graves of the
recently dead were meant as food for Atuas of lower
rank. A kava root was a constant form of offering
all over Polynesia. In the excellent work of the
Rev. George Turner, entitled *Nineteen Years in
Polynesia* (p. 241), I find it said of the Samoans
(near neighbours of the Tongans) :—

The *offerings* were principally cooked food. As in ancient
Greece so in Samoa, the first cup was in honour of the god. It
was either poured out on the ground or *waved* towards the
heavens, reminding us again of the Mosaic ceremonies. The
chiefs all drank a portion out of the same cup, according to
rank ; and after that, the food brought as an offering was divided
and eaten " *there before the Lord.*"

In Tonga, when they consulted a god who had a

priest, the latter, as representative of the god, had the first cup ; but if the god, like Ta-li-y-Tooboó, had no priest, then the chief place was left vacant, and was supposed to be occupied by the god himself. When the first cup of kava was filled, the mataboole who acted as master of the ceremonies said, " Give it to your god," and it was offered, though only as a matter of form. In Tonga and Samoa there were many sacred places or *morais*, with houses of the ordinary construction, but which served as temples in consequence of being dedicated to various gods ; and there were altars on which the sacrifices were offered ; nevertheless there were few or no images. Mariner mentions none in Tonga, and the Samoans seem to have been regarded as no better than atheists by other Polynesians because they had none. It does not appear that either of these peoples had images even of their family or ancestral gods.

In Tahiti and the adjacent islands, Moerenhout (t. i. p. 471) makes the very interesting observation, not only that idols were often absent, but that, where they existed, the images of the gods served merely as depositories for the proper representatives of the divinity. Each of these was called a *maro aurou*, and was a kind of girdle artistically adorned with red, yellow, blue, and black feathers—the red feathers being especially important—which were consecrated and kept as sacred objects within the idols. They were worn by great personages on solemn occasions, and conferred upon their wearers a sacred and almost divine character. There is no distinct evidence that the *maro aurou* was supposed to have any special efficacy in divination, but one cannot fail to see a certain parallelism between this holy girdle, which endowed its wearer with a particular sanctity, and the ephod.

According to the Rev. R. Taylor, the New Zealanders formerly used the word *karakia* (now employed for "prayer") to signify a "spell, charm, or incantation," and the utterance of these karakias constituted the chief part of their cult. In the south, the officiating priest had a small image, "about eighteen inches long, resembling a peg with a carved head," which reminds one of the form commonly attributed to the teraphim.

The priest first bandaged a fillet of red parrot feathers under the god's chin, which was called his pahau or beard; this bandage was made of a certain kind of sennet, which was tied on in a peculiar way. When this was done it was taken possession of by the Atua, whose spirit entered it. The priest then either held it in the hand and vibrated it in the air, whilst the powerful karakia was repeated, or he tied a piece of string (formed of the centre of a flax leaf) round the neck of the image and stuck it in the ground. He sat at a little distance from it, leaning against a tuahu, a short stone pillar stuck in the ground in a slanting position, and holding the string in his hand, he gave the god a jerk to arrest his attention, lest he should be otherwise engaged, like Baal of old, either hunting, fishing, or sleeping, and therefore must be awaked. . . . The god is supposed to make use of the priest's tongue in giving a reply. Image-worship appears to have been confined to one part of the island. The Atua was supposed only to enter the image for the occasion. The natives declare they did not worship the image itself, but only the Atua it represented, and that the image was merely used as a way of approaching him.[1]

[1] *Te Ika a Maui : New Zealand and its Inhabitants*, p. 72.

This is the excuse for image-worship which the more intelligent idolaters make all the world over: but it is more interesting to observe that, in the present case, we seem to have the equivalents of divination by teraphim, with the aid of something like an ephod (which, however, is used to sanctify the image and not the priest) mixed up together. Many Hebrew archæologists have supposed that the term " ephod " is sometimes used for an image (particularly in the case of Gideon's ephod), and the story of Micah, in the book of Judges, shows that images were, at any rate, employed in close association with the ephod. If the pulling of the string to call the attention of the god seems as absurd to us as it appears to have done to the worthy missionary, who tells us of the practice, it should be recollected that the high priest of Jahveh was ordered to wear a garment fringed with golden bells.

And it shall be upon Aaron to minister; and the sound thereof shall be heard when he goeth in unto the holy place before Jahveh, and when he cometh out, that he die not (Exod. xxviii. 35).

An escape from the obvious conclusion suggested by this passage has been sought in the supposition that these bells rang for the sake of the worshippers, as at the elevation of the host in the Roman Catholic ritual; but then why should the priest be threatened with the well-known penalty for inadvisedly beholding the divinity ?

In truth, the intermediate step between the Maori practice and that of the old Israelites is furnished by the Kami temples in Japan. These are provided with bells which the worshippers who present themselves ring, in order to call the attention of the ancestor-god to their presence. Grant the fundamental assumption of the essentially human character of the spirit, whether Atua, Kami, or Elohim, and all these practices are equally rational.

The sacrifices to the gods in Tonga, and elsewhere in Polynesia, were ordinarily social gatherings, in which the god, either in his own person or in that of his priestly representative, was supposed to take part. These sacrifices were offered on every occasion of importance, and even the daily meals were prefaced by oblations and libations of food and drink, exactly answering to those offered by the old Romans to their manes, penates, and lares. The sacrifices had no moral significance, but were the necessary result of the theory that the god was either a deified ghost of an ancestor or chief, or, at any rate, a being of like nature to these. If one wanted to get anything out of him, therefore, the first step was to put him in good humour by gifts; and if one desired to escape his wrath, which might be excited by the most trifling neglect or unintentional disrespect, the great thing was to pacify him by costly presents. King Finow appears to have been somewhat of a freethinker (to the great horror of his subjects), and it was only his untimely death which prevented him from dealing with the priest of a god, who had not returned a favourable answer to his supplications, as Saul dealt with the priests of the sanctuary of Jahveh at Nob.

Nevertheless, Finow showed his practical belief in the gods during the sickness of a daughter, to whom he was fondly attached, in a fashion which has a close parallel in the history of Israel.

If the gods have any resentment against us, let the whole weight of vengeance fall on my head. I fear not their vengeance—but spare my child ; and I earnestly entreat you, Toobo Totái [the god whom he had invoked], to exert all your influence with the other gods that I alone may suffer all the punishment they desire to inflict (vol. i. p. 354).

So when the king of Israel has sinned by "numbering the people," and they are punished for his fault by a pestilence which slays seventy thousand innocent men, David cries to Jahveh :—

Lo, I have sinned, and I have done perversely : but these sheep, what have they done ? let thine hand, I pray thee, be against me, and against my father's house (2 Sam. xxiv. 17).

Human sacrifices were extremely common in Polynesia ; and, in Tonga, the "devotion" of a child by strangling was a favourite method of averting the wrath of the gods. The well - known instances of Jephthah's sacrifice of his daughter and of David's giving up the seven sons of Saul to be sacrificed by the Gibeonites "before Jahveh," appear to me to leave no doubt that the old Israelites, even when devout worshippers of Jahveh, considered human sacrifices, under certain circumstances, to be not only permissible but laudable. Samuel's hewing to pieces of the miserable captive, sole survivor of his nation, Agag, "before Jahveh," can hardly be viewed in any other light. The life of Moses is redeemed from Jahveh, who "sought to slay him," by Zipporah's symbolical sacrifice of her child, by the bloody operation of circumcision. Jahveh expressly affirms that the first-born males of men and beasts are devoted to him; in accordance with that claim, the first-born males of the beasts are duly sacrificed; and it is only by special permission that the claim to the first-born of men is waived, and it is enacted that they may be redeemed (Exod. xiii. 12-15). Is it possible to avoid the conclusion that immolation of their first-born sons would have been incumbent on the worshippers of Jahveh, had they not been thus specially excused? Can any other conclusion be drawn from the history of Abraham and Isaac? Does Abraham exhibit any indication of surprise when he receives the astounding order to sacrifice his son? Is there the slightest evidence that there was anything in his intimate and personal acquaintance with the character of the Deity, who had eaten the meat and drunk the milk which Abraham set before him under the oaks of Mamre, to lead him to hesitate — even to wait twelve or fourteen hours for a repetition of the command? Not a whit. We are told that "Abraham rose early in the morning" and led his only child to the slaughter, as if it were the most ordinary business imaginable. Whether the story has any historical foundation or not, it is valuable as showing that the writer of it conceived Jahveh as a deity whose requirement of such a sacrifice need excite neither astonishment, nor suspicion of mistake, on the part of his devotee. Hence, when the incessant human sacrifices in Israel, during the age of the kings, are put down to the influence of foreign idolatries, we may fairly inquire whether editorial Bowdlerising has not prevailed over historical truth.

An attempt to compare the ethical standards of two nations, one of which has a written code, while the other has not, is beset with difficulties. With all that is strange and, in many cases, repulsive to us in the social arrangements and opinions respecting moral obligation among the Tongans, as they are placed before us, with perfect candour, in Mariner's account, there is much that indicates a strong ethical sense. They showed great kindliness to one another, and faithfulness in standing by their comrades in war. No people could have better observed either the third or the fifth commandment; for they had a particular horror of blasphemy, and their respectful tenderness towards their parents and, indeed, towards old people in general, was remarkable.

It cannot be said that the eighth commandment was generally observed, especially where Europeans were concerned; but nevertheless a well-bred Tongan looked upon theft as a meanness to which he would not condescend. As to the seventh commandment, any breach of it was considered scandalous in women and as something to be avoided in self-respecting men, but among unmarried and widowed people chastity was held very cheap. Nevertheless the women were extremely well treated, and often showed themselves capable of great devotion and entire faithfulness. In the matter of cruelty, treachery, and

bloodthirstiness, these islanders were neither better
nor worse than most peoples of antiquity.   It is to
the credit of the Tongans that they particularly
objected to slander; nor can covetousness be regarded as their characteristic; for Mariner says :—

> When any one is about to eat, he always shares out what he
> has to those about him, without any hesitation, and a contrary
> conduct would be considered exceedingly vile and selfish (vol.
> ii. p. 145).

In fact, they thought very badly of the English when
Mariner told them that his countrymen did not act
exactly on that principle.   It further appears that
they decidedly belonged to the school of intuitive
moral philosophers, and believed that virtue is its
own reward ; for

> Many of the chiefs, on being asked by Mr. Mariner what
> motives they had for conducting themselves with propriety,
> besides the fear of misfortunes in this life, replied, the agreeable
> and happy feeling which a man experiences within himself when
> he does any good action or conducts himself nobly and gener-
> ously as a man ought to do ; and this question they answered as
> if they wondered such a question should be asked (vol. ii.
> p. 161).

One may read from the beginning of the book of
Judges to the end of the books of Samuel without
discovering that the old Israelites had a moral standard which differs, in any essential respect (except
perhaps in regard to the chastity of unmarried
women), from that of the Tongans.   Gideon, Jephthah, Samson, and David are strong-handed men,
some of whom are not outdone by any Polynesian
chieftain in the matter of murder and treachery;

while Deborah's jubilation over Jael's violation of the primary duty of hospitality, proffered and accepted under circumstances which give a peculiarly atrocious character to the murder of the guest; and her witch-like gloating over the picture of the disappointment of the mother of the victim—

> The mother of Sisera cried through the lattice,
> Why is his chariot so long in coming? (Jud. v. 28).

—would not have been out of place in the choral service of the most sanguinary god in the Polynesian pantheon.

With respect to the cannibalism which the Tongans occasionally practised, Mariner says :—

> Although a few young ferocious warriors chose to imitate what they considered a mark of courageous fierceness in a neighbouring nation, it was held in disgust by everybody else (vol. ii. p. 171).

That the moral standard of Tongan life was less elevated than that indicated in the "Book of the Covenant" (Exod. xxi.-xxiii.) may be freely admitted. But then the evidence that this Book of the Covenant, and even the ten commandments as given in Exodus, were known to the Israelites of the time of Samuel and Saul, is (to say the least) by no means conclusive. The Deuteronomic version of the fourth commandment is hopelessly discrepant from that which stands in Exodus. Would any later writer have ventured to alter the commandments as given from Sinai, if he had had before him that which professed to be an accurate statement of the "ten words" in Exodus? And if the writer of

Deuteronomy had not Exodus before him, what is the value of the claim of the version of the ten commandments therein contained to authenticity? From one end to the other of the books of Judges and Samuel, the only " commandments of Jahveh " which are specially adduced refer to the prohibition of the worship of other gods, or are orders given *ad hoc*, and have nothing to do with questions of morality.

In Polynesia, the belief in witchcraft, in the appearance of spiritual beings in dreams, in possession as the cause of diseases, and in omens, prevailed universally. Mariner tells a story of a woman of rank who was greatly attached to King Finow, and who, for the space of six months after his death, scarcely ever slept elsewhere than on his grave, which she kept carefully decorated with flowers :—

One day she went, with the deepest affliction, to the house of Mo-oonga Toobó, the widow of the deceased chief, to communicate what had happened to her at the *fytoca* [grave] during several nights, and which caused her the greatest anxiety. She related that she had dreamed that the late How [king] appeared to her and, with a countenance full of disappointment, asked why there yet remained at Vavaoo so many evil-designing persons : for he declared that, since he had been at Bolotoo, his spirit had been disturbed[1] by the evil machinations of wicked men conspiring against his son ; but he declared that "the youth" should not be molested nor his power shaken by the spirit of rebellion ; that he therefore came to her with a warning voice to prevent such disastrous consequences (vol. i. p. 424).

On inquiry it turned out that the charm of *tattao* had been performed on Finow's grave, with the view

[1] Compare : "And Samuel said unto Saul, Why hast thou disquieted me ? " (1 Sam. xxviii. 15).

of injuring his son, the reigning king, and it is to be presumed that it was this sorcerer's work which had "disturbed" Finow's spirit. The Rev. Richard Taylor says in the work already cited: "The account given of the witch of Endor agrees most remarkably with the witches of New Zealand" (p. 45).

The Tongans also believed in a mode of divination (essentially similar to the casting of lots) by the twirling of a cocoa-nut.

> The object of inquiry . . . is chiefly whether a sick person will recover; for this purpose the nut being placed on the ground, a relation of the sick person determines that, if the nut, when again at rest, points to such a quarter, the east for example, that the sick man will recover; he then prays aloud to the patron god of the family that he will be pleased to direct the nut so that it may indicate the truth; the nut being next spun, the result is attended to with confidence, at least with a full conviction that it will truly declare the intentions of the gods at the time (vol. ii. p. 227).

Does not the action of Saul, on a famous occasion, involve exactly the same theological presuppositions?

> Therefore Saul said unto Jahveh, the Elohim of Israel, Shew the right. And Jonathan and Saul were taken *by lot:* but the people escaped. And Saul said, Cast *lots* between me and Jonathan my son. And Jonathan was taken. And Saul said to Jonathan, Tell me what thou hast done. . . . And the people rescued Jonathan so that he died not (1 Sam. xiv. 41-45).

As the Israelites had great yearly feasts, so had the Polynesians; as the Israelites practised circumcision, so did many Polynesian people; as the Israelites had a complex and often arbitrary-seeming multitude of distinctions between clean and unclean things, and clean and unclean states of men, to which

they attached great importance, so had the Polynesians their notions of ceremonial purity and their
tabu, an equally extensive and strange system of
prohibitions, violation of which was visited by death.
These doctrines of cleanness and uncleanness no
doubt may have taken their rise in the real or fancied
utility of the prescriptions, but it is probable that
the origin of many is indicated in the curious habit
of the Samoans to make fetishes of living animals.
It will be recollected that these people had no " gods
made with hands," but they substituted animals for
them.

     At his birth

every Samoan was supposed to be taken under the care of some
tutelary god or *aitu* [= Atua] as it was called. The help of perhaps half a dozen different gods was invoked in succession on the
occasion, but the one who happened to be addressed just as the
child was born was marked and declared to be the child's god
for life.

     These gods were supposed to appear in some *visible incarnation*,
and the particular thing in which his god was in the habit of
appearing was, to the Samoan, an object of veneration. It was
in fact his idol, and he was careful never to injure it or treat
it with contempt. One, for instance, saw his god in the eel,
another in the shark, another in the turtle, another in the dog,
another in the owl, another in the lizard ; and so on, throughout
all the fish of the sea and birds and four-footed beasts and
creeping things. In some of the shell-fish even, gods were
supposed to be present. A man would eat freely of what was
regarded as the incarnation of the god of another man, but the
incarnation of his own particular god he would consider it death
to injure or eat.[1]

     We have here that which appears to be the origin,
or one of the origins, of food prohibitions, on the one

[1] Turner, *Nineteen Years in Polynesia*, p. 238.

hand, and of totemism on the other. When it is remembered that the old Israelities sprang from ancestors who are said to have resided near, or in, one of the great seats of ancient Babylonian civilisation, the city of Ur; that they had been, it is said for centuries, in close contact with the Egyptians; and that, in the theology of both the Babylonians and the Egyptians there is abundant evidence, notwithstanding their advanced social organisation, of the belief in spirits, with sorcery, ancestor-worship, the deification of animals, and the converse animalisation of gods—it obviously needs very strong evidence to justify the belief that the rude tribes of Israel did not share the notions from which their far more civilised neighbours had not emancipated themselves.

But it is surely needless to carry the comparison further. Out of the abundant evidence at command, I think that sufficient has been produced to furnish ample grounds for the belief, that the old Israelites of the time of Samuel entertained theological conceptions which were on a level with those current among the more civilised of the Polynesian islanders, though their ethical code may possibly, in some respects, have been more advanced.[1]

A theological system of essentially similar character, exhibiting the same fundamental conceptions respecting the continued existence and incessant interference in human affairs of disembodied spirits, prevails, or formerly prevailed, among the whole of

[1] See Lippert's excellent remarks on this subject, *Der Seelencult*, p. 89.

the inhabitants of the Polynesian and Melanesian islands, and among the people of Australia, notwithstanding the wide differences in physical character and in grade of civilisation which obtain among them. And the same proposition is true of the people who inhabit the riverain shores of the Pacific Ocean, whether Dyaks, Malays, Indo-Chinese, Chinese, Japanese, the wild tribes of America, or the highly civilised old Mexicans and Peruvians. It is no less true of the Mongolic nomads of Northern Asia, of the Asiatic Aryans, and of the ancient Greeks and Romans, and it holds good among the Dravidians of the Dekhan and the negro tribes of Africa. No tribe of savages, which has yet been discovered, has been conclusively proved to have so poor a theological equipment as to be devoid of a belief in ghosts, and in the utility of some form of witchcraft in influencing those ghosts. And there is no nation, modern or ancient, which, even at this moment, has wholly given up the belief; and in which it has not, at one time or other, played a great part in practical life.

This *sciotheism*,[1] as it might be called, is found in several degrees of complexity, in rough correspondence with the stages of social organisation, and, like these, separated by no sudden breaks.

In its simplest condition, such as may be met with among the Australian savages, theology is a mere

[1] *Sciography* has the authority of Cudworth, *Intellectual System*, vol. ii. p. 836. Sciomancy (σκιομαντεία), which, in the sense of divination by ghosts, may be found in Bailey's *Dictionary* (1751), also furnishes a precedent for my coinage.

belief in the existence, powers, and disposition (usually malignant) of ghostlike entities who may be propitiated or scared away ; but no cult can properly be said to exist. And, in this stage, theology is wholly independent of ethics. The moral code, such as is implied by public opinion, derives no sanction from the theological dogmas, and the influence of the spirits is supposed to be exerted out of mere caprice or malice.

As a next stage, the fundamental fear of ghosts and the consequent desire to propitiate them acquire an organised ritual in simple forms of ancestor-worship, such as the Rev. Mr. Turner describes among the people of Tanna (*l.c.* p. 88); and this line of development may be followed out until it attains its acme in the State-theology of China and the Kami-theology[1] of Japan. Each of these is essentially ancestor-worship, the ancestors being reckoned back through family groups of higher and higher order, sometimes with strict reference to the principle of agnation, as in old Rome ; and, as in the latter, it is intimately bound up with the whole organisation of the State. There are no idols ; inscribed tablets in China, and strips of paper lodged in a peculiar portable shrine in Japan, represent the souls of the deceased, or the special seats which they occupy when sacrifices are offered by their descendants. In Japan it is interesting to observe that a national

---

[1] " Kami " is used in the sense of Elohim ; and is also, like our word " Lord," employed as a title of respect among men, as indeed Elohim was.

Kami—Ten-zio-dai-zin—is worshipped as a sort of
Jahveh by the nation in general, and (as Lippert has
observed) it is singular that his special seat is a portable litter-like shrine, termed the Mikosi, in some
sort analogous to the Israelitic ark.   In China, the
emperor is the representative of the primitive ancestors, and stands, as it were, between them and the
supreme cosmic deities—Heaven and Earth—who
are superadded to them, and who answer to the
Tangaloa and the Maui of the Polynesians.

Sciotheism, under the form of the deification of
ancestral ghosts, in its most pronounced form, is
therefore the chief element in the theology of a great
moiety, possibly of more than half, of the human
race.   I think this must be taken to be a matter of
fact—though various opinions may be held as to how
this ancestor-worship came about.   But, on the other
hand, it is no less a matter of fact that there are very
few people without additional gods, who cannot, with
certainty, be accounted for as deified ancestors.

With all respect for the distinguished authorities
on the other side, I cannot find good reasons for
accepting the theory that the cosmic deities—who
are superadded to deified ancestors even in China;
who are found all over Polynesia, in Tangaloa and
Maui, and in old Peru, in the Sun—are the product
either of the "search after the infinite," or of mistakes arising out of the confusion of a great chief's
name with the thing signified by the name.   But,
however this may be, I think it is again merely
matter of fact that, among a large portion of man-

kind, ancestor-worship is more or less thrown into
the background either by such cosmic deities, or
by tribal gods of uncertain origin, who have been
raised to eminence by the superiority in warfare, or
otherwise, of their worshippers.

Among certain nations, the polytheistic theology,
thus constituted, has become modified by the selection
of some one cosmic or tribal god, as the only god
to whom worship is due on the part of that nation
(though it is by no means denied that other nations
have a right to worship other gods), and thus results
a worship of one God—*monolatry*, as Wellhausen calls
it—which is very different from genuine monotheism.[1]
In ancestral sciotheism, and in this *monolatry*, the
ethical code, often of a very high order, comes into
closer relation with the theological creed. Morality is taken under the patronage of the god or gods,
who reward all morally good conduct and punish all
morally evil conduct in this world or the next. At
the same time, however, they are conceived to be
thoroughly human, and they visit any shadow of
disrespect to themselves, shown by disobedience to
their commands, or by delay, or carelessness, in
carrying them out, as severely as any breach of the
moral laws. Piety means minute attention to the
due performance of all sacred rites, and covers any
number of lapses in morality, just as cruelty, treachery, murder, and adultery did not bar David's claim
to the title of the man after God's own heart among
the Israelites; crimes against men may be expiated,

[1 The Assyrians thus raised Assur to a position of pre-eminence.]

but blasphemy against the gods is an unpardonable sin. Men forgive all injuries but those which touch their self-esteem; and they make their gods after their own likeness, in their own image make they them.

It is in the category of monolatry that I conceive the theology of the old Israelites must be ranged. They were polytheists, in so far as they admitted the existence of other Elohim of divine rank beside Jahveh; they differed from ordinary polytheists, in so far as they believed that Jahveh was the supreme god and the one proper object of their own national worship. But it will doubtless be objected that I have been building up a fictitious Israelitic theology on the foundation of the recorded habits and customs of the people, when they had lapsed from the ordinances of their great lawgiver and prophet Moses, and that my conclusions may be good for the perverts to Canaanitish theology, but not for the true observers of the Sinaitic legislation. The answer to the objection is that—so far as I can form a judgment of that which is well ascertained in the history of Israel—there is very little ground for believing that we know much, either about the theological and social value of the influence of Moses, or about what happened during the wanderings in the Desert.

The account of the Exodus and of the occurrences in the Sinaitic peninsula; in fact, all the history of Israel before the invasion of Canaan, is full of wonderful stories which may be true, in so far as they are conceivable occurrences, but which are certainly not probable, and which I, for one, decline to accept until evidence, which deserves that name, is offered of their historical truth.   Up to this time I know of none.[1] Furthermore, I see no answer to the argument that one has no right to pick out of an obviously unhistorical statement the assertions which happen to be probable and discard the rest.   But it is also certain that a primitively veracious tradition may be smothered under subsequent mythical additions, and that one has no right to cast away the former along with the latter.   Thus, perhaps the fairest way of stating the case may be as follows.

There can be no *à priori* objection to the supposition that the Israelites were delivered from their Egyptian bondage by a leader called Moses, and that he exerted a great influence over their subsequent organisation in the desert.   There is no reason to doubt that, during their residence in the land of Goshen, the Israelites knew nothing of Jahveh ; but, as their own prophets declare (see Ezek. xx.), were polytheistic idolaters, sharing in the worst practices of their neighbours.   As to their conduct in other respects, nothing is known.   But it may fairly be suspected that their ethics were not of a higher order than those of Jacob their progenitor, in which case they might derive great profit from contact with Egyptian society, which held honesty and truthful-

---

[1] I refer those who wish to know the reasons which lead me to take up this position to the works of Reuss and Wellhausen, [and especially to Stade's *Geschichte des Volkes Israel.*]

ness in the highest esteem.   Thanks to the Egyptolo-
gers, we now know, with all requisite certainty, the
moral standard of that society in the time, and long
before the time, of Moses.   It can be determined
from the scrolls buried with the mummified dead and
from the inscriptions on the tombs and memorial
statues of that age.   For, though the lying of epi-
taphs is proverbial, so far as their subject is concerned, they give an unmistakable insight into that
which the writers and the readers of them think
praiseworthy.

In the famous tombs at Beni Hassan there is a
record of the life of Prince Nakht, who served Osertasen II., a Pharaoh of the twelfth dynasty, as
governor of a province.   The inscription speaks in
his name : " I was a benevolent and kindly governor
who loved his country. . . . Never was a little child
distressed nor a widow ill-treated by me.   I have
never repelled a workman or hindered a shepherd.
I gave alike to the widow and to the married woman,
and have not preferred the great to the small in my
gifts."   And we have the high authority of the late
Dr. Samuel Birch for the statement that the inscriptions of the twelfth dynasty abound in injunctions of
a high ethical character.   "To feed the hungry, give
drink to the thirsty, clothe the naked, bury the dead,
loyally serve the king, formed the first duty of a
pious man and faithful subject."[1]   The people for
whom these inscriptions embodied their ideal of
praiseworthiness assuredly had no imperfect concep-

[1] Bunsen, *Egypt's Place*, vol. v. p. 129, note.

tion of either justice or mercy.   But there is a document which gives still better evidence of the moral standard of the Egyptians.   It is the "Book of the Dead," a sort of "Guide to Spiritland," the whole, or a part, of which was buried with the mummy of every well-to-do Egyptian, while extracts from it are found in innumerable inscriptions.   Portions of this work are of extreme antiquity, evidence of their existence occurring as far back as the fifth and sixth dynasties; · while the 125th chapter, which constitutes a sort of book by itself, and is known as the "Book of Redemption in the Hall of the two Truths," is frequently inscribed upon coffins and other monuments of the nineteenth dynasty (that under which, there is reason to believe, the Israelites were oppressed and the Exodus took place), and it occurs, more than once, in the famous tombs of the kings of this and the preceding dynasty at Thebes.[1]   This "Book of Redemption" is chiefly occupied by the so-called "negative confession" made to the forty-two Divine Judges, in which the soul of the dead denies that he has committed faults of various kinds.   It is, therefore, obvious that the Egyptians conceived that their gods commanded them not to do the deeds which are here denied.   The "Book of Redemption," in fact, implies the existence in the mind of the Egyptians, if not in a formal writing, of a series of ordinances couched, like the majority of the ten command-

[1] See Birch, in *Egypt's Place*, vol. v. ; and Brugsch, *History of Egypt*.

ments, in negative terms. And it is easy to prove the implied existence of a series which nearly answers to the "ten words." Of course a polytheistic and image-worshipping people, who observed a great many holy days, but no Sabbaths, could have nothing analogous to the first or the second and the fourth commandments of the Decalogue; but, answering to the third, is "I have not blasphemed;" to the fifth, "I have not reviled the face of the king or my father;" to the sixth, "I have not murdered;" to the seventh, "I have not committed adultery;" to the eighth, "I have not stolen," "I have not done fraud to man;" to the ninth, "I have not told falsehoods in the tribunal of truth," and, further, "I have not calumniated the slave to his master." I find nothing exactly similar to the tenth commandment; but that the inward disposition of mind was held to be of no less importance than the outward act is to be gathered from the praises of kindliness already cited and the cry of "I am pure," which is repeated by the soul on trial. Moreover, there is a minuteness of detail in the confession which shows no little delicacy of moral appreciation — "I have not privily done evil against mankind," "I have not afflicted men," "I have not withheld milk from the mouths of sucklings," "I have not been idle," "I have not played the hypocrite," "I have not told falsehoods," "I have not corrupted woman or man," "I have not caused fear," "I have not multiplied words in speaking."

Would that the moral sense of the nineteenth

century A.D. were as far advanced as that of the Egyptians in the nineteenth century B.C. in this last particular! What incalculable benefit to mankind would flow from strict observance of the commandment, "Thou shalt not multiply words in speaking!" Nothing is more remarkable than the stress which the old Egyptians, here and elsewhere, lay upon this and other kinds of truthfulness, as compared with the absence of any such requirement in the Israelitic Decalogue, in which only a specific kind of untruthfulness is forbidden.

If, as the story runs, Moses was adopted by a princess of the royal house, and was instructed in all the wisdom of the Egyptians, it is surely incredible that he should not have been familiar, from his youth up, with the high moral code implied in the "Book of Redemption." It is surely impossible that he should have been less familiar with the complete legal system, and with the method of administration of justice, which, even in his time, had enabled the Egyptian people to hold together, as a complex social organisation, for a period far longer than the duration of old Roman society, from the building of the city to the death of the last Cæsar. Nor need we look to Moses alone for the influence of Egypt upon Israel. It is true that the Hebrew nomads who came into contact with the Egyptians of Osertasen, or of Ramses, stood in much the same relation to them, in point of culture, as a Germanic tribe did to the Romans of Tiberius or of Marcus Antoninus, or as Captain Cook's Omai did to the English of George the Third. But,

o

at the same time, any difficulty of communication which might have arisen out of this circumstance was removed by the long pre-existing intercourse of other Semites, of every grade of civilisation, with the Egyptians. In Mesopotamia and elsewhere, as in Phenicia, Semitic people had attained to a social organisation as advanced as that of the Egyptians; Semites had conquered and occupied Lower Egypt for centuries. So extensively had Semitic influences penetrated Egypt that the Egyptian language, during the period of the nineteenth dynasty, is said by Brugsch to be as full of Semitisms as German is of Gallicisms; while Semitic deities had supplanted the Egyptian gods at Heliopolis and elsewhere. On the other hand, the Semites, as far as Phenicia, were extensively influenced by Egypt.

It is generally admitted[1] that Moses, Phinehas (and perhaps Aaron), are names of Egyptian origin, and there is excellent authority for the statement that the name *Abir*, which the Israelites gave to their golden calf, and which is also used to signify the strong, the heavenly, and even God,[2] is simply the Egyptian Apis. Brugsch points out that the god Tum, or Tom, who was the special object of worship in the city of Pi-Tom, with which the Israelites were only too familiar, was called Ānkh and the "great god," and had no image. Ānkh means "He who lives," "the living one," a name the resemblance of which to the

[1] Even by Graetz, who, though a fair enough historian, cannot be accused of any desire to over-estimate the importance of Egyptian influence upon his people.

[2] Graetz, *Geschichte der Juden*, Bd. i. p. 370.

" I am that I am " of Exodus is unmistakable, whatever may be the value of the fact. Every discussion
of Israelitic ritual seeks and finds the explanation of
its details in the portable sacred chests, the altars,
the priestly dress, the breastplate, the incense, and
the sacrifices depicted on the monuments of Egypt.
But it must be remembered that these signs of the
influence of Egypt upon Israel are not necessarily
evidence that such influence was exerted before the
Exodus. It may have come much later, through the
close connection of the Israel of David and Solomon,
first with Phenicia and then with Egypt.

If we suppose Moses to have been a man of the
stamp of Calvin, there is no difficulty in conceiving
that he may have constructed the substance of the
ten words, and even of the Book of the Covenant,
which curiously resembles parts of the Book of the
Dead, from the foundation of Egyptian ethics and
theology which had filtered through to the Israelites
in general, or had been furnished specially to himself
by his early education ; just as the great Genevese
reformer built up a puritanic social organisation on so
much as remained of the ethics and theology of the
Roman Church, after he had trimmed them to his
liking.

Thus, I repeat, I see no *à priori* objection to the
assumption that Moses may have endeavoured to give
his people a theologico-political organisation based on
the ten commandments (though certainly not quite
in their present form) and the Book of the Covenant,
contained in our present book of Exodus. But

whether there is such evidence as amounts to proof,
or, I had better say, to probability, that even this
much of the Pentateuch owes its origin to Moses is
another matter.   The mythical character of the accessories of the Sinaitic history is patent, and it would
take a good deal more evidence than is afforded by the
bare assertion of an unknown writer to justify the
belief that the people who " saw the thunderings and
the lightnings and the voice of the trumpet and
the mountain smoking" (Exod. xx. 18); to whom
Jahveh orders Moses to say, " Ye yourselves have seen
that I have talked with you from heaven.   Ye shall
not make other gods with me; gods of silver and
gods of gold ye shall not make unto you" (*ibid.* 22,
23), should, less than six weeks afterwards, have done
the exact thing they were thus awfully forbidden to
do.   Nor is the credibility of the story increased by
the statement that Aaron, the brother of Moses, the
witness and fellow-worker of the miracles before
Pharaoh, was their leader and the artificer of the
idol.   And yet, at the same time, Aaron was apparently so ignorant of wrongdoing that he made
proclamation, " To-morrow shall be a feast to Jahveh,"
and the people proceeded to offer their burnt-offerings
and peace-offerings, as if everything in their proceedings must be satisfactory to the Deity with whom
they had just made a solemn covenant to abolish
image-worship.   It seems to me that, on a survey of
all the facts of the case, only a very cautious and
hypothetical judgment is justifiable.   It may be that
Moses profited by the opportunities afforded him of

access to what was best in Egyptian society to become acquainted, not only with its advanced ethical and legal code, but with the more or less pantheistic unification of the Divine to which the speculations of the Egyptian thinkers, like those of all polytheistic philosophers, from Polynesia to Greece, tend ; if indeed the theology of the period of the nineteenth dynasty was not, as some Egyptologists think, a modification of an earlier, more distinctly monotheistic doctrine of a long antecedent age. It took only half a dozen centuries for the theology of Paul to become the theology of Gregory the Great ; and it is possible that twenty centuries lay between the theology of the first worshippers in the sanctuary of the Sphinx and that of the priests of Ramses Maimun.

It may be that the ten commandments and the Book of the Covenant are based upon faithful traditions of the efforts of a great leader to raise his followers to his own level. For myself, as a matter of pious opinion, I like to think so ; as I like to imagine that, between Moses and Samuel, there may have been many a seer, many a herdsman such as him of Tekoah, lonely amidst the hills of Ephraim and Judah, who cherished and kept alive these traditions. In the present results of Biblical criticism, however, I can discover no justification for the common assumption that, between the time of Joshua and that of Rehoboam, the Israelites were familiar with either the Deuteronomic or the Levitical legislation ; or that the theology of the Israelites, from the king who sat on the throne to the lowest of his subjects, was in any important respect different from that which might naturally be expected from their previous history and the conditions of their existence. But there is excellent evidence to the contrary effect. And, for my part, I see no reason to doubt that, like the rest of the world, the Israelites had passed through a period of mere ghost-worship, and had advanced through Ancestor-worship and Fetishism and Totemism to the theological level at which we find them in the books of Judges and Samuel.

All the more remarkable, therefore, is the extraordinary change which is to be noted in the eighth century B.C. The student who is familiar with the theology implied, or expressed, in the books of Judges, Samuel, and the first book of Kings, finds himself in a new world of thought, in the full tide of a great reformation, when he reads Joel, Amos, Hosea, Isaiah, Micah, and Jeremiah.

The essence of this change is the reversal of the position which, in primitive society, ethics holds in relation to theology. Originally, that which men worship is a theological hypothesis, not a moral ideal. The prophets, in substance, if not always in form, preach the opposite doctrine. They are constantly striving to free the moral ideal from the stifling embrace of the current theology and its concomitant ritual. Theirs was not an intellectual criticism, argued on strictly scientific grounds; the image-worshippers and the believers in the efficacy of sacrifices and ceremonies might logically have held their own against anything the prophets have to say; it was an ethical criticism. From the height of his moral intuition—that the whole duty of man is to do justice and love mercy and to bear himself as humbly as befits his insignificance in face of the Infinite—the prophet simply laughs at the idolaters of stocks and stones and the idolaters of ritual. Idols of the first kind, in his experience, were inseparably united with the practice of immorality, and they were to be ruthlessly destroyed. As for sacrifices and ceremonies, whatever their intrinsic value might be, they might be tolerated on condition of ceasing to be idols; they might even be praiseworthy on condition of being made to subserve the worship of the true Jahveh—the moral ideal.

If the realm of David had remained undivided, if the Assyrian and the Chaldean and the Egyptian had left Israel to the ordinary course of development of an Oriental kingdom, it is possible that the effects of the reforming zeal of the prophets of the eighth and seventh centuries might have been effaced by the growth, according to its inevitable tendencies, of the theology which they combated. But the captivity made the fortune of the ideas which it was the privilege of these men to launch upon an endless career. With the abolition of the Temple-services for more than half a century, the priest must have lost and the scribe gained influence. The puritanism of a vigorous minority among the Babylonian Jews rooted out polytheism from all its hiding-places in the theology which they had inherited; they created the first

consistent, remorseless, naked monotheism, which, so
far as history records, appeared in the world (for
Zoroastrism is practically ditheism, and Buddhism
any-theism or no-theism); and they inseparably
united therewith an ethical code, which, for its purity
and for its efficiency as a bond of social life, was and
is, unsurpassed. So I think we must not judge Ezra
and Nehemiah and their followers too hardly, if they
exemplified the usual doom of poor humanity to
escape from one error only to fall into another; if
they failed to free themselves as completely from the
idolatry of ritual as they had from that of images and
dogmas; if they cherished the new fetters of the
Levitical legislation which they had fitted upon
themselves and their nation, as though such bonds
had the sanctity of the obligations of morality; and
if they led succeeding generations to spend their best
energies in building that "hedge round the Torah"
which was meant to preserve both ethics and theology, but which too often had the effect of pampering
the latter and starving the former. The world being
what it was, it is to be doubted whether Israel would
have preserved intact the pure ore of religion, which
the prophets had extracted for the use of mankind as
well as for their nation, had not the leaders of the
nation been zealous, even to death, for the dross
of the law in which it was embedded. The
struggle of the Jews, under the Maccabean house,
against the Seleucidæ was as important for mankind
as that of the Greeks against the Persians. And, of
all the strange ironies of history, perhaps the strangest

is that " Pharisee " is current, as a term of reproach, among the theological descendants of that sect of Nazarenes who, without the martyr spirit of those primitive Puritans, would never have come into existence. They, like their historical successors, our own Puritans, have shared the general fate of the poor wise men who save cities.

A criticism of theology from the side of science is not thought of by the prophets, and is at most indicated in the books of Job and Ecclesiastes, in both of which the problem of vindicating the ways of God to man is given up, though on different grounds, as a hopeless one. But with the extensive introduction of Greek thought among the Jews, which took place, not only during the domination of the Seleucidæ in Palestine, but in the great Judaic colony which flourished in Egypt under the Ptolemies, criticism, on both ethical and scientific grounds, took a new departure.

In the hands of the Alexandrian Jews, as represented by Philo, the fundamental axiom of later Jewish, as of Christian monotheism, that the Deity is infinitely perfect and infinitely good, worked itself out into its logical consequence—agnostic theism. Philo will allow of no point of contact between God and a world in which evil exists. For him God has no relation to space or to time, and, as infinite, suffers no predicate beyond that of existence. It is, therefore, absurd to ascribe to Him mental faculties and affections comparable in the remotest degree to those of men ; He is in no way an object of cognition ; He is ἄποιος and ἀκατάληκτος[1]—without quality and incomprehensible. That is to say, the Alexandrian Jew of the first century had anticipated the reasonings of Hamilton and Mansell in the nineteenth, and, for him, God is the Unknowable in the sense in which that term is used by Mr. Herbert Spencer. Moreover, Philo's definition of the Supreme Being would not be inconsistent with that "substantia constans infinitis attributis, quorum unumquodque æternam et infinitam essentiam exprimit," given by another great Israelite, were it not that Spinoza's doctrine of the immanence of the Deity in the world puts him, at any rate formally, at the antipodes of theological speculation. But the conception of the essential incognoscibility of the Deity is the same in each case. However, Philo was too thorough an Israelite and too much the child of his time to be content with this agnostic position. With the help of the Platonic and Stoic philosophy, he constructed an apprehensible, if not comprehensible, quasi-deity out of the Logos ; while other more or less personified divine powers, or attributes, bridged over the interval between God and man ; between the sacred existence, too pure to be called by any name which implied a conceivable quality, and the gross and evil world of matter. In order to get over the ethical difficulties presented by the naïve naturalism of many parts of

---

[1] See the careful analysis of the work of the Alexandrian philosopher and theologian (who, it should be remembered, was a most devout Jew, held in the highest esteem by his countrymen) in Siegfried's *Philo von Alexandrien*, 1875. [Also Dr. J. Drummond's *Philo Judæus*, 1888.]

those Scriptures, in the divine authority of which he firmly believed, Philo borrowed from the Stoics (who had been in like straits in respect of Greek mythology), that great Excalibur which they had forged with infinite pains and skill—the method of allegorical interpretation. This mighty "two-handed engine at the door" of the theologian is warranted to make a speedy end of any and every moral or intellectual difficulty, by showing that, taken allegorically or, as it is otherwise said, "poetically," or, "in a spiritual sense," the plainest words mean whatever a pious interpreter desires they should mean. In Biblical phrase, Zeno (who probably had a strain of Semitic blood in him) was the "father of all such as reconcile." No doubt Philo and his followers were eminently religious men; but they did endless injury to the cause of religion by laying the foundations of a new theology, while equipping the defenders of it with the subtlest of all weapons of offence and defence, and with an inexhaustible store of sophistical arguments of the most plausible aspect.

The question of the real bearing upon theology of the influence exerted by the teaching of Philo's contemporary, Jesus of Nazareth, is one upon which it is not germane to my present purpose to enter. I take it simply as an unquestionable fact that his immediate disciples, known to their countrymen as " Nazarenes," were regarded as, and considered themselves to be, perfectly orthodox Jews belonging to the puritanic or pharisaic section of their people, and differing from the rest only in their belief that the Messiah had already come.   Christianity, it is said, first became clearly differentiated at Antioch, and it separated itself from orthodox Judaism by denying the obligation of the rite of circumcision and of the food prohibitions, prescribed by the law. Henceforward theology became relatively stationary among the Jews,[1] and the history of its rapid progress in a new course of evolution is the history of the Christian Churches, orthodox and heterodox.   The steps in this evolution are obvious.   The first is the birth of a new theological scheme arising out of the union of elements derived from Greek philosophy with elements derived from Israelitic theology.   In the fourth Gospel, the Logos, raised to a somewhat higher degree of personification than in the Alexandrian theosophy, is identified with Jesus of Nazareth. In the Epistles, especially the later of those attributed to Paul, the Israelitic ideas of the Messiah and of sacrificial atonement coalesce with one another and with the embodiment of the Logos in Jesus, until the apotheosis of the Son of man is almost, or quite, effected.   The history of Christian dogma, from Justin to Athanasius, is a record of continual progress in the same direction, until the fair body of religion, revealed in almost naked purity by the prophets, is

---

[1] I am not unaware of the existence of many and widely divergent sects and schools among the Jews at all periods of their history, since the dispersion.   But I imagine that orthodox Judaism is now pretty much what it was in Philo's time ; while Peter and Paul, if they could return to life, would certainly have to learn the catechism of either the Roman, Greek, or Anglican Churches, if they desired to be considered orthodox Christians.

once more hidden under a new accumulation of
dogmas and of ritual practices of which the primitive
Nazarene knew nothing; and which he would probably have regarded as blasphemous if he could have
been made to understand them.

As, century after century, the ages roll on, poly-
theism comes back under the disguise of Mariolatry
and the adoration of saints; image-worship becomes
as rampant as in old Egypt; adoration of relics takes
the place of the old fetish-worship; the virtues of the
ephod pale before those of holy coats and handker-
chiefs; shrines and calvaries make up for the loss of
the ark and of the high places; and even the lustral
fluid of paganism is replaced by holy water at the
porches of the temples. A touching ceremony—the
common meal originally eaten in pious memory of a
loved teacher—was metamorphosed into a flesh-and-
blood sacrifice, supposed to possess exactly that re-
deeming virtue which the prophets denied to the
flesh-and-blood sacrifices of their day; while the
minute observance of ritual was raised to a degree of
punctilious refinement which Levitical legislators
might envy. And with the growth of this theology,
grew its inevitable concomitant, the belief in evil
spirits, in possession, in sorcery, in charms and
omens, until the Christians of the twelfth century
after our era were sunk in more debased and brutal
superstitions than are recorded of the Israelites in the
twelfth century before it.

The greatest men of the Middle Ages are unable to
escape the infection. Dante's "Inferno" would be

revolting if it were not so often sublime, so often exquisitely tender. The hideous pictures which cover a vast space on the south wall of the Campo Santo of Pisa convey information, as terrible as it is indisputable, of the theological conceptions of Dante's countrymen in the fourteenth century, whose eyes were addressed by the painters of those disgusting scenes, and whose approbation they knew how to win. A candid Mexican of the time of Cortez, could he have seen this Christian burial-place, would have taken it for an appropriately adorned Teocalli. The professed disciple of the God of justice and of mercy might there gloat over the sufferings of his fellow-men depicted as undergoing every extremity of atrocious and sanguinary torture to all eternity, for theological errors no less than for moral delinquencies; while, in the central figure of Satan,[1] occupied in champing up souls in his capricious and well-toothed jaws, to void them again for the purpose of undergoing fresh suffering, we have the counterpart of the strange Polynesian and Egyptian dogma that there were certain gods who employed themselves in devouring the ghostly flesh of the spirits of the dead.

---

[1] Dante's description of Lucifer engaged in the eternal mastication of Brutus, Cassius, and Judas Iscariot—

> "Da ogni bocca dirompea co' denti
> Un peccatore, a guisa di maciulla,
> Sì che tre ne facea così dolenti.
> A quel dinanzi il mordere era nulla,
> Verso 'l graffiar, chè tal volta la schiena
> Rimanea della pelle tutta brulla "—

is quite in harmony with the Pisan picture and perfectly Polynesian in conception.

But, in justice to the Polynesians, it must be recollected that, after three such operations, they thought the soul was purified and happy. In the view of the Christian theologian the operation was only a preparation for new tortures continued for ever and aye.

With the growth of civilisation in Europe, and with the revival of letters and of science in the fourteenth and fifteenth centuries, the ethical and intellectual criticism of theology once more recommenced, and arrived at a temporary resting-place in the confessions of the various reformed Protestant sects in the sixteenth century; almost all of which, as soon as they were strong enough, began to persecute those who carried criticism beyond their own limit. But the movement was not arrested by these ecclesiastical barriers, as their constructors fondly imagined it would be; it was continued, tacitly or openly, by Galileo, by Hobbes, by Descartes, and especially by Spinoza, in the seventeenth century; by the English Freethinkers, by Rousseau, by the French Encyclopædists, and by the German Rationalists, among whom Lessing stands out a head and shoulders taller than the rest, throughout the eighteenth century; by the historians, the philologers, the Biblical critics, the geologists, and the biologists in the nineteenth century, until it is obvious to all who can see that the moral sense and the really scientific method of seeking for truth are once more predominating over false science. Once more ethics and theology are parting company.

It is my conviction that, with the spread of true scientific culture, whatever may be the medium, his-

torical, philological, philosophical, or physical, through
which that culture is conveyed, and with its necessary concomitant, a constant elevation of the standard
of veracity, the end of the evolution of theology will
be like its beginning—it will cease to have any relation to ethics.   I suppose that, so long as the human
mind exists, it will not escape its deep-seated instinct
to personify its intellectual conceptions.   The science
of the present day is as full of this particular form of
intellectual shadow-worship as is the nescience of
ignorant ages.   The difference is that the philosopher
who is worthy of the name knows that his personified
hypotheses, such as law, and force, and ether, and the
like, are merely useful symbols, while the ignorant
and the careless take them for adequate expressions
of reality.   So, it may be, that the majority of mankind may find the practice of morality made easier by
the use of theological symbols.   And unless these are
converted from symbols into idols, I do not see that
science has anything to say to the practice, except to
give an occasional warning of its dangers.   But, when
such symbols are dealt with as real existences, I think
the highest duty which is laid upon men of science is
to show that these dogmatic idols have no greater
value than the fabrications of men's hands, the stocks
and the stones, which they have replaced.

V

SCIENCE AND MORALS

In spite of long and, perhaps, not unjustifiable hesitation, I begin to think that there must be something in telepathy. For evidence, which I may not disregard, is furnished by the last number of the *Fortnightly Review* that, among the hitherto undiscovered endowments of the human species, there may be a power even more wonderful than the mystic faculty by which the esoterically Buddhistic sage "upon the farthest mountain in Cathay" reads the inmost thoughts of a dweller within the homely circuit of the London postal district. Great indeed is the insight of such a seer; but how much greater is his who combines the feat of reading, not merely the thoughts of which the thinker is aware, but those of which he knows nothing; who sees him unconsciously drawing the conclusions which he repudiates, and supporting the doctrines which he detests. To reflect upon the confusion which the working of such a power as this may introduce into one's ideas of personality and responsibility is perilous—madness lies that way. But truth is truth, and I am almost fain to believe in this magical visibility of the non-existent when the only alternative is the supposition that the writer of the article on "Materialism and Morality" in vol. xl. (1886) of the *Fortnightly Review*, in spite of his manifest ability and honesty, has pledged himself, so far as I am concerned, to what, if I may trust my own knowledge of my own thoughts, must be called a multitude of errors of the first magnitude.

I so much admire Mr. Lilly's outspokenness, I am so completely satisfied of the uprightness of his intentions, that it is repugnant to me to quarrel with anything he may say; and I sympathise so warmly with his manly scorn of the vileness of much that passes under the name of literature in these times, that I would willingly be silent under his by no means unkindly exposition of his theory of my own tenets, if I thought that such personal abnegation would serve the interest of the cause we both have at heart. But I cannot think so. My creed may be an ill-favoured thing, but it is mine own, as Touchstone says of his lady-love; and I have so high an opinion of the solid virtues of the object of my affections that I cannot calmly see her personated by a wench who is much uglier and has no virtue worth speaking of. I hope I should be ready to stand by a falling cause if I had ever adopted it; but suffering for a falling cause, which one has done one's best to bring to the ground, is a kind of martyrdom for which I have no taste. In my opinion, the philosophical theory which Mr. Lilly attributes to me— but which I have over and over again disclaimed

—is untenable and destined to extinction; and I not unreasonably demur to being counted among its defenders.

After the manner of a mediæval disputant, Mr. Lilly posts up three theses, which, as he conceives, embody the chief heresies propagated by the late Professor Clifford, Mr. Herbert Spencer, and myself. He says that we agree "(1) in putting aside, as unverifiable, everything which the senses cannot verify; (2) everything beyond the bounds of physical science; (3) everything which cannot be brought into a laboratory and dealt with chemically" (p. 578).

My lamented young friend Clifford, sweetest of natures though keenest of disputants, is out of reach of our little controversies, but his works speak for him, and those who run may read a refutation of Mr. Lilly's assertions in them.   Mr. Herbert Spencer, hitherto, has shown no lack either of ability or of inclination to speak for himself; and it would be a superfluity, not to say an impertinence, on my part, to take up the cudgels for him.   But, for myself, if my knowledge of my own consciousness may be assumed to be adequate (and I make not the least pretension to acquaintance with what goes on in my "Unbewusstsein"), I may be permitted to observe that the first proposition appears to me to be not true; that the second is in the same case; and that, if there be gradations in untrueness, the third is so monstrously untrue that it hovers on the verge of absurdity, even if it does not actually flounder in that logical limbo.   Thus, to all three theses, I reply in

appropriate fashion, *Nego*—I say No ; and I proceed
to state the grounds of that negation, which the proprieties do not permit me to make quite so emphatic
as I could desire.

Let me begin with the first assertion, that I "put
aside, as unverifiable, everything which the senses
cannot verify." Can such a statement as this be
seriously made in respect of any human being? But
I am not appointed apologist for mankind in general;
and confining my observations to myself, I beg leave
to point out that, at this present moment, I entertain
an unshakable conviction that Mr. Lilly is the victim
of a patent and enormous misunderstanding, and that
I have not the slightest intention of putting that conviction aside because I cannot " verify " it either by
touch, or taste, or smell, or hearing, or sight, which
(in the absence of any trace of telepathic faculty)
make up the totality of my senses.

Again, I may venture to admire the clear and
vigorous English in which Mr. Lilly embodies his
views ; but the source of that admiration does not lie
in anything which my five senses enable me to discover in the pages of his article, and of which an
orang-outang might be just as acutely sensible. No,
it lies in an appreciation of literary form and logical
structure by æsthetic and intellectual faculties which
are not senses, and which are not unfrequently sadly
wanting where the senses are in full vigour. My poor
relation may beat me in the matter of sensation ; but
I am quite confident that, when style and syllogisms
are to be dealt with, he is nowhere.

If there is anything in the world which I do firmly believe in, it is the universal validity of the law of causation ; but that universality cannot be proved by any amount of experience, let alone that which comes to us through the senses. And when an effort of volition changes the current of my thoughts, or when an idea calls up another associated idea, I have not the slightest doubt that the process to which the first of the phenomena, in each case, is due stands in the relation of cause to the second. Yet the attempt to verify this belief by sensation would be sheer lunacy. Now I am quite sure that Mr. Lilly does not doubt my sanity ; and the only alternative seems to be the admission that his first proposition is erroneous.

The second thesis charges me with putting aside " as unverifiable " " everything beyond the bounds of physical science." Again I say, No. Nobody, I imagine, will credit me with a desire to limit the empire of physical science, but I really feel bound to confess that a great many very familiar and, at the same time, extremely important phenomena lie quite beyond its legitimate limits. I cannot conceive, for example, how the phenomena of consciousness, as such and apart from the physical process by which they are called into existence, are to be brought within the bounds of physical science. Take the simplest possible example, the feeling of redness. Physical science tells us that it commonly arises as a consequence of molecular changes propagated from the eye to a certain part of the substance of the brain, when vibrations of the luminiferous ether of a certain character fall upon the retina. Let us suppose the process of physical analysis pushed so far that one could view the last link of this chain of molecules, watch their movements as if they were billiard balls, weigh them, measure them, and know all that is physically knowable about them. Well, even in that case, we should be just as far from being able to include the resulting phenomenon of consciousness, the feeling of redness, within the bounds of physical science, as we are at present. It would remain as unlike the phenomena we know under the names of matter and motion as it is now. If there is any plain truth upon which I have made it my business to insist over and over again it is this —and whether it is a truth or not, my insistence upon it leaves not a shadow of justification for Mr. Lilly's assertion.

But I ask in this case also, how is it conceivable that any man, in possession of all his natural faculties, should hold such an opinion? I do not suppose that I am exceptionally endowed because I have all my life enjoyed a keen perception of the beauty offered us by nature and by art. Now physical science may and probably will, some day, enable our posterity to set forth the exact physical concomitants and conditions of the strange rapture of beauty. But if ever that day arrives, the rapture will remain, just as it is now, outside and beyond the physical world; and, even in the mental world, something superadded to mere sensation. I do not wish to crow unduly over my humble cousin the orang, but in the æsthetic province, as in that of the intellect, I am afraid he is nowhere. I doubt not he would detect a fruit amidst a wilderness of leaves where I could see nothing; but I am tolerably confident that he has never been awe-struck, as I have been, by the dim religious gloom, as of a temple devoted to the earthgods, of the tropical forest which he inhabits. Yet I doubt not that our poor long-armed and short-legged friend, as he sits meditatively munching his durian fruit, has something behind that sad Socratic face of his which is utterly " beyond the bounds of physical science." Physical science may know all about his clutching the fruit and munching it and digesting it, and how the physical titillation of his palate is transmitted · to some microscopic cells of the gray matter of his brain. But the feelings of sweetness and of satisfac-tion which, for a moment, hang out their signal lights in his melancholy eyes, are as utterly outside the bounds of physics as is the " fine frenzy " of a human rhapsodist.

Does Mr. Lilly really believe that, putting me aside, there is any man with the feeling of music in him who disbelieves in the reality of the delight which he derives from it, because that delight lies outside the bounds of physical science, not less than outside the region of the mere sense of hearing? But, it may be, that he includes music, painting, and sculpture under the head of physical science, and in that case I can only regret I am unable to follow him in his ennoblement of my favourite pursuits.

The third thesis runs that I put aside " as unveri-fiable " " everything which cannot be brought into a laboratory and dealt with chemically;" and, once more, I say No. This wondrous allegation is no novelty; it has not unfrequently reached me from that region where gentle (or ungentle) dulness so often holds unchecked sway — the pulpit. But I marvel to find that a writer of Mr. Lilly's intelligence and good faith is willing to father such a wastrel. If I am to deal with the thing seriously, I find myself met by one of the two horns of a dilemma. Either some meaning, as unknown to usage as to the dictionaries, attaches to "laboratory" and "chemical," or the proposition is (what am I to say in my sore need for a gentle and yet appropriate word?)—well—unhistorical.

Does Mr. Lilly suppose that I put aside "as unverifiable" all the truths of mathematics, of philology, of history? And if I do not, will he have the great goodness to say how the binomial theorem is to be dealt with "chemically," even in the best appointed "laboratory"; or where the balances and crucibles are kept by which the various theories of the nature of the Basque language may be tested; or what reagents will extract the truth from any given History of Rome, and leave the errors behind as a residual calx?

I really cannot answer these questions, and unless Mr. Lilly can, I think he would do well hereafter to think more than twice before attributing such preposterous notions to his fellow-men, who, after all, as a learned counsel said, are vertebrated animals.

The whole thing perplexes me much; and I am

sure there must be an explanation which will leave
Mr. Lilly's reputation for common sense and fair
dealing untouched. Can it be—I put this forward
quite tentatively—that Mr. Lilly is the victim of a
confusion, common enough among thoughtless people,
and into which he has fallen unawares? Obviously,
it is one thing to say that the logical methods of
physical science are of universal applicability, and
quite another to affirm that all subjects of thought lie
within the province of physical science. I have often
declared my conviction that there is only one method
by which intellectual truth can be reached, whether
the subject-matter of investigation belongs to the
world of physics or to the world of consciousness;
and one of the arguments in favour of the use of
physical science as an instrument of education which
I have oftenest used is that, in my opinion, it exercises
young minds in the appreciation of inductive evidence
better than any other study. But while I repeat my
conviction that the physical sciences probably furnish
the best and most easily appreciable illustrations of
the one and indivisible mode of ascertaining truth by
the use of reason, I beg leave to add that I have
never thought of suggesting that other branches of
knowledge may not afford the same discipline; and
assuredly I have never given the slightest ground for
the attribution to me of the ridiculous contention that
there is nothing true outside the bounds of physical
science. Doubtless people who wanted to say something damaging, without too nice a regard to its
truth or falsehood, have often enough misrepresented

my plain meaning. But Mr. Lilly is not one of
these folks at whom one looks and passes by, and I
can but sorrowfully wonder at finding him in such
company.

So much for the three theses which Mr. Lilly has
nailed on to a page of this Review. I think I have
shown that the first is inaccurate, that the second is
inaccurate, and that the third is inaccurate; and that
these three inaccurates constitute one prodigious,
though I doubt not unintentional, misrepresentation.
If Mr. Lilly and I were dialectic gladiators, fighting
in the arena of the *Fortnightly*, under the eye of an
editorial lanista, for the delectation of the public, my
best tactics would now be to leave the field of battle.
For the question whether I do, or do not, hold certain
opinions is a matter of fact, with regard to which my
evidence is likely to be regarded as conclusive—at
least until such time as the telepathy of the unconscious is more generally recognised.

However, some other assertions are made by Mr.
Lilly which more or less involve matters of opinion
whereof the rights and wrongs are less easily settled,
but in respect of which he seems to me to err quite
as seriously as about the topics we have been hitherto
discussing. And the importance of these subjects
leads me to venture upon saying something about
them, even though I am thereby compelled to leave
the safe ground of personal knowledge.

Before launching the three torpedoes which have
so sadly exploded on board his own ship, Mr. Lilly
says that with whatever " rhetorical ornaments I may

gild my teaching," it is "Materialism." Let me observe, in passing, that rhetorical ornament is not in my way, and that gilding refined gold would, to my mind, be less objectionable than varnishing the fair face of truth with that pestilent cosmetic, rhetoric. If I believed that I had any claim to the title of "Materialist," as that term is understood in the language of philosophy and not in that of abuse, I should not attempt to hide it by any sort of gilding. I have not found reason to care much for hard names in the course of the last thirty years, and I am too old to develop a new sensitiveness. But, to repeat what I have more than once taken pains to say in the most unadorned of plain language, I repudiate, as philosophical error, the doctrine of Materialism as I understand it, just as I repudiate the doctrine of Spiritualism as Mr. Lilly presents it, and my reason for thus doing is, in both cases, the same; namely, that, whatever their differences, Materialists and Spiritualists agree in making very positive assertions about matters of which I am certain I know nothing, and about which I believe they are, in truth, just as ignorant. And further, that, even when their assertions are confined to topics which lie within the range of my faculties, they often appear to me to be in the wrong. And there is yet another reason for objecting to be identified with either of these sects; and that is that each is extremely fond of attributing to the other, by way of reproach, conclusions which are the property of neither, though they infallibly flow from the logical

development of the first principles of both.   Surely
a prudent man is not to be reproached because he
keeps clear of the squabbles of these philosophical
Bianchi and Neri, by refusing to have anything to do
with either ?

I understand the main tenet of Materialism to be
that there is nothing in the universe but matter and
force ; and that all the phenomena of nature are ex-
plicable by deduction from the properties assignable
to these two primitive factors.   That great champion
of Materialism whom Mr. Lilly appears to consider
to be an authority in physical science, Dr. Büchner,
embodies this article of faith on his title-page.
Kraft und Stoff—force and matter—are paraded as
the Alpha and Omega of existence.   This I apprehend is the fundamental article of the faith materialistic ; and whosoever does not hold it is condemned
by the more zealous of the persuasion (as I have
some reason to know) to the Inferno appointed for
fools or hypocrites.   But all this I heartily disbelieve ; and at the risk of being charged with wearisome repetition of an old story, I will briefly give my
reasons for persisting in my infidelity.   In the first
place, as I have already hinted, it seems to me pretty
plain that there is a third thing in the universe, to
wit, consciousness, which, in the hardness of my
heart or head, I cannot see to be matter or force, or
any conceivable modification of either, however intimately the manifestations of the phenomena of
consciousness may be connected with the phenomena
known as matter and force.   In the second place,

the arguments used by Descartes and Berkeley to show that our certain knowledge does not extend beyond our states of consciousness, appear to me to be as irrefragable now as they did when I first became acquainted with them some half century ago. All the materialistic writers I know of who have tried to bite that file have simply broken their teeth. But, if this is true, our one certainty is the existence of the mental world, and that of *Kraft und Stoff* falls into the rank of, at best, a highly probable hypothesis.

Thirdly, when I was a mere boy, with a perverse tendency to think when I ought to have been playing, my mind was greatly exercised by this formidable problem, What would become of things if they lost their qualities? As the qualities had no objective existence, and the thing without qualities was nothing, the solid world seemed whittled away—to my great horror. As I grew older, and learned to use the terms matter and force, the boyish problem was revived, *mutato nomine*. On the one hand, the notion of matter without force seemed to resolve the world into a set of geometrical ghosts, too dead even to jabber. On the other hand, Boscovich's hypothesis, by which matter was resolved into centres of force, was very attractive. But when one tried to think it out, what in the world became of force considered as an objective entity? Force, even the most materialistic of philosophers will agree with the most idealistic, is nothing but a name for the cause of motion. And if, with Boscovich, I resolved things into centres of force, then matter vanished altogether and left immaterial entities in its place. One might as well frankly accept Idealism and have done with it.

I must make a confession, even if it be humiliating. I have never been able to form the slightest conception of those "forces" which the Materialists talk about, as if they had samples of them many years in bottle. They tell me that matter consists of atoms, which are separated by mere space devoid of contents; and that, through this void, radiate the attractive and repulsive forces whereby the atoms affect one another. If anybody can clearly conceive the nature of these things which not only exist in nothingness, but pull and push there with great vigour, I envy him for the possession of an intellect of larger grasp, not only than mine, but than that of Leibnitz or of Newton.[1] To me the "chimæra, bombinans in vacuo quia comedit secundas intentiones" of the schoolmen is a familiar and domestic creature compared with such "forces." Besides, by the hypothesis, the forces are not matter; and thus all that is of any particular consequence in the world turns out to be not matter on the Materialist's own showing. Let it not be supposed that I am casting

---

[1] See the famous *Collection of Papers*, published by Clarke in 1717. Leibnitz says: "'Tis also a supernatural thing that bodies should *attract* one another at a distance without any intermediate means." And Clarke, on behalf of Newton, caps this as follows: "That one body should attract another without any intermediate *means* is, indeed, not a *miracle*, but a contradiction; for 'tis supposing something to act where it is not."

a doubt upon the propriety of the employment of the terms " atom " and " force," as they stand among the working hypotheses of physical science. As formulæ which can be applied, with perfect precision and great convenience, in the interpretation of nature, their value is incalculable ; but, as real entities, having an objective existence, an indivisible particle which nevertheless occupies space is surely inconceivable ; and with respect to the operation of that atom, where it is not, by the aid of a " force " resident in nothingness, I am as little able to imagine it as I fancy any one else is.

Unless and until anybody will resolve all these doubts and difficulties for me, I think I have a right to hold aloof from Materialism. As to Spiritualism, it lands me in even greater difficulties when I want to get change for its notes-of-hand in the solid coin of reality. For the assumed substantial entity, spirit, which is supposed to underlie the phenomena of consciousness, as matter underlies those of physical nature, leaves not even a geometrical ghost when these phenomena are abstracted. And, even if we suppose the existence of such an entity apart from qualities—that is to say, a bare existence—for mind ; how does anybody know that it differs from that other entity, apart from qualities, which is the supposed substratum of matter ? Spiritualism is, after all, little better than Materialism turned upside down. And if I try to think of the " spirit " which a man, by this hypothesis, carries about under his hat, as something devoid of relation to space, and as something indivisible, even in thought, while it is, at the same time, supposed to be in that place and to be possessed of half a dozen different faculties, I confess I get quite lost.

As I have said elsewhere, if I were forced to choose between Materialism and Idealism, I should elect for the latter; and I certainly would have nothing to do with the effete mythology of Spiritualism. But I am not aware that I am under any compulsion to choose either the one or the other. I have always entertained a strong suspicion that the sage who maintained that man is the measure of the universe was sadly in the wrong; and age and experience have not weakened that conviction. In following these lines of speculation I am reminded of the quarter-deck walks of my youth. In taking that form of exercise you may perambulate through all points of the compass with perfect safety, so long as you keep within certain limits : forget those limits, in your ardour, and mere smothering and spluttering, if not worse, await you. I stick by the deck and throw a life-buoy now and then to the struggling folk who have gone overboard; and all I get for my humanity is the abuse of all whenever they leave off abusing one another.

Tolerably early in life I discovered that one of the unpardonable sins, in the eyes of most people, is for a man to presume to go about unlabelled. The world regards such a person as the police do an unmuzzled dog, not under proper control. I could find no label that would suit me, so, in my desire to range myself and be respectable, I invented one; and, as the chief thing I was sure of was that I did not know a great

many things that the —ists and the —ites about me
professed to be familiar with, I called myself an
Agnostic.　Surely no denomination could be more
modest or more appropriate; and I cannot imagine
why I should be every now and then haled out of my
refuge and declared sometimes to be a Materialist,
sometimes an Atheist, sometimes a Positivist; and
sometimes, alas and alack, a cowardly or reactionary
Obscurantist.

I trust that I have, at last, made my case clear, and
that henceforth I shall be allowed to rest in peace—at
least, after a further explanation or two, which Mr.
Lilly proves to me may be necessary.　It has been
seen that my excellent critic has original ideas respecting the meaning of the words "laboratory" and
"chemical"; and, as it appears to me, his definition
of "Materialist" is quite as much peculiar to himself.
For, unless I misunderstand him, and I have taken
pains not to do so, he puts me down as a Materialist
(over and above the grounds which I have shown to
have no foundation); firstly, because I have said
that consciousness is a function of the brain; and,
secondly, because I hold by determinism.　With
respect to the first point, I am not aware that there
is any one who doubts that, in the proper physiological
sense of the word function, consciousness, in certain
forms at any rate, is a cerebral function.　In physiology
we call function that effect, or series of effects, which
results from the activity of an organ.　Thus, it is the
function of muscle to give rise to motion; and the
muscle gives rise to motion when the nerve which

supplies it is stimulated.   If one of the nerve-bundles in a man's arm is laid bare and a stimulus is applied to certain of the nervous filaments, the result will be production of motion in that arm.   If others are stimulated, the result will be the production of the state of consciousness called pain.   Now, if I trace these last nerve-filaments, I find them to be ultimately connected with part of the substance of the brain, just as the others turn out to be connected with muscular substance.   If the production of motion in the one case is properly said to be the function of the muscular substance, why is the production of a state of consciousness in the other case not to be called a function of the cerebral substance ?   Once upon a time, it is true, it was supposed that a certain "animal spirit" resided in muscle and was the real active agent.   But we have done with that wholly superfluous fiction so far as the muscular organs are concerned.   Why are we to retain a corresponding fiction for the nervous organs ?

If it is replied that no physiologist, however spiritual his leanings, dreams of supposing that simple sensations require a "spirit" for their production, then I must point out that we are all agreed that consciousness is a function of matter, and that particular tenet must be given up as a mark of Materialism.   Any further argument will turn upon the question, not whether consciousness is a function of the brain, but whether all forms of consciousness are so.   Again, I hold it would be quite correct to say that material changes are the causes of psychical

phenomena (and, as a consequence, that the organs
in which these changes take place have the production
of such phenomena for their function), even if the
spiritualistic hypothesis had any foundation.   For
nobody hesitates to say that an event A is the cause
of an event Z, even if there are as many intermediate
terms, known and unknown, in the chain of causation
as there are letters between A and Z.   The man who
pulls the trigger of a loaded pistol placed close to
another's head certainly is the cause of that other's
death, though, in strictness, he " causes " nothing
but the movement of the finger upon the trigger.
And, in like manner, the molecular change which
is brought about in a certain portion of the cerebral
substance by the stimulation of a remote part of the
body would be properly said to be the cause of the
consequent feeling, whatever unknown terms were
interposed between the physical agent and the actual
psychical product.   Therefore, unless Materialism
has the monopoly of the right use of language, I
see nothing materialistic in the phraseology which
I have employed.

The only remaining justification which Mr. Lilly
offers for dubbing me a Materialist, *malgré moi*, arises
out of a passage which he quotes, in which I say that
the progress of science means the extension of the
province of what we call matter and force, and the
concomitant gradual banishment from all regions of
human thought of what we call spirit and spontaneity.
I hold that opinion now, if anything, more firmly
than I did when I gave utterance to it a score of

years ago, for it has been justified by subsequent
events. But what that opinion has to do with
Materialism I fail to discover. In my judgment, it is
consistent with the most thorough-going Idealism, and
the grounds of that judgment are really very plain
and simple.

The growth of science, not merely of physical
science, but of all science, means the demonstration of
order and natural causation among phenomena which
had not previously been brought under those conceptions. Nobody who is acquainted with the progress
of scientific thinking in every department of human
knowledge, in the course of the last two centuries,
will be disposed to deny that immense provinces have
been added to the realm of science ; or to doubt that
the next two centuries will be witnesses of a vastly
greater annexation. More particularly in the region
of the physiology of the nervous system, is it justifiable to conclude from the progress that has been
made in analysing the relations between material and
psychical phenomena, that vast further advances will
be made ; and that, sooner or later, all the so-called
spontaneous operations of the mind will have, not
only their relations to one another, but their relations
to physical phenomena, connected in natural series of
causes and effects, strictly defined. In other words,
while, at present, we know only the nearer moiety of
the chain of causes and effects, by which the phenomena we call material give rise to those which we
call mental ; hereafter, we shall get to the further end
of the series.

In my innocence, I have been in the habit of supposing that this is merely a statement of facts, and that the good Bishop Berkeley, if he were alive, would find such facts fit into his system without the least difficulty. That Mr. Lilly should play into the hands of his foes, by declaring that unmistakable facts make for them, is an exemplification of ways that are dark, quite unintelligible to me. Surely Mr. Lilly does not hold that the disbelief in spontaneity —which term, if it has any meaning at all, means uncaused action—is a mark of the beast Materialism? If so, he must be prepared to tackle many of the Cartesians (if not Descartes himself), Spinoza and Leibnitz among the philosophers, Augustine, Thomas Aquinas, Calvin and his followers among theologians, as Materialists—and that surely is a sufficient *reductio ad absurdum* of such a classification.

The truth is, that in his zeal to paint "Materialism," in large letters, on everything he dislikes, Mr. Lilly forgets a very important fact, which, however, must be patent to every one who has paid attention to the history of human thought; and that fact is, that every one of the speculative difficulties which beset Kant's three problems, the existence of a Deity, the freedom of the will, and immortality, existed ages before anything that can be called physical science, and would continue to exist if modern physical science were swept away. All that physical science has done has been to make, as it were, visible and tangible some difficulties that formerly were more hard of apprehension. Moreover, these difficulties exist just as much on the hypothesis of Idealism as on that of Materialism.

The student of nature, who starts from the axiom of the universality of the law of causation, cannot refuse to admit an eternal existence; if he admits the conservation of energy, he cannot deny the possibility of an eternal energy; if he admits the existence of immaterial phenomena in the form of consciousness, he must admit the possibility, at any rate, of an eternal series of such phenomena; and, if his studies have not been barren of the best fruit of the investigation of nature, he will have enough sense to see that when Spinoza says, "Per Deum intelligo ens absolute infinitum, hoc est substantiam constantem infinitis attributis," the God so conceived is one that only a very great fool would deny, even in his heart. Physical science is as little Atheistic as it is Materialistic.

So with respect to immortality. As physical science states this problem, it seems to stand thus: "Is there any means of knowing whether the series of states of consciousness, which has been casually associated for threescore years and ten with the arrangement and movements of innumerable millions of successively different material molecules, can be continued, in like association, with some substance which has not the properties of matter and force?" As Kant said, on a like occasion, if anybody can answer that question, he is just the man I want to see. If he says that consciousness cannot exist, except in relation of cause and effect with certain organic molecules, I must ask how he knows that; and if he says it can, I must put the same question. And I am afraid that, like jesting Pilate, I shall not think it worth while (having but little time before me) to wait for an answer.

Lastly, with respect to the old riddle of the freedom of the will. In the only sense in which the word freedom is intelligible to me—that is to say, the absence of any restraint upon doing what one likes within certain limits—physical science certainly gives no more ground for doubting it than the common sense of mankind does. And if physical science, in strengthening our belief in the universality of causation and abolishing chance as an absurdity, leads to the conclusions of determinism, it does no more than follow the track of consistent and logical thinkers in philosophy and in theology, before it existed or was thought of. Whoever accepts the universality of the law of causation as a dogma of philosophy, denies the existence of uncaused phenomena. And the essence of that which is improperly called the freewill doctrine is that occasionally, at any rate, human volition is self-caused, that is to say, not caused at all; for to cause oneself one must have anteceded oneself—which is, to say the least of it, difficult to imagine.

Whoever accepts the existence of an omniscient Deity as a dogma of theology, affirms that the order of things is fixed from eternity to eternity; for the fore-knowledge of an occurrence means that the occurrence will certainly happen; and the certainty of an event happening is what is meant by its being fixed or fated.[1]

Whoever asserts the existence of an omnipotent Deity, that he made and sustains all things, and is the *causa causarum*, cannot, without a contradiction in terms, assert that there is any cause independent of him ; and it is a mere subterfuge to assert that the cause of all things can " permit " one of these things to be an independent cause.

[1] I may cite in support of this obvious conclusion of sound reasoning, two authorities who will certainly not be regarded lightly by Mr. Lilly. These are Augustine and Thomas Aquinas. The former declares that " Fate " is only an ill-chosen name for Providence.

" Prorsus divina providentia regna constituuntur humana. Quæ si propterea quisquam fato tribuit, quia ipsam Dei voluntatem vel potestatem fati nomine appellat, *sententiam teneat, linguam corrigat* " (Augustinus *De Civitate Dei*, V. c. i.)

The other great doctor of the Catholic Church, " Divus Thomas," as Suarez calls him, whose marvellous grasp and subtlety of intellect seem to me to be almost without a parallel, puts the whole case into a nutshell, when he says that the ground for doing a thing in the mind of the doer is as it were the pre-existence of the thing done :

" Ratio autem alicujus fiendi in mente actoris existens est quædam præ-existentia rei fiendæ in eo " (*Summa*, Qu. xxiii. Art. i.)

If this is not enough, I may further ask what " Materialist " has ever given a better statement of the case for determinism, on theistic grounds, than is to be found in the following passage of the *Summa*, Qu. xiv. Art. xiii.

" Omnia quæ sunt in tempore, sunt Deo ab æterno præsentia, non solum ea ex ratione quâ habet rationes rerum apud se presentes, ut quidam dicunt, sed quia ejus intuitus fertur ab æterno supra omnia, prout sunt in sua præsentialitate. *Unde manifestum est quod contingentia infallibiliter a Deo cognoscuntur*, in quantum subduntur divino conspectui secundum suam præsentialitatem ; et tamen sunt futura contingentia, suis causis proximis comparata."

[As I have not said that Thomas Aquinas is professedly a determinist, I do not see the bearing of citations from him which may be more or less inconsistent with the foregoing.]

Whoever asserts the combination of omniscience and omnipotence as attributes of the Deity, does implicitly assert predestination. For he who knowingly makes a thing and places it in circumstances the operation of which on that thing he is perfectly acquainted with, does predestine that thing to whatever fate may befall it.

Thus, to come, at last, to the really important part of all this discussion, if the belief in a God is essential to morality, physical science offers no obstacle thereto; if the belief in immortality is essential to morality, physical science has no more to say against the probability of that doctrine than the most ordinary experience has, and it effectually closes the mouths of those who pretend to refute it by objections deduced from merely physical data. Finally, if the belief in the uncausedness of volition is essential to morality, the student of physical science has no more to say against that absurdity than the logical philosopher or theologian. Physical science, I repeat, did not invent determinism, and the deterministic doctrine would stand on just as firm a foundation as it does if there were no physical science. Let any one who doubts this read Jonathan Edwards, whose demonstrations are derived wholly from philosophy and theology.

Thus, when Mr. Lilly, like another Solomon Eagle, goes about proclaiming " Woe to this wicked city," and denouncing physical science as the evil genius of modern days—mother of materialism, and fatalism, and all sorts of other condemnable isms—I

venture to beg him to lay the blame on the right
shoulders; or, at least, to put in the dock, along
with Science, those sinful sisters of hers, Philosophy and Theology, who, being so much older,
should have known better than the poor Cinderella of the schools and universities over which they
have so long dominated.   No doubt modern society
is diseased enough; but then it does not differ from
older civilisations in that respect.   Societies of men are
fermenting masses, and as beer has what the Germans
call " Oberhefe " and " Unterhefe," so every society
that has existed has had its scum at the top and its
dregs at the bottom; but I doubt if any of the " ages
of faith " had less scum or less dregs, or even showed a
proportionally greater quantity of sound wholesome
stuff in the vat.   I think it would puzzle Mr. Lilly,
or any one else, to adduce convincing evidence that,
at any period of the world's history, there was a more
widespread sense of social duty, or a greater sense of
justice, or of the obligation of mutual help, than in
this England of ours.   Ah ! but, says Mr. Lilly, these
are all products of our Christian inheritance ; when
Christian dogmas vanish virtue will disappear too, and
the ancestral ape and tiger will have full play.   But
there are a good many people who think it obvious that
Christianity also inherited a good deal from Paganism
and from Judaism; and that, if the Stoics and the
Jews revoked their bequest, the moral property of
Christianity would realise very little.   And, if morality has survived the stripping off of several sets
of clothes which have been found to fit badly, why

should it not be able to get on very well in the light and handy garments which Science is ready to provide ?

But this by the way. If the diseases of society consist in the weakness of its faith in the existence of the God of the theologians, in a future state, and in uncaused volitions, the indication, as the doctors say, is to suppress Theology and Philosophy, whose bickerings about things of which they know nothing have been the prime cause and continual sustenance of that evil scepticism which is the Nemesis of meddling with the unknowable.

Cinderella is modestly conscious of her ignorance of these high matters. She lights the fire, sweeps the house, and provides the dinner ; and is rewarded by being told that she is a base creature, devoted to low and material interests. But in her garret she has fairy visions out of the ken of the pair of shrews who are quarrelling downstairs. She sees the order which pervades the seeming disorder of the world ; the great drama of evolution, with its full share of pity and terror, but also with abundant goodness and beauty, unrolls itself before her eyes ; and she learns, in her heart of hearts, the lesson, that the foundation of morality is to have done, once and for all, with lying ; to give up pretending to believe that for which there is no evidence, and repeating unintelligible propositions about things beyond the possibilities of knowledge.

She knows that the safety of morality lies neither in the adoption of this or that philosophical specula-

tion, or this or that theological creed, but in a
real and living belief in that fixed order of nature
which sends social disorganisation upon the track
of immorality, as surely as it sends physical
disease after physical trespasses. And of that firm
and lively faith it is her high mission to be the
priestess.

VI

SCIENTIFIC AND PSEUDO-SCIENTIFIC REALISM

NEXT to undue precipitation in anticipating the results of pending investigations, the intellectual sin which is commonest and most hurtful to those who devote themselves to the increase of knowledge is the omission to profit by the experience of their predecessors recorded in the history of science and philosophy. It is true that, at the present day, there is more excuse than at any former time for such neglect. No small labour is needed to raise oneself to the level of the acquisitions already made ; and able men, who have achieved thus much, know that, if they devote themselves body and soul to the increase of their store, and avoid looking back, with as much care as if the injunction laid on Lot and his family were binding upon them, such devotion is sure to be richly repaid by the joys of the discoverer and the solace of fame, if not by rewards of a less elevated character.

So, following the advice of Francis Bacon, we refuse *inter mortuos quærere vivum;* we leave the past to bury its dead, and ignore our intellectual ancestry. Nor are we content with that. We follow the evil example set us, not only by Bacon but by almost all the men of the Renaissance, in pouring scorn upon the work of our immediate spiritual fore-fathers, the schoolmen of the middle ages. It is accepted as a truth which is indisputable, that, for seven or eight centuries, a long succession of able men —some of them of transcendent acuteness and ency-clopædic knowledge — devoted laborious lives to the grave discussion of mere frivolities and the arduous pursuit of intellectual will-o'-the-wisps. To say nothing of a little modesty, a little impartial pondering over personal experience might suggest a doubt as to the adequacy of this short and easy method of dealing with a large chapter of the history of the human mind. Even an acquaintance with popular literature which had extended so far as to include that part of the contributions of Sam Slick which contains his weighty aphorism that " there is a great deal of human nature in all mankind," might raise a doubt whether, after all, the men of that epoch, who, take them all round, were endowed with wisdom and folly in much the same proportion as ourselves, were likely to display nothing better than the qualities of energetic idiots, when they devoted their faculties to the elucidation of problems which were to them, and indeed are to us, the most serious which life has to offer. Speaking for myself, the longer I live the more I am disposed to think that there is much less either of pure folly, or of pure wickedness, in the world than is commonly supposed. It may be doubted if any sane man ever said to himself, " Evil be thou my good," and I have never yet had the good fortune to meet with a perfect

fool.   When I have brought to the inquiry the patience
and long-suffering which become a scientific investigator, the most promising specimens have turned out
to have a good deal to say for themselves from their
own point of view.   And, sometimes, calm reflection
has taught the humiliating lesson, that their point of
view was not so different from my own as I had
fondly imagined.   Comprehension is more than half-
way to sympathy, here as elsewhere.

If we turn our attention to scholastic philosophy in
the frame of mind suggested by these prefatory remarks,
it assumes a very different character from that which
it bears in general estimation.   No doubt it is surrounded by a dense thicket of thorny logomachies and
obscured by the dust-clouds of a barbarous and perplexing terminology.   But suppose that, undeterred
by much grime and by many scratches, the explorer
has toiled through this jungle, he comes to an open
country which is amazingly like his dear native land.
The hills which he has to climb, the ravines he has to
avoid, look very much the same ; there is the same
infinite space above, and the same abyss of the unknown below ; the means of travelling are the same,
and the goal is the same.

That goal for the schoolmen, as for us, is the
settlement of the question how far the universe is the
manifestation of a rational order ; in other words, how
far logical deduction from indisputable premisses will
account for that which has happened and does happen.
That was the object of scholasticism, and, so far as I
am aware, the object of modern science may be ex-

pressed in the same terms. In pursuit of this end, modern science takes into account all the phenomena of the universe which are brought to our knowledge by observation or by experiment. It admits that there are two worlds to be considered, the one physical and the other psychical; and that though there is a most intimate relation and interconnection between the two, the bridge from one to the other has yet to be found; that their phenomena run, not in one series, but along two parallel lines.

To the schoolmen the duality of the universe appeared under a different aspect. How this came about will not be intelligible unless we clearly apprehend the fact that they did really believe in dogmatic Christianity as it was formulated by the Roman Church. They did not give a mere dull assent to anything the Church told them on Sundays, and ignore her teachings for the rest of the week; but they lived and moved and had their being in that supersensible theological world which was created, or rather grew up, during the first four centuries of our reckoning, and which occupied their thoughts far more than the sensible world in which their earthly lot was cast.

For the most part, we learn history from the colourless compendiums or partisan briefs of mere scholars, who have too little acquaintance with practical life, and too little insight into speculative problems, to understand that about which they write. In historical science, as in all sciences which have to do with concrete phenomena, laboratory practice is indispensable; and the laboratory practice of historical science is afforded, on the one hand, by active social and political life, and, on the other, by the study of those tendencies and operations of the mind which embody themselves in philosophical and theological systems. Thucydides and Tacitus, and, to come nearer our own time, Hume and Grote, were men of affairs, and had acquired, by direct contact with social and political history in the making, the secret of understanding how such history is made. Our notions of the intellectual history of the middle ages are, unfortunately, too often derived from writers who have never seriously grappled with philosophical and theological problems : and hence that strange myth of a millennium of moonshine to which I have adverted.

However, no very profound study of the works of contemporary writers who, without devoting themselves specially to theology or philosophy, were learned and enlightened—such men, for example, as Eginhard or Dante—is necessary to convince oneself that, for them, the world of the theologian was an ever-present and awful reality. From the centre of that world, the Divine Trinity, surrounded by a hierarchy of angels and saints, contemplated and governed the insignificant sensible world in which the inferior spirits of men, burdened with the debasement of their material embodiment and continually solicited to their perdition by a no less numerous and almost as powerful hierarchy of devils, were constantly struggling on the edge of the pit of everlasting damnation.[1]

[1] There is no exaggeration in this brief and summary view of the Catholic cosmos. But it would be unfair to leave it to be supposed that

The men of the middle ages believed that through the Scriptures, the traditions of the Fathers, and the authority of the Church, they were in possession of far more, and more trustworthy, information with respect to the nature and order of things in the theological world than they had in regard to the nature and order of things in the sensible world. And, if the two sources of information came into conflict, so much the worse for the sensible world, which, after all, was more or less under the dominion of Satan. Let us suppose that a telescope powerful enough to show us what is going on in the nebula of the sword of Orion, should reveal a world in which stones fell upwards, parallel lines met, and the fourth dimension of space was quite obvious. Men of science would have only two alter-

the Reformation made any essential alteration, except perhaps for the worse, in that cosmology which called itself "Christian." The protagonist of the Reformation, from whom the whole of the Evangelical sects are lineally descended, states the case with that plainness of speech, not to say brutality, which characterised him. Luther says that man is a beast of burden who only moves as his rider orders; sometimes God rides him, and sometimes Satan. "Sic voluntas humana in medio posita est, ceu jumentum ; si insederit Deus, vult et vadit, quo vult Deus. . . . Si insederit Satan, vult et vadit, quo vult Satan ; nec est in ejus arbitrio ad utrum sessorem currere, aut eum quærere, sed ipsi sessores certant ob ipsum obtinendum et possidendum " (*De Servo Arbitrio*, M. Lutheri Opera, ed. 1546, t. ii. p. 468). One may hear substantially the same doctrine preached in the parks and at street-corners by zealous volunteer missionaries of Evangelicism, any Sunday, in modern London. Why these doctrines, which are conspicuous by their absence in the four Gospels, should arrogate to themselves the title of Evangelical, in contradistinction to Catholic, Christianity, may well perplex the impartial inquirer, who, if he were obliged to choose between the two, might naturally prefer that which leaves the poor beast of burden a little freedom of choice.

natives before them.   Either the terrestrial and the nebular facts must be brought into harmony by such feats of subtle sophistry as the human mind is always capable of performing when driven into a corner; or science must throw down its arms in despair, and commit suicide, either by the admission that the universe is, after all, irrational, inasmuch as that which is truth in one corner of it is absurdity in another, or by a declaration of incompetency.

In the middle ages, the labours of those great men who endeavoured to reconcile the system of thought which started from the data of pure reason, with that which started from the data of Roman theology, produced the system of thought· which is known as scholastic philosophy; the alternative of surrender and suicide is exemplified by Avicenna and his followers when they declared that that which is true in theology may be false in philosophy, and *vice versâ;* and by Sanchez in his famous defence of the thesis " *Quod nil scitur.*"

To those who deny the validity of one of the primary assumptions of the disputants—who decline, on the ground of the utter insufficiency of the evidence, to put faith in the reality of that other world, the geography and the inhabitants of which are so confidently described in the so-called[1] Christianity of Catholicism—the long and bitter contest, which

[1] I say "so-called" not by way of offence, but as a protest against the monstrous assumption that Catholic Christianity is explicitly or implicitly contained in any trustworthy record of the teaching of Jesus of Nazareth.

engaged the best intellects for so many centuries, may
seem a terrible illustration of the wasteful way in
which the struggle for existence is carried on in the
world of thought, no less than in that of matter.
But there is a more cheerful mode of looking at the
history of scholasticism.   It ground and sharpened
the dialectic implements of our race as perhaps
nothing but discussions, in the result of which men
thought their eternal no less than their temporal
interests were at stake, could have done.   When a
logical blunder may ensure combustion, not only in
the next world but in this, the construction of syllogisms acquires a peculiar interest.   Moreover, the
schools kept the thinking faculty alive and active,
when the disturbed state of civil life, the mephitic
atmosphere engendered by the dominant ecclesiasticism, and the almost total neglect of natural knowledge, might well have stifled it.   And, finally, it
should be remembered that scholasticism really did
thresh out pretty effectually certain problems which
have presented themselves to mankind ever since
they began to think, and which, I suppose, will
present themselves so long as they continue to think.
Consider, for example, the controversy of the Realists
and the Nominalists, which was carried on with
varying fortunes, and under various names, from the
time of Scotus Erigena to the end of the scholastic
period.   Has it now a merely antiquarian interest?
Has Nominalism, in any of its modifications, so completely won the day that Realism may be regarded
as dead and buried without hope of resurrection?

Many people seem to think so, but it appears to me that, without taking Catholic philosophy into consideration, one has not to look about far to find evidence that Realism is still to the fore, and indeed extremely lively.[1]

The other day I happened to meet with a report of a sermon recently preached in St. Paul's Cathedral. From internal evidence I am inclined to think that the report is substantially correct. But as I have not the slightest intention of finding fault with the eminent theologian and eloquent preacher to whom the discourse is attributed, for employment of scientific language in a manner for which he could find only too many scientific precedents, the accuracy of the report in detail is not to the purpose. I may safely take it as the embodiment of views which are thought to be quite in accordance with science by many excellent, instructed, and intelligent people.

The preacher further contended that it was yet more difficult to realise that our earthly home would become the scene of a vast physical catastrophe. Imagination recoils from the idea that the

---

[1] It may be desirable to observe that, in modern times, the term " Realism " has acquired a signification wholly different from that which attached to it in the middle ages. We commonly use it as the contrary of Idealism. The Idealist holds that the phenomenal world has only a subjective existence, the Realist that it has an objective existence. I am not aware that any mediæval philosopher was an Idealist in the sense in which we apply the term to Berkeley. In fact, the cardinal defect of their speculations lies in their oversight of the considerations which lead to Idealism. If many of them regarded the material world as a negation, it was an active negation ; not zero, but a minus quantity.

course of nature—the phrase helps to disguise the truth—so unvarying and regular, the ordered sequence of movement and life, should suddenly cease. Imagination looks more reasonable when it assumes the air of scientific reason. Physical law, it says, will prevent the occurrence of catastrophes only anticipated by an apostle in an unscientific age. Might not there, however, be a suspension of a lower law by the intervention of a higher? Thus every time we lifted our arms we defied the laws of gravitation, and in railways and steamboats powerful laws were held in check by others. The flood and the destruction of Sodom and Gomorrah were brought about by the operations of existing laws, and may it not be that in His illimitable universe there are more important laws than those which surround our puny life— moral and not merely physical forces? Is it inconceivable that the day will come when these royal and ultimate laws shall wreck the natural order of things which seems so stable and so fair? Earthquakes were not things of remote antiquity, as an island off Italy, the Eastern Archipelago, Greece, and Chicago bore witness. . . . In presence of a great earthquake men feel how powerless they are, and their very knowledge adds to their weakness. The end of human probation, the final dissolution of organised society, and the destruction of man's home on the surface of the globe, were none of them violently contrary to our present experience, but only the extension of present facts. The presentiment of death was common; there were felt to be many things which threatened the existence of society; and as our globe was a ball of fire, at any moment the pent-up forces which surge and boil beneath our feet might be poured out (*Pall Mall Gazette*, December 6, 1886).

The preacher appears to entertain the notion that the occurrence of a " catastrophe "[1] involves a breach of the present order of nature—that it is an event incompatible with the physical laws which at present

---

[1] At any rate a catastrophe greater than the flood, which, as I observe with interest, is as calmly assumed by the preacher to be an historical event as if science had never had a word to say on that subject!

obtain.   He seems to be of opinion that " scientific
reason " lends its authority to the imaginative supposition that physical law will prevent the occurrence of the
" catastrophes " anticipated by an unscientific apostle.

Scientific reason, like Homer, sometimes nods ; but
I am not aware that it has ever dreamed dreams of
this sort.   The fundamental axiom of scientific thought
is that there is not, never has been, and never will be,
any disorder in nature.   The admission of the occurrence of any event which was not the logical consequence of the immediately antecedent events, according
to these definite, ascertained, or unascertained rules
which we call the " laws of nature," would be an act
of self-destruction on the part of science.

" Catastrophe " is a relative conception.   For ourselves it means an event which brings about very
terrible consequences to man, or impresses his mind
by its magnitude relatively to him.   But events which
are quite in the natural order of things to us, may be
frightful catastrophes to other sentient beings.   Surely
no interruption of the order of nature is involved if,
in the course of descending through an Alpine pine-
wood, I jump upon an anthill and in a moment wreck
a whole city and destroy a hundred thousand of its
inhabitants.   To the ants the catastrophe is worse
than the earthquake of Lisbon.   To me it is the
natural and necessary consequence of the laws of
matter in motion.   A redistribution of energy has
taken place, which is perfectly in accordance with
natural order, however unpleasant its effects may be
to the ants.

Imagination, inspired by scientific reason, and not merely assuming the airs thereof, as it unfortunately too often does in the pulpit, so far from having any right to repudiate catastrophes and deny the possibility of the cessation of motion and life, easily finds justification for the exactly contrary course. Kant in his famous *Theory of the Heavens* declares the end of the world and its reduction to a formless condition to be a necessary consequence of the causes to which it owes its origin and continuance. And, as to catastrophes of prodigious magnitude and frequent occurrence, they were the favourite *asylum ignorantiæ* of geologists, not a quarter of a century ago. If modern geology is becoming more and more disinclined to call in catastrophes to its aid, it is not because of any *à priori* difficulty in reconciling the occurrence of such events with the universality of order, but because the *à posteriori* evidence of the occurrence of events of this character in past times has more or less completely broken down.

It is, to say the least, highly probable that this earth is a mass of extremely hot matter, invested by a cooled crust, through which the hot interior still continues to cool, though with extreme slowness. It is no less probable that the faults and dislocations, the foldings and fractures, everywhere visible in the stratified crust, its large and slow movements through miles of elevation and depression, and its small and rapid movements which give rise to the innumerable perceived and unperceived earthquakes which are constantly occurring, are due to the shrinkage of the crust on its cooling and contracting nucleus.

Without going beyond the range of fair scientific analogy, conditions are easily conceivable which should render the loss of heat far more rapid than it is at present; and such an occurrence would be just as much in accordance with ascertained laws of nature as the more rapid cooling of a redhot bar, when it is thrust into cold water, than when it remains in the air. But much more rapid cooling might entail a shifting and rearrangement of the parts of the crust of the earth on a scale of unprecedented magnitude, and bring about " catastrophes " to which the earth-quake of Lisbon is but a trifle. It is conceivable that man and his works and all the higher forms of ani-mal life should be utterly destroyed; that mountain regions should be converted into ocean depths and the floor of oceans raised into mountains; and the earth become a scene of horror which even the lurid fancy of the writer of the Apocalypse would fail to portray. And yet, to the eye of science, there would be no more disorder here than in the Sabbatical peace of a summer sea. Not a link in the chain of natural causes and effects would be broken, nowhere would there be the slightest indication of the " sus-pension of a lower law by a higher." If a sober scientific thinker is inclined to put little faith in the wild vaticinations of universal ruin which, in a less saintly person than the seer of Patmos, might seem to be dictated by the fury of a revengeful fanatic rather than by the spirit of the teacher who bid men love

their enemies, it is not on the ground that they contradict scientific principles ; but because the evidence
of their scientific value does not fulfil the conditions
on which weight is attached to evidence. The
imagination which supposes that it does, simply does
not "assume the air of scientific reason."

I repeat that, if imagination is used within the
limits laid down by science, disorder is unimaginable.
If a being endowed with perfect intellectual and
æsthetic faculties, but devoid of the capacity for
suffering pain, either physical or moral, were to devote
his utmost powers to the investigation of nature, the
universe would seem to him to be a sort of kaleidoscope, in which, at every successive moment of time, a
new arrangement of parts of exquisite beauty and
symmetry would present itself; and each of them
would show itself to be the logical consequence of the
preceding arrangement, under the conditions which
we call the laws of nature. Such a spectator might
well be filled with that *Amor intellectualis Dei*, the
beatific vision of the *vita contemplativa*, which some
of the greatest thinkers of all ages, Aristotle, Aquinas,
Spinoza, have regarded as the only conceivable eternal
felicity ; and the vision of illimitable suffering, as if
sensitive beings were unregarded animalcules which
had got between the bits of glass of the kaleidoscope,
which mars the prospect to us poor mortals, in no
wise alters the fact that order is lord of all, and disorder only a name for that part of the order which
gives us pain.

The other fallacious employment of the names of

scientific conceptions which pervades the preacher's utterance, brings me back to the proper topic of the present paper. It is the use of the word "law" as if it denoted a thing—as if a "law of nature," as science understands it, were a being endowed with certain powers, in virtue of which the phenomena expressed by that law are brought about. The preacher asks, "Might not there be a suspension of a lower law by the intervention of a higher?" He tells us that every time we lift our arms we defy the law of gravitation. He asks whether some day certain "royal and ultimate laws" may not come and "wreck" those laws which are at present, it would appear, acting as nature's police. It is evident, from these expressions, that "laws," in the mind of the preacher, are entities having an objective existence in a graduated hierarchy. And it would appear that the "royal laws" are by no means to be regarded as constitutional royalties: at any moment, they may, like Eastern despots, descend in wrath among the middle-class and plebeian laws, which have hitherto done the drudgery of the world's work, and, to use phraseology not unknown in our seats of learning—"make hay" of their belongings. Or perhaps a still more familiar analogy has suggested this singular theory; and it is thought that high laws may "suspend" low laws, as a bishop may suspend a curate.

Far be it from me to controvert these views, if any one likes to hold them. All I wish to remark is that such a conception of the nature of "laws" has nothing to do with modern science. It is scholastic

realism—realism as intense and unmitigated as that
of Scotus Erigena a thousand years ago. The essence
of such realism is that it maintains the objective existence of universals, or, as we call them nowadays,
general propositions. It affirms, for example, that
" man " is a real thing, apart from individual men,
having its existence, not in the sensible, but in the
intelligible world, and clothing itself with the accidents of sense to make the Jack and Tom and Harry
whom we know. Strange as such a notion may
appear to modern scientific thought, it really pervades ordinary language. There are few people who
would, at once, hesitate to admit that colour, for
example, exists apart from the mind which conceives
the idea of colour. They hold it to be something
which resides in the coloured object ; and so far they
are as much Realists as if they had sat at Plato's feet.
Reflection on the facts of the case must, I imagine,
convince every one that " colour " is—not a mere
name, which was the extreme Nominalist position—
but a name for that group of states of feeling which
we call blue, red, yellow, and so on, and which we
believe to be caused by luminiferous vibrations which
have not the slightest resemblance to colour ; while
these again are set afoot by states of the body to
which we ascribe colour, but which are equally devoid
of likeness to colour.

In the same way, a law of nature, in the scientific
sense, is the product of a mental operation upon the
facts of nature which come under our observation,
and has no more existence outside the mind than

colour has.   The law of gravitation is a statement of
the manner in which experience shows that bodies,
which are free to move, do, in fact, move towards one
another.   But the other facts of observation, that
bodies are not always moving in this fashion, and
sometimes move in a contrary direction, are implied
in the words " free to move."   If it is a law of nature
that bodies tend to move towards one another in a
certain way ; it is another and no less true law of
nature that, if bodies are not free to move as they
tend to do, either in consequence of an obstacle, or of
a contrary impulse from some other source of energy
than that to which we give the name of gravitation,
they either stop still, or go another way.

Scientifically speaking, it is the acme of absurdity to
talk of a man defying the law of gravitation when he lifts
his arm.   The general store of energy in the universe
working through terrestrial matter is doubtless tending
to bring the man's arm down ; but the particular
fraction of that energy which is working through
certain of his nervous and muscular organs is tending
to drive it up, and more energy being expended on
the arm in the upward than in the downward direction,
the arm goes up accordingly.   But the law of gravitation is no more defied in this case than when a grocer
throws so much sugar into the empty pan of his
scales that the one which contains the weight kicks
the beam.

The tenacity of the wonderful fallacy that the
laws of nature are agents, instead of being, as they
really are, a mere record of experience, upon which we

base our interpretations of that which does happen, and our anticipation of that which will·happen, is an interesting psychological fact; and would be unintelligible if the tendency of the human mind towards realism were less strong.

Even at the present day, and in the writings of men who would at once repudiate scholastic realism in any form, " law " is often inadvertently employed in the sense of cause, just as, in common life, a man will say that he is compelled by the law to do so and so, when, in point of fact, all he means is that the law orders him to do it, and tells him what will happen if he does not do it. We commonly hear of bodies falling to the ground by reason of the law of gravitation, whereas that law is simply the record of the fact that, according to all experience, they have so fallen (when free to move), and of the grounds of a reasonable expectation that they will so fall. If it should be worth anybody's while to seek for examples of such misuse of language on my own part, I am not at all sure he might not succeed, though I have usually been on my guard against such looseness of expression. If I am guilty, I do penance beforehand, and only hope that I may thereby deter others from committing the like fault. And I venture on this personal observation by way of showing that I have no wish to bear hardly on the preacher for falling into an error for which he might find good precedents. But it is one of those errors which, in the case of a person engaged in scientific pursuits, do little harm, because it is corrected as soon as its consequences become obvious; while those who know physical science only by name are, as has been seen, easily led to build a mighty fabric of unrealities on this fundamental fallacy. In fact, the habitual use of the word "law," in the sense of an active thing, is almost a mark of pseudo-science; it characterises the writings of those who have appropriated the forms of science without knowing anything of its substance.

There are two classes of these people : those who are ready to believe in any miracle so long as it is guaranteed by ecclesiastical authority; and those who are ready to believe in any miracle so long as it has some different guarantee. The believers in what are ordinarily called miracles — those who accept the miraculous narratives which they are taught to think are essential elements of religious doctrine—are in the one category; the spirit-rappers, table-turners, and all the other devotees of the occult sciences of our day are in the other : and, if they disagree in most things they agree in this, namely, that they ascribe to science a dictum that is not scientific; and that they endeavour to upset the dictum thus foisted on science by a realistic argument which is equally unscientific.

It is asserted, for example, that, on a particular occasion, water was turned into wine; and, on the other hand, it is asserted that a man or a woman "levitated" to the ceiling, floated about there, and finally sailed out by the window. And it is assumed that the pardonable scepticism, with which most scientific men receive these statements, is due to the fact that they feel themselves justified in denying the possibility of any such metamorphosis of water or of any such levitation, because such events are contrary to the laws of nature. So the question of the preacher is triumphantly put: How do you know that there are not "higher" laws of nature than your chemical and physical laws, and that these higher laws may not intervene and "wreck" the latter?

The plain answer to this question is, Why should anybody be called upon to say how he knows that which he does not know? You are assuming that laws are agents—efficient causes of that which happens—and that one law can interfere with another. To us, that assumption is as nonsensical as if you were to talk of a proposition of Euclid being the cause of the diagram which illustrates it, or of the integral calculus interfering with the rule of three. Your question really implies that we pretend to complete knowledge not only of all past and present phenomena, but of all that are possible in the future, and we leave all that sort of thing to the adepts of esoteric Buddhism. Our pretensions are infinitely more modest. We have succeeded in finding out the rules of action of a little bit of the universe; we call these rules "laws of nature," not because anybody knows whether they bind nature or not, but because we find it is obligatory on us to take them into account, both as actors' under nature, and as interpreters of nature. We have any quantity of genuine miracles of our own, and if you will furnish us with as good evidence of your miracles as we have of ours, we shall be quite happy to accept them and to amend our expression of the laws of nature in accordance with the new facts.

As to the particular cases adduced, we are so perfectly fair-minded as to be willing to help your case as far as we can. You are quite mistaken in supposing that anybody who is acquainted with the possibilities of physical science will undertake categorically to deny that water may be turned into wine. Many very competent judges are already inclined to think that the bodies, which we have hitherto called elementary, are really composite arrangements of the particles of a uniform primitive matter. Supposing that view to be correct, there would be no more theoretical difficulty about turning water into alcohol, ethereal and colouring matters, than there is, at this present moment, any practical difficulty in working other such miracles; as when we turn sugar into alcohol, carbonic acid, glycerine, and succinic acid; or transmute gas-refuse into perfumes rarer than musk and dyes richer than Tyrian purple. If the so-called "elements," oxygen and hydrogen, which compose water, are aggregates of the same ultimate particles, or physical units, as those which enter into the structure of the so-called element "carbon," it is obvious that alcohol and other substances, composed of carbon, hydrogen, and oxygen, may be produced by a rearrangement of some of the units of oxygen and hydrogen into the "element" carbon, and their synthesis with the rest of the oxygen and hydrogen.

s

Theoretically, therefore, we can have no sort of objection to your miracle. And our reply to the levitators is just the same. Why should not your friend "levitate"? Fish are said to rise and sink in the water by altering the volume of an internal air-receptacle; and there may be many ways science, as yet, knows nothing of, by which we, who live at the bottom of an ocean of air, may do the same thing. Dialectic gas and wind appear to be by no means wanting among you, and why should not long practice in pneumatic philosophy have resulted in the internal generation of something a thousand times rarer than hydrogen, by which, in accordance with the most ordinary natural laws, you would not only rise to the ceiling and float there in quasi-angelic posture, but perhaps, as one of your feminine adepts is said to have done, flit swifter than train or telegram to "still-vexed Bermoothes," and twit Ariel, if he happens to be there, for a sluggard? We have not the presumption to deny the possibility of anything you affirm; only, as our brethren are particular about evidence, do give us as much to go upon as may save us from being roared down by their inextinguishable laughter.

Enough of the realism which clings about "laws." There are plenty of other exemplifications of its vitality in modern science, but I will cite only one of them.

This is the conception of "vital force" which comes straight from the philosophy of Aristotle. It is a fundamental proposition of that philosophy that

VI   SCIENTIFIC AND PSEUDO-SCIENTIFIC REALISM   259

a natural object is composed of two constituents—the one its matter, conceived as inert or even, to a certain extent, opposed to orderly and purposive motion ; the other its form, conceived as a quasi-spiritual something, containing or conditioning the actual activities of the body and the potentiality of its possible activities.

I am disposed to think that the prominence of this conception in Aristotle's theory of things arose from the circumstance that he was, to begin with and throughout his life, devoted to biological studies. In fact it is a notion which must force itself upon the mind of any one who studies biological phenomena, without reference to general physics, as they now stand. Everybody who observes the obvious pheno-mena of the development of a seed into a tree, or of an egg into an animal, will note that a relatively form-less mass of matter gradually grows, takes a definite shape and structure, and, finally, begins to perform actions which contribute towards a certain end, namely, the maintenance of the individual in the first place, and of the species in the second. Starting from the axiom that every event has a cause, we have here the *causa finalis* manifested in the last set of phenomena, the *causa materialis* and *formalis* in the first, while the existence of a *causa efficiens* within the seed or egg and its product, is a corollary from the phenomena of growth and metamorphosis, which proceed in unbroken succession and make up the life of the animal or plant.

Thus, at starting, the egg or seed is matter having

a "form" like all other material bodies. But this form has the peculiarity, in contradistinction to lower substantial "forms," that it is a power which constantly works towards an end by means of living organisation.

So far as I know, Leibnitz is the only philosopher (at the same time a man of science, in the modern sense, of the first rank) who has noted that the modern conception of Force, as a sort of atmosphere enveloping the particles of bodies, and having potential or actual activity, is simply a new name for the Aristotelian Form.[1] In modern biology, up till within quite recent times, the Aristotelian conception held undisputed sway; living matter was endowed with "vital force," and that accounted for everything. Whosoever was not satisfied with that explanation was treated to that very "plain argument"—"confound you eternally"—wherewith Lord Peter overcomes the doubts of his brothers in the *Tale of a Tub*. "Materialist" was the mildest term applied to him—fortunate if he escaped pelting with "infidel" and "atheist." There may be scientific Rip Van Winkles about, who still hold by vital force; but among those biologists who have not been asleep for the last quarter of a century "vital force" no longer figures in the vocabulary of science. It is a patent survival of realism; the generalisation from experience that all living bodies exhibit certain activities of a definite character is made the basis of the notion

---

[1] "Les formes des anciens ou Entéléchies ne sont autre chose que les forces" (Leibnitz, *Lettre au Père Bouvet*, 1697).

that every living body contains an entity, " vital force," which is assumed to be the cause of those activities.

It is remarkable, in looking back, to notice to what an extent this and other survivals of scholastic realism arrested or, at any rate, impeded the application of sound scientific principles to the investigation of bio- logical phenomena. When I was beginning to think about these matters, the scientific world was occa- sionally agitated by discussions respecting the nature of the " species " and " genera " of Naturalists, of a different order from the disputes of a later time. I think most were agreed that a " species " was some- thing which existed objectively, somehow or other, and had been created by a Divine fiat. As to the objective reality of genera, there was a good deal of difference of opinion. On the other hand, there were a few who could see no objective reality in anything but individuals, and looked upon both species and genera as hypostatised universals. As for myself, I seem to have unconsciously emulated William of Occam, inasmuch as almost the first public discourse I ever ventured upon, dealt with " Animal Individual- ity," and its tendency was to fight the Nominalist battle even in that quarter.

Realism appeared in still stranger forms at the time to which I refer. The community of plan which is observable in each great group of animals was hypostatised into a Platonic idea with the appropriate name of " archetype," and we were told, as a disciple of Philo-Judæus might have told us, that this realistic figment was " the archetypal light " by which Nature has been guided amidst the " wreck of worlds." So, again, another naturalist, who had no less earned a well-deserved reputation by his contributions to positive knowledge, put forward a theory of the production of living things which, as nearly as the increase of knowledge allowed, was a reproduction of the doctrine inculcated by the Jewish Cabbala.

Annexing the archetype notion, and carrying it to its full logical consequence, the author of this theory conceived that the species of animals and plants were so many incarnations of the thoughts of God—material representations of Divine ideas—during the particular period of the world's history at which they existed. But, under the influence of the embryological and palæontological discoveries of modern times, which had already lent some scientific support to the revived ancient theories of cosmical evolution or emanation, the ingenious author of this speculation, while denying and repudiating the ordinary theory of evolution by successive modification of individuals, maintained and endeavoured to prove the occurrence of a progressive modification in the Divine ideas of successive epochs.

On the foundation of a supposed elevation of organisation in the whole living population of any epoch as compared with that of its predecessor, and a supposed complete difference in species between the populations of any two epochs (neither of which suppositions has stood the test of further inquiry), the author of this speculation based his conclusion that the Creator had, so to speak, improved upon his thoughts as time went on; and that, as each such

amended scheme of creation came up, the embodiment
of the earlier divine thoughts was swept away by a
universal catastrophe, and an incarnation of the improved ideas took its place. Only after the last such
" wreck " thus brought about, did the embodiment of
a divine thought, in the shape of the first man, make
its appearance as the *ne plus ultra* of the cosmogonical
process.

I imagine that Louis Agassiz, the genial back-
woodsman of the science of my young days, who did
more to open out new tracks in the scientific forest
than most men, would have been much surprised to
learn that he was preaching the doctrine of the
Cabbala, pure and simple. According to this modification of Neoplatonism by contact with Hebrew
speculation, the divine essence is unknowable—without form or attribute; but the interval between it
and the world of sense is filled by intelligible entities,
which are nothing but the familiar hypostatised
abstractions of the realists. These have emanated,
like immense waves of light, from the divine centre,
and, as ten consecutive zones of Sephiroth, form the
universe. The farther away from the centre, the
more the primitive light wanes, until the periphery
ends in those mere negations, darkness and evil,
which are the essence of matter. On this, the divine
agency transmitted through the Sephiroth operates
after the fashion of the Aristotelian forms, and, at
first, produces the lowest of a series of worlds. After
a certain duration the primitive world is demolished
and its fragments used up in making a better; and

this process is repeated, until at length a final world, with man for its crown and finish, makes its appearance. It is needless to trace the process of retrogressive metamorphosis by which, through the agency of the Messiah, the steps of the process of evolution here sketched are retraced. Sufficient has been said to prove that the extremest realism current in the philosophy of the thirteenth century can be fully matched by the speculations of our own time.

VII

SCIENCE AND PSEUDO-SCIENCE

In the opening sentences of a contribution to the last number of this Review,[1] the Duke of Argyll has favoured me with a lecture on the proprieties of controversy, to which I should be disposed to listen with more docility if his Grace's precepts appeared to me to be based upon rational principles, or if his example were more exemplary.

With respect to the latter point, the Duke has thought fit to entitle his article " Professor Huxley on Canon Liddon," and thus forces into prominence an element of personality, which those who read the paper which is the object of the Duke's animadversions will observe I have endeavoured, most carefully, to avoid. My criticisms dealt with a report of a sermon, published in a newspaper, and thereby addressed to all the world. Whether that sermon was preached by A or B was not a matter of the smallest consequence ; and I went out of my way to absolve the learned divine to whom the discourse was attributed, from the responsibility for statements which, for anything I knew to the contrary, might contain

[1] *Nineteenth Century*, March 1887.

imperfect, or inaccurate, representations of his views. The assertion that I had the wish or was beset by any "temptation to attack" Canon Liddon is simply contrary to fact.

But suppose that if, instead of sedulously avoiding even the appearance of such attack, I had thought fit to take a different course; suppose that, after satisfying myself that the eminent clergyman whose name is paraded by the Duke of Argyll had really uttered the words attributed to him from the pulpit of St. Paul's, what right would any one have to find fault with my action on grounds either of justice, expediency, or good taste?

Establishment has its duties as well as its rights. The clergy of a State Church enjoy many advantages over those of unprivileged and unendowed religious persuasions; but they lie under a correlative responsibility to the State, and to every member of the body politic. I am not aware that any sacredness attaches to sermons. If preachers stray beyond the doctrinal limits set by lay lawyers, the Privy Council will see to it; and, if they think fit to use their pulpits for the promulgation of literary, or historical, or scientific errors, it is not only the right, but the duty, of the humblest layman, who may happen to be better informed, to correct the evil effects of such perversion of the opportunities which the State affords them and such misuse of the authority which its support lends them. Whatever else it may claim to be, in its relations with the State, the Established Church is a branch of the Civil Service; and, for those who repudiate the ecclesiastical authority of the clergy, they are merely civil servants, as much responsible to the English people for the proper performance of their duties as any others.

The Duke of Argyll tells us that the "work and calling" of the clergy prevent them from "pursuing disputation as others can." I wonder if his Grace ever reads the so-called religious newspapers. It is not an occupation which I should commend to any one who wishes to employ his time profitably; but a very short devotion to this exercise will suffice to convince him that the "pursuit of disputation," carried to a degree of acrimony and vehemence unsurpassed in lay controversies, seems to be found quite compatible with the "work and calling" of a remarkably large number of the clergy.

Finally, it appears to me that nothing can be in worse taste than the assumption that a body of English gentlemen can, by any possibility, desire that immunity from criticism which the Duke of Argyll claims for them. Nothing would be more personally offensive to me than the supposition that I shirked criticism, just or unjust, of any lecture I ever gave. I should be utterly ashamed of myself if, when I stood up as an instructor of others, I had not taken every pains to assure myself of the truth of that which I was about to say; and I should feel myself bound to be even more careful with a popular assembly, who would take me more or less on trust, than with an audience of competent and critical experts.

I decline to assume that the standard of morality,

in these matters, is lower among the clergy than it is among scientific men. I refuse to think that the priest who stands up before a congregation, as the minister and interpreter of the Divinity, is less careful in his utterances, less ready to meet adverse comment, than the layman who comes before his audience, as the minister and interpreter of nature. Yet what should we think of the man of science who, when his ignorance or his carelessness was exposed, whined about the want of delicacy of his critics, or pleaded his "work and calling" as a reason for being let alone ?

No man, nor any body of men, is good enough, or wise enough, to dispense with the tonic of criticism. Nothing has done more harm to the clergy than the practice, too common among laymen, of regarding them, when in the pulpit, as a sort of chartered libertines, whose divagations are not to be taken seriously. And I am well assured that the distinguished divine, to whom the sermon is attributed, is the last person who would desire to avail himself of the dishonouring protection which has been superfluously thrown over him.

So much for the lecture on propriety. But the Duke of Argyll, to whom the hortatory style seems to come naturally, does me the honour to make my sayings the subjects of a series of other admonitions, some on philosophical, some on geological, some on biological topics. I can but rejoice that the Duke's authority in these matters is not always employed to show that I am ignorant of them ; on the contrary, I meet with an amount of agreement, even of approbation, for which I proffer such gratitude as may be due, even if that gratitude is sometimes almost over-shadowed by surprise.

I am unfeignedly astonished to find that the Duke of Argyll, who professes to intervene on behalf of the preacher, does really, like another Balaam, bless me altogether in respect of the main issue.

I denied the justice of the preacher's ascription to men of science of the doctrine that miracles are in-credible, because they are violations of natural law ; and the Duke of Argyll says that he believes my " denial to be well founded. The preacher was answering an objection which has now been generally abandoned." Either the preacher knew this or he did not know it. It seems to me, as a mere lay teacher, to be a pity that the " great dome of St. Paul's " should have been made to " echo " (if so be that such stentorian effects were really produced) a statement which, admitting the first alternative, was unfair, and, admitting the second, was ignorant.[1]

[1] The Duke of Argyll speaks of the recent date of the demon-stration of the fallacy of the doctrine in question. "Recent" is a relative term, but I may mention that the question is fully discussed in my book on "Hume"; which, if I may believe my publishers, has been read by a good many people since it appeared in 1879. Moreover, I observe, from a note at page 89 of *The Reign of Law*, a work to which I shall have occasion to advert by and by, that the Duke of Argyll draws attention to the circumstance that, so long ago as 1866, the views which I hold on this subject were well known. The Duke, in fact, writing about this time, says, after quoting a phrase of mine : " The question of miracles seems now to be admitted on all hands to be simply a question of evidence." In science we think that a teacher who ignores views which have been discussed *coram populo* for twenty years, is hardly up to the mark.

Having thus sacrificed one half of the preacher's arguments, the Duke of Argyll proceeds to make equally short work with the other half. It appears that he fully accepts my position that the occurrence of those events, which the preacher speaks of as catastrophes, is no evidence of disorder, inasmuch as such catastrophes may be necessary occasional consequences of uniform changes. Whence I conclude, his Grace agrees with me, that the talk about royal laws " wrecking" ordinary laws may be eloquent metaphor, but is also nonsense.

And now comes a further surprise. After having given these superfluous stabs to the slain body of the preacher's argument, my good ally remarks, with magnificent calmness : " So far, then, the preacher and the professor are at one." " Let them smoke the calumet." By all means : smoke would be the most appropriate symbol of this wonderful attempt to cover a retreat. After all, the Duke has come to bury the preacher, not to praise him ; only he makes the funeral obsequies look as much like a triumphal procession as possible.

So far as the questions between the preacher and myself are concerned, then, I may feel happy. The authority of the Duke of Argyll is ranged on my side. But the Duke has raised a number of other questions, with respect to which I fear I shall have to dispense with his support—nay, even be compelled to differ from him as much, or more, than I have done about his Grace's new rendering of the " benefit of clergy."

In discussing catastrophes, the Duke indulges in statements, partly scientific, partly anecdotic, which appear to me to be somewhat misleading. We are told, to begin with, that Sir Charles Lyell's doctrine respecting the proper mode of interpreting the facts of geology (which is commonly called uniformitarianism) "does not hold its head quite so high as it once did." That is great news indeed. But is it true? All I can say is that I am aware of nothing that has happened of late that can in any way justify it; and my opinion is, that the body of Lyell's doctrine, as laid down in that great work, *The Principles of Geology*, whatever may have happened to its head, is a chief and permanent constituent of the foundations of geological science.

But this question cannot be advantageously discussed, unless we take some pains to discriminate between the essential part of the uniformitarian doctrine and its accessories; and it does not appear that the Duke of Argyll has carried his studies of geological philosophy so far as this point. For he defines uniformitarianism to be the assumption of the "extreme slowness and perfect continuity of all geological changes."

What "perfect continuity" may mean in this definition, I am by no means sure; but I can only imagine that it signifies the absence of any break in the course of natural order during the millions of years, the lapse of which is recorded by geological phenomena.

Is the Duke of Argyll prepared to say that any

geologist of authority, at the present day, believes
that there is the slightest evidence of the occurrence
of supernatural intervention, during the long ages of
which the monuments are preserved to us in the crust
of the earth ?   And if he is not, in what sense has
this part of the uniformitarian doctrine, as he defines
it, lowered its pretensions to represent scientific
truth ?

As to the " extreme slowness of all geological
changes," it is simply a popular error to regard that
as, in any wise, a fundamental and necessary dogma
of uniformitarianism.   It is extremely astonishing to
me that any one who has carefully studied Lyell's
great work can have so completely failed to appreciate
its purport, which yet is " writ large " on the very
title-page : " *The Principles of Geology, being an
attempt to explain the former changes of the earth's
surface by reference to causes now in operation.*"
The essence of Lyell's doctrine is here written so that
those who run may read ; and it has nothing to do
with the quickness or slowness of the past changes of the
earth's surface ; except in so far as existing analogous
changes may go on slowly, and therefore create a
presumption in favour of the slowness of past changes.

With that epigrammatic force which characterises
his style, Buffon wrote, nearly a hundred and fifty
years ago, in his famous *Théorie de la Terre* : " Pour
juger de ce qui est arrivé, et même de ce qui arrivera,
nous n'avons qu'à examiner ce qui arrive."   The key
of the past, as of the future, is to be sought in the
present, and only when known causes of change have

been shown to be insufficient have we any right to have recourse to unknown causes. Geology is as much a historical science as archæology; and I apprehend that all sound historical investigation rests upon this axiom. It underlay all Hutton's work and animated Lyell and Scrope in their successful efforts to revolutionise the geology of half a century ago.

There is no antagonism whatever, and there never was, between the belief in the views which had their chief and unwearied advocate in Lyell and the belief in the occurrence of catastrophes. The first edition of Lyell's *Principles*, published in 1830, lies before me; and a large part of the first volume is occupied by an account of volcanic, seismic, and diluvial catastrophes which have occurred within the historical period. Moreover, the author, over and over again, expressly draws the attention of his readers to the consistency of catastrophes with his doctrine.

Notwithstanding, therefore, that we have not witnessed within the last three thousand years the devastation by deluge of a large continent, yet, as we may predict the future occurrence of such catastrophes, we are authorised to regard them as part of the present order of nature, and they may be introduced into geological speculations respecting the past, provided that we do not imagine them to have been more frequent or general than we expect them to be in time to come (vol. i. p. 89).

Again :—

If we regard each of the causes separately, which we know to be at present the most instrumental in remodelling the state of the surface, we shall find that we must expect each to be in action for thousands of years, without producing any extensive alter-

T

ations in the habitable surface, and then to give rise, during a very brief period, to important revolutions (vol. ii. p. 161).[1]

Lyell quarrelled with the catastrophists then, by no means because they assumed that catastrophes occur and have occurred, but because they had got into the habit of calling on their god Catastrophe to help them, when they ought to have been putting their shoulders to the wheel of observation of the present course of nature, in order to help themselves out of their difficulties. And geological science has become what it is, chiefly because geologists have gradually accepted Lyell's doctrine and followed his precepts.

So far as I know anything about the matter, there is nothing that can be called proof, that the causes of geological phenomena operated more intensely or more rapidly, at any time between the older tertiary and the oldest palæozoic epochs than they have done between the older tertiary epoch and the present day. And if that is so, uniformitarianism, even as limited by Lyell,[2] has no call to lower its crest. But if the facts were otherwise, the position Lyell took up remains

---

[1] See also vol. i. p. 460. In the ninth edition (1853), published twenty-three years after the first, Lyell deprives even the most careless reader of any excuse for misunderstanding him : " So in regard to subterranean movements, the theory of the perpetual uniformity of the force which they exert on the earth-crust is quite consistent with the admission of their alternate development and suspension for indefinite periods within limited geographical areas" (p. 187).

[2] A great many years ago (Presidential Address to the Geological Society, 1869) I ventured to indicate that which seemed to me to be the weak point, not in the fundamental principles of uniformitarianism, but in uniformitarianism as taught by Lyell. It lay, to my mind, in

impregnable. He did not say that the geological
operations of nature were never more rapid, or more
vast, than they are now ; what he did maintain is the
very different proposition that there is no good evidence of anything of the kind. And that proposition
has not yet been shown to be incorrect.

I owe more than I can tell to the careful study of
the *Principles of Geology* in my young days ; and,
long before the year 1856, my mind was familiar with
the truth that "the doctrine of. uniformity is not
incompatible with great and sudden changes," which,
as I have shown, is taught *totidem verbis* in that work.
Even had it been possible for me to shut my eyes to
the sense of what I had read in the *Principles*,
Whewell's *Philosophy of the Inductive Sciences*,
published in 1840, a work with which I was also
tolerably familiar, must have opened them. For the
always acute, if not always profound, author, in
arguing against Lyell's uniformitarianism, expressly
points out that it does not in any way contravene the
occurrence of catastrophes.

With regard to such occurrences [earthquakes, deluges, etc.],
terrible as they appear at the time, they may not much affect the

---

the refusal by Hutton, and in a less degree by Lyell, to look beyond
the limits of the time recorded by the stratified rocks. I said : " This
attempt to limit, at a particular point, the progress of inductive and
deductive reasoning from the things which are to the things which
were—this faithlessness to its own logic, seems to me to have cost
uniformitarianism the place as the permanent form of geological
speculation which it might otherwise have held " (*Lay Sermons*, p. 260).
The context shows that "uniformitarianism" here means that doctrine,
as limited in application by Hutton and Lyell, and that what I mean
by "evolutionism" is consistent and thoroughgoing uniformitarianism.

average rate of change : there may be a *cycle*, though an irregular one, of rapid and slow change : and if such cycles go on succeeding each other, we may still call the order of nature uniform, notwithstanding the periods of violence which it involves.[1]

The reader who has followed me through this brief chapter of the history of geological philosophy will probably find the following passage in the paper of the Duke of Argyll to be not a little remarkable :—

> Many years ago, when I had the honour of being President of the British Association,[2] I ventured to point out, in the presence and in the hearing of that most distinguished man [Sir C. Lyell] that the doctrine of uniformity was not incompatible with great and sudden changes, since cycles of these and other cycles of comparative rest might well be constituent parts of that uniformity which he asserted. Lyell did not object to this extended interpretation of his own doctrine, and indeed expressed to me his entire concurrence.

I should think he did ; for, as I have shown, there was nothing in it that Lyell himself had not said, six-and-twenty years before, and enforced, three years before ; and it is almost verbally identical with the view of uniformitarianism taken by Whewell, sixteen years before, in a work with which, one would think, that any one who undertakes to discuss the philosophy of science should be familiar.

Thirty years have elapsed since the beginner of 1856 persuaded himself that he enlightened the foremost geologist of his time, and one of the most acute and far-seeing men of science of any time, as to the scope of the doctrines which the veteran philosopher

---

[1] *Philosophy of the Inductive Sciences*, vol. i. p. 670. New edition, 1847.   [2] At Glasgow in 1856.

had grown gray in promulgating; and the Duke of Argyll's acquaintance with the literature of geology has not, even now, become sufficiently profound to dissipate that pleasant delusion.

If the Duke of Argyll's guidance in that branch of physical science, with which alone he has given evidence of any practical acquaintance, is thus unsafe, I may breathe more freely in setting my opinion against the authoritative deliverances of his Grace about matters which lie outside the province of geology.

And here the Duke's paper offers me such a wealth of opportunities that choice becomes embarrassing. I must bear in mind the good old adage, "Non multa sed multum." Tempting as it would be to follow the Duke through his labyrinthine misunderstandings of the ordinary terminology of philosophy, and to comment on the curious unintelligibility which hangs about his frequent outpourings of fervid language, limits of space oblige me to restrict myself to those points, the discussion of which may help to enlighten the public in respect of matters of more importance than the competence of my Mentor for the task which he has undertaken.

I am not sure when the employment of the word Law, in the sense in which we speak of laws of nature, commenced, but examples of it may be found in the works of Bacon, Descartes, and Spinoza. Bacon employs "Law" as the equivalent of "Form," and I am inclined to think that he may be responsible for a good deal of the confusion that has subsequently arisen; but I am not aware that the term is used by other authorities, in the seventeenth and eighteenth centuries, in any other sense than that of "rule" or "definite order" of the coexistence of things or succession of events in nature. Descartes speaks of "règles, que je nomme les lois de la nature." Leibnitz says "loi ou règle générale," as if he considered the terms interchangeable.

The Duke of Argyll, however, affirms that the "law of gravitation" as put forth by Newton was something more than the statement of an observed order. He admits that Kepler's three laws "were an observed order of facts and nothing more." As to the law of gravitation, "it contains an element which Kepler's laws did not contain, even an element of causation, the recognition of which belongs to a higher category of intellectual conceptions than that which is concerned in the mere observation and record of separate and apparently unconnected facts." There is hardly a line in these paragraphs which appears to me to be indisputable. But, to confine myself to the matter in hand, I cannot conceive that any one who had taken ordinary pains to acquaint himself with the real nature of either Kepler's or Newton's work could have written them. That the labours of Kepler, of all men in the world, should be called " mere observation and record," is truly wonderful. And any one who will look into the *Principia*, or the *Optics*, or the *Letters to Bentley*, will see, even if he has no more special knowledge of the topics discussed than I have, that Newton over and over again insisted that he had nothing to do with gravitation as a physical cause, and that when he used the terms attraction, force, and the like, he employed them, as he says, "*mathematicè*" and not "*physicè*."

> How these attractions [of gravity, magnetism, and electricity] may be performed, I do not here consider. What I call attraction may be performed by impulse or by some other means unknown to me. I use that word here to signify only in a general way any force by which bodies tend towards one another, whatever be the cause.[1]

According to my reading of the best authorities upon the history of science, Newton discovered neither gravitation, nor the law of gravitation; nor did he pretend to offer more than a conjecture as to the causation of gravitation. Moreover, his assertion that the notion of a body acting where it is not, is one that no competent thinker could entertain, is antagonistic to the whole current conception of attractive and repulsive forces, and therefore of "the attractive force of gravitation." What, then, was that labour of unsurpassed magnitude and excellence and immortal influence which Newton did perform? In the first place, Newton defined the laws, rules, or observed order of the phenomena of motion, which come under our daily observation, with greater precision than had been before attained; and, by following out with marvellous power and subtlety the mathematical consequences of these rules, he almost created the modern science of pure mechanics. In the second place, applying exactly the same method

---

[1] *Optics*, query 31.

to the explication of the facts of astronomy as that
which was applied a century and a half later to the
facts of geology by Lyell, he set himself to solve the
following problem.   Assuming that all bodies, free to
move, tend to approach one another as the earth and
the bodies on it do ; assuming that the strength of
that tendency is directly as the mass and inversely as
the squares of the distances ; assuming that the laws
of motion, determined for terrestrial bodies, hold good
throughout the universe ; assuming that the planets
and their satellites were created and placed at their
observed mean distances, and that each received a
certain impulse from the Creator ; will the form of
the orbits, the varying rates of motion of the planets,
and the ratio between those rates and their distances
from the sun which must follow by mathematical
reasoning from these premisses, agree with the order
of facts determined by Kepler and others, or not ?

Newton, employing mathematical methods which
are the admiration of adepts, but which no one but
himself appears to have been able to use with ease, not
only answered this question in the affirmative, but
stayed not his constructive genius before it had
founded modern physical astronomy.

The historians of mechanical and of astronomical
science appear to be agreed that he was the first
person who clearly and distinctly put forth the hypothesis that the phenomena comprehended under the
general name of " gravity " follow the same order
throughout the universe, and that all material bodies
exhibit these phenomena ; so that, in this sense, the

idea of universal gravitation may, doubtless, be properly ascribed to him.

Newton proved that the laws of Kepler were particular consequences of the laws of motion and the law of gravitation—in other words, the reason of the first lay in the two latter. But to talk of the law of gravitation alone as the reason of Kepler's laws, and still more as standing in any causal relation to Kepler's laws, is simply a misuse of language. It would really be interesting if the Duke of Argyll would explain how he proposes to set about showing that the elliptical form of the orbits of the planets, the constant area described by the radius vector, and the proportionality of the squares of the periodic times to the cubes of the distances from the sun, are either caused by the " force of gravitation " or deducible from the "law of gravitation." I conceive that it would be about as apposite to say that the various compounds of nitrogen with oxygen are caused by chemical attraction and deducible from the atomic theory.

Newton assuredly lent no shadow of support to the modern pseudo-scientific philosophy which confounds laws with causes. I have not taken the trouble to trace out this commonest of fallacies to its first beginning ; but I was familiar with it in full bloom, more than thirty years ago, in a work which had a great vogue in its day—the *Vestiges of the Natural History of Creation*—of which the first edition was published in 1844.

It is full of apt and forcible illustrations of pseudo-scientific realism. Consider, for example, this gem serene. When a boy who has climbed a tree loses his hold of the branch, "the law of gravitation unrelentingly pulls him to the ground, and then he is hurt," whereby the Almighty is quite relieved from any responsibility for the accident. Here is the " law of gravitation " acting as a cause in a way quite in accordance with the Duke of Argyll's conception of it. In fact, in the mind of the author of the *Vestiges*, " laws " are existences intermediate between the Creator and his works, like the " ideas " of the Platonisers or the Logos of the Alexandrians.[1] I may cite a passage which is quite in the vein of Philo :—

> We have seen powerful evidences that the construction of this globe and its associates ; and, inferentially, that of all the other globes in space, was the result, not of any immediate or personal exertion on the part of the Deity, but of natural laws which are the expression of his will. What is to hinder our supposing that the organic creation is also a result of natural laws which are in like manner an expression of his will ? (p. 154, 1st edition).

And creation " operating by law " is constantly cited as relieving the Creator from trouble about insignificant details.

I am perplexed to picture to myself the state of mind which accepts these verbal juggleries. It is intelligible that the Creator should operate according to such rules as he might think fit to lay down for himself (and therefore according to law); but that would leave the operation of his will just as much a direct personal act as it would be under any other circum-

---

[1] The author recognises this in his *Explanations*.

stances. I can also understand that (as in Leibnitz's caricature of Newton's views) the Creator might have made the cosmical machine, and, after setting it going, have left it to itself till it needed repair. But then, by the supposition, his personal responsibility would have been involved in all that it did, just as much as a dynamiter is responsible for what happens when he has set his machine going and left it to explode.

The only hypothesis which gives a sort of mad consistency to the Vestigiarian's views is the supposition that laws are a kind of angels or demiurgoi, who, being supplied with the Great Architect's plan, were permitted to settle the details among themselves. Accepting this doctrine, the conception of royal laws and plebeian laws, and of those more than Homeric contests in which the big laws "wreck" the little ones, becomes quite intelligible. And, in fact, the honour of the paternity of those remarkable ideas which come into full flower in the preacher's discourse, must, so far as my imperfect knowledge goes, be attributed to the author of the *Vestiges*.

But the author of the *Vestiges* is not the only writer who is responsible for the current pseudo-scientific mystifications which hang about the term "law." When I wrote my paper about "Scientific and Pseudo-Scientific Realism," I had not read a work by the Duke of Argyll, *The Reign of Law*, which, I believe, has enjoyed, possibly still enjoys, a widespread popularity. But the vivacity of the Duke's attack led me to think it possible that criticisms directed elsewhere might have come home to him. And, in fact, I find that the second chapter of the work in question, which is entitled "Law; its definitions," is, from my point of view, a sort of "summa" of pseudo-scientific philosophy. It will be worth while to examine it in some detail.

In the first place, it is to be noted that the author of the *Reign of Law* admits that "law," in many cases, means nothing more than the statement of the order in which facts occur, or, as he says, "an observed order of facts" (p. 66). But his appreciation of the value of accuracy of expression does not hinder him from adding, almost in the same breath, "In this sense the laws of nature are simply those facts of nature which recur according to rule" (p. 66). Thus "laws," which were rightly said to be the statement of an order of facts in one paragraph, are declared to be the facts themselves in the next.

We are next told that, though it may be customary and permissible to use "law" in the sense of a statement of the order of facts, this is a low use of the word; and indeed, two pages farther on, the writer, flatly contradicting himself, altogether denies its admissibility.

An observed order of facts, to be entitled to the rank of a law, must be an order so constant and uniform as to indicate necessity, and necessity can only arise out of the action of some compelling force (p. 68).

This is undoubtedly one of the most singular propositions that I have ever met with in a professedly scientific work, and its rarity is embellished by another direct self-contradiction which it implies. For on the preceding page (67), when the Duke of Argyll is speaking of the laws of Kepler, which he admits to be laws, and which are types of that which men of science understand by "laws," he says that they are "simply and purely an order of facts." Moreover, he adds: "A very large proportion of the laws of every science are laws of this kind and in this sense."

If, according to the Duke of Argyll's admission, law is understood, in this sense, thus widely and constantly by scientific authorities, where is the justification for his unqualified assertion that such statements of the observed order of facts are not "entitled to the rank" of laws?

But let us examine the consequences of the really interesting proposition I have just quoted. I presume that it is a law of nature that "a straight line is the shortest distance between two points." This law affirms the constant association of a certain fact of form with a certain fact of dimension. Whether the notion of necessity which attaches to it has an *à priori* or an *à posteriori* origin is a question not relevant to the present discussion. But I would beg to be informed, if it is necessary, where is the "compelling force" out of which the necessity arises; and further, if it is not necessary, whether it loses the character of a law of nature?

I take it to be a law of nature, based on unexceptionable evidence, that the mass of matter remains unchanged, whatever chemical or other modifications it may undergo. This law is one of the foundations of chemistry. But it is by no means necessary. It is quite possible to imagine that the mass of matter should vary according to circumstances, as we know its weight does. Moreover, the determination of the " force" which makes mass constant (if there is any intelligibility in that form of words) would not, so far as I can see, confer any more validity on the law than it has now.

There is a law of nature, so well vouched by experience, that all mankind, from pure logicians in search of examples to parish sextons in search of fees, confide in it. This is the law that "all men are mortal." It is simply a statement of the observed order of facts that all men sooner or later die. I am not acquainted with any law of nature which is more "constant and uniform" than this. But will any one tell me that death is "necessary"? Certainly there is no *à priori* necessity in the case, for various men have been imagined to be immortal. And I should be glad to be informed of any "necessity" that can be deduced from biological considerations. It is quite conceivable, as has recently been pointed out, that some of the lowest forms of life may be immortal, after a fashion. However this may be, I would further ask, supposing "all men are mortal" to be a real law of nature, where and what is that to which, with any propriety, the title of "compelling force" of the law can be given?

On page 69, the Duke of Argyll asserts that the law of gravitation "is a law in the sense, not merely of a rule, but of a cause." But this revival of the teaching of the *Vestiges* has already been examined

and disposed of ; and when the Duke of Argyll states
that the "observed order" which Kepler had discovered was simply a necessary consequence of the
force of "gravitation," I need not recapitulate the
evidence which proves such a statement to be wholly
fallacious. But it may be useful to say, once more,
that, at this present moment, nobody knows anything
about the existence of a "force" of gravitation apart
from the fact ; that Newton declared the ordinary
notion of such force to be inconceivable ; that various
attempts have been made to account for the order of
facts we call gravitation, without recourse to the
notion of attractive force ; that, if such a force exists,
it is utterly incompetent to account for Kepler's laws,
without taking into the reckoning a great number of
other considerations ; and, finally, that all we know
about the "force" of gravitation, or any other so-
called "force," is that it is a name for the hypothetical cause of an observed order of facts.

Thus, when the Duke of Argyll says : " Force, ascertained according to some measure of its operation
—this is indeed one of the definitions, but only one,
of a scientific law " (p. 71), I reply that it is a definition which must be repudiated by every one who
possesses an adequate acquaintance with either the
facts, or the philosophy, of science and relegated to
the limbo of pseudo-scientific fallacies. If the human
mind had never entertained this notion of "force,"
nay, if it substituted bare invariable succession for
the ordinary notion of causation, the idea of law, as
the expression of a constantly-observed order, which

generates a corresponding intensity of expectation in our minds, would have exactly the same value, and play its part in real science, exactly as it does now.

It is needless to extend further the present excursus on the origin and history of modern pseudo-science. Under such high patronage as it has enjoyed, it has grown and flourished until, nowadays, it is becoming somewhat rampant. It has its weekly " Ephemerides," in which every hew pseudo-scientific mare's-nest is hailed and belauded with the unconscious unfairness of ignorance ; and an army of " reconcilers," enlisted in its service, whose business seems to be to mix the black of dogma and the white of science into the neutral tint of what they call liberal theology.

I remember that, not long after the publication of the *Vestiges,* a shrewd and sarcastic countryman of the author defined it as " cauld kail made het again." A cynic might find amusement in the reflection that, at the present time, the principles and the methods of the much-vilified Vestigiarian are being " made het again " ; and are not only " echoed by the dome of St. Paul's," but thundered from the castle of Inveraray. But my turn of mind is not cynical, and I can but regret the waste of time and energy bestowed on the endeavour to deal with the most difficult problems of science, by those who have neither undergone the discipline, nor possess the information, which are indispensable to the successful issue of such an enterprise.

I have already had occasion to remark that the

Duke of Argyll's views of the conduct of controversy are different from mine ; and this much-to-be-lamented discrepancy becomes yet more accentuated when the Duke reaches biological topics. Anything that was good enough for Sir Charles Lyell, in his department of study, is certainly good enough for me in mine ; and I by no means demur to being pedagogically instructed about a variety of matters with which it has been the business of my life to try to acquaint myself. But the Duke of Argyll is not content with favouring me with his opinions about my own business ; he also answers for mine ; and, at that point, really the worm must turn. I am told that "no one knows better than Professor Huxley" a variety of things which I really do not know ; and I am said to be a disciple of that "Positive Philosophy" which I have, over and over again, publicly repudiated in language which is certainly not lacking in intelligibility, whatever may be its other defects.

I am told that I have been amusing myself with a "metaphysical exercitation or logomachy" (may I remark incidentally that these are not quite convertible terms ?), when, to the best of my belief, I have been trying to expose a process of mystification, based upon the use of scientific language by writers who exhibit no sign of scientific training, of accurate scientific knowledge, or of clear ideas respecting the philosophy of science, which is doing very serious harm to the public. Naturally enough, they take the lion's skin of scientific phraseology for evidence that the voice which issues from beneath it is the voice of

science, and I desire to relieve them from the consequences of their error.

The Duke of Argyll asks, apparently with sorrow
that it should be his duty to subject me to reproof—

> What shall we say of a philosophy which confounds the or-
> ganic with the inorganic, and, refusing to take note of a difference
> so profound, assumes to explain under one common abstraction,
> the movements due to gravitation and the movements due to the
> mind of man ?

To which I may fitly reply by another question :
What shall we say to a controversialist who attributes
to the subject of his attack opinions which are notoriously not his ; and expresses himself in such a
manner that it is obvious he is unacquainted with
even the rudiments of that knowledge which is
necessary to the discussion into which he has rushed ?

What line of my writing can the Duke of Argyll
produce which confounds the organic with the inorganic ?

As to the latter half of the paragraph, I have to
confess a doubt whether it has any definite meaning.
But I imagine that the Duke is alluding to my assertion that the law of gravitation is nowise " suspended "
or " defied " when a man lifts his arm ; but that,
under such circumstances, part of the store of energy
in the universe operates on the arm at a mechanical
advantage as against the operation of another part.
I was simple enough to think that no one who had as
much knowledge of physiology as is to be found in
an elementary primer, or who had ever heard of the
greatest physical generalisation of modern times—

the doctrine of the conservation of energy—would dream of doubting my statement; and I was further simple enough to think that no one who lacked these qualifications would feel tempted to charge me with error. It appears that my simplicity is greater than my powers of imagination.

The Duke of Argyll may not be aware of the fact, but it is nevertheless true, that when a man's arm is raised, in sequence to that state of consciousness we call a volition, the volition is not the immediate cause of the elevation of the arm. On the contrary, that operation is effected by a certain change of form, technically known as "contraction" in sundry masses of flesh, technically known as muscles, which are fixed to the bones of the shoulder in such a manner that, if these muscles contract, they must raise the arm. Now each of these muscles is a machine comparable, in a certain sense, to one of the donkey-engines of a steamship, but more complete, inasmuch as the source of its ability to change its form, or contract, lies within itself. Every time that, by contracting, the muscle does work, such as that involved in raising the arm, more or less of the material which it contains is used up, just as more or less of the fuel of a steam-engine is used up, when it does work. And I do not think there is a doubt in the mind of any competent physicist or physiologist that the work done in lifting the weight of the arm is the mechanical equivalent of a certain proportion of the energy set free by the molecular changes which take place in the muscle. It is further a tolerably well-based belief that this, and all other forms of energy. are mutually convertible ; and, therefore, that they all come under that general law or statement of the order of facts, called the conservation of energy.   And, as that certainly is an abstraction, so the view which the Duke of Argyll thinks so extremely absurd is really one of the commonplaces of physiology.   But this Review is hardly an appropriate place for giving instruction in the elements of that science, and I content myself with recommending the Duke of Argyll to devote some study to Book II. chap. v. section 4 of my friend Dr. Foster's excellent text-book of Physiology (1st edition, 1877, p. 321), which begins thus :—

Broadly speaking, the animal body is a machine for converting potential into actual energy.   The potential energy is supplied by the food ; this the metabolism of the body converts into the actual energy of heat and mechanical labour.

There is no more difficult problem in the world than that of the relation of the state of consciousness, termed volition, to the mechanical work which frequently follows upon it.   But no one can even comprehend the nature of the problem, who has not carefully studied the long series of modes of motion which, without a break, connect the energy which does that work with the general store of energy. The ultimate form of the problem is this : Have we any reason to believe that a feeling, or state of consciousness, is capable of directly affecting the motion of even the smallest conceivable molecule of matter ? Is such a thing even conceivable ?   If we answer these questions in the negative, it follows that volition may be a sign, but cannot be a cause, of bodily motion. If we answer them in the affirmative, then states of consciousness become undistinguishable from material things ; for it is the essential nature of matter to be the vehicle or substratum of mechanical energy.

There is nothing new in all this. I have merely put into modern language the issue raised by Descartes more than two centuries ago. The philosophies of the Occasionalists, of Spinoza, of Malebranche, of modern idealism and modern materialism, have all grown out of the controversies which Cartesianism evoked. Of all this the pseudo-science of the present time appears to be unconscious ; otherwise it would hardly content itself with "making het again" the pseudo-science of the past.

In the course of these observations I have already had occasion to express my appreciation of the copious and perfervid eloquence which enriches the Duke of Argyll's pages. I am almost ashamed that a constitutional insensibility to the Sirenian charms of rhetoric has permitted me, in wandering through these flowery meads, to be attracted, almost exclusively, to the bare places of fallacy and the stony grounds of deficient information, which are disguised, though not concealed, by these floral decorations. But, in his concluding sentences, the Duke soars into a Tyrtæan strain which roused even my dull soul.

It was high time, indeed, that some revolt should be raised against that Reign of Terror which had come to be established in the scientific world under the abuse of a great name. Professor Huxley has not joined this revolt openly, for as yet, indeed, it is only beginning to raise its head. But more than once—and very lately—he has uttered a warning voice against the shallow dogmatism that has provoked it. The time is coming when that revolt will be carried further. Higher interpretations will be established. Unless I am much mistaken, they are already coming in sight (p. 339).

I have been living very much out of the world for the last two or three years, and when I read this denunciatory outburst, as of one filled with the spirit of prophecy, I said to myself, "Mercy upon us, what has happened? Can it be that X. and Y. (it would be wrong to mention the names of the vigorous young friends which occurred to me) are playing Danton and Robespierre; and that a guillotine is erected in the courtyard of Burlington House for the benefit of all anti-Darwinian Fellows of the Royal Society? Where are the secret conspirators against this tyranny, whom I am supposed to favour, and yet not have the courage to join openly? And to think of my poor oppressed friend, Mr. Herbert Spencer, 'compelled to speak with bated breath' (p. 338) certainly for the first time in my thirty-odd years' acquaintance with him!" My alarm and horror at the supposition that, while I had been fiddling (or at any rate physicking), my beloved Rome had been burning, in this fashion, may be imagined.

I am sure the Duke of Argyll will be glad to hear that the anxiety he created was of extremely short duration. It is my privilege to have access to the best sources of information, and nobody in the scientific world can tell me anything about either the " Reign of

Terror " or "the Revolt."   In fact, the scientific world
laughs most indecorously at the notion of the existence
of either; and some are so lost to the sense of the
scientific dignity, that they descend to the use of
transatlantic slang, and call it a " bogus scare."   As
to my friend Mr. Herbert Spencer, I have every reason
to know that, in the *Factors of Organic Evolution,*
he has said exactly what was in his mind, without any
particular deference to the opinions of the person
whom he is pleased to regard as his most dangerous
critic and Devil's Advocate-General, and still less of
any one else.

I do not know whether the Duke of Argyll pictures himself as the Tallien of this imaginary revolt
against a no less imaginary Reign of Terror.   But if
so, I most respectfully but firmly decline to join his
forces.   It is only a few weeks since I happened to
read over again the first article which I ever wrote
(now twenty-seven years ago) on the *Origin of Species,*
and I found nothing that I wished to modify in the
opinions that are there expressed, though the subsequent vast accumulation of evidence in favour of Mr.
Darwin's views would give me much to add.   As is
the case with all new doctrines, so with that of
Evolution, the enthusiasm of advocates has sometimes
tended to degenerate into fanaticism ; and mere speculation has, at times, threatened to shoot beyond its
legitimate bounds.   I have occasionally thought it
wise to warn the more adventurous spirits among us
against these dangers, in sufficiently plain language ;
and I have sometimes jestingly said that I expected,

if I lived long enough, to be looked on as a reactionary by some of my more ardent friends. But nothing short of midsummer madness can account for the fiction that I am waiting till it is safe to join openly a revolt, hatched by some person or persons unknown, against an intellectual movement with which I am in the most entire and hearty sympathy. It is a great many years since, at the outset of my career, I had to think seriously what life had to offer that was worth having. I came to the conclusion that the chief good, for me, was freedom to learn, think, and say what I pleased, when I pleased. I have acted on that conviction, and have availed myself of the " rara temporum felicitas ubi sentire quæ velis, et quæ sentias dicere licet," which is now enjoyable, to the best of my ability ; and though strongly, and perhaps wisely, warned that I should probably come to grief, I am entirely satisfied with the results of the line of action I have adopted.

My career is at an end. I have

> Warmed both hands before the fire of life ;

and nothing is left me, before I depart, but to help, or at any rate to abstain from hindering, the younger generation of men of science in doing better service to the cause we have at heart than I have been able to render.

And yet, forsooth, I am supposed to be waiting for the signal of " revolt," which some fiery spirits among these young men are to raise before I dare express my real opinions concerning questions about which we older men had to fight, in the teeth of fierce public opposition and obloquy—of something which might almost justify even the grandiloquent epithet of a Reign of Terror—before our excellent .successors had left school.

It would appear that the spirit of pseudo-science has impregnated even the imagination of the Duke of Argyll. The scientific imagination always restrains itself within the limits of probability.

VIII

AN EPISCOPAL TRILOGY

IF there is any truth in the old adage that a burnt child dreads the fire, I ought to be very loath to touch a sermon, while the memory of what befell me on a recent occasion, possibly not yet forgotten by the readers of this Review, is uneffaced. But I suppose that even the distinguished censor of that unheard-of audacity to which not even the newspaper report of a sermon is sacred, can hardly regard a man of science as either indelicate or presumptuous, if he ventures to offer some comments upon three discourses, specially addressed to the great assemblage of men of science which recently gathered at Manchester, by three bishops of the State Church. On my return to England not long ago, I found a pamphlet[1] containing a version, which I presume to be authorised, of these sermons, among the huge mass of letters and papers which had accumulated during two months' absence ; and I have read them not only

---

[1] " The Advance of Science." Three sermons preached in Manchester Cathedral on Sunday, September 4, 1887, during the meeting of the British Association for the Advancement of Science, by the Bishop of Carlisle, the Bishop of Bedford, and the Bishop of Manchester.

with attentive interest, but with a feeling of satisfaction which is quite new to me as a result of hearing, or reading, sermons. These excellent discourses, in fact, appear to me to signalise a new departure in the course adopted by theology towards science, and to indicate the possibility of bringing about an honourable *modus vivendi* between the two. How far the three bishops speak as accredited representatives of the Church is a question to be considered by and by. Most assuredly, I am not authorised to represent any one but myself. But I suppose that there must be a good many people in the Church of the bishops' way of thinking; and I have reason to believe that, in the ranks of science, there are a good many persons who, more or less, share my views. And it is to these sensible people on both sides, as the bishops and I must needs think those who agree with us, that my present observations are addressed. They will probably be astonished to learn how insignificant, in principle, their differences are.

It is impossible to read the discourses of the three prelates without being impressed by the knowledge which they display, and by the spirit of equity, I might say of generosity, towards science which pervades them. There is no trace of that tacit or open assumption that the rejection of theological dogmas, on scientific grounds, is due to moral perversity, which is the ordinary note of ecclesiastical homilies on this subject, and which makes them look so supremely silly to men whose lives have been spent in wrestling with these questions. There is no attempt to hide

away real stumbling-blocks under rhetorical stucco ;
no resort to the *tu quoque* device of setting scientific
blunders against theological errors ; no suggestion
that an honest man may keep contradictory beliefs
in separate pockets of his brain ; no question that
the method of scientific investigation is valid, whatever the results to which it may lead ; and that the
search after truth, and truth only, ennobles the
searcher and leaves no doubt that his life, at any rate,
is worth living.   The Bishop of Carlisle declares himself pledged to the belief that " the advancement of
science, the progress of human knowledge, is in itself a
worthy aim of the greatest effort of the greatest minds."

How often was it my fate, a quarter of a century
ago, to see the whole artillery of the pulpit brought
to bear upon the doctrine of evolution and its sup-
porters !   Any one unaccustomed to the amenities of
ecclesiastical controversy would have thought we
were too wicked to be permitted to live.   But let us
hear the Bishop of Bedford.   After a perfectly frank
statement of the doctrine of evolution and some of
its obvious consequences, that learned prelate pleads,
with all earnestness, against

a hasty denunciation of what *may* be proved to have at least
some elements of truth in it, a contemptuous rejection of theories
which we *may* some day learn to accept as freely and with as
little sense of inconsistency with God's word as we now accept
the theory of the earth's motion round the sun, or the long
duration of the geological epochs (p. 28).

I do not see that the most convinced evolutionist
could ask any one, whether cleric or layman, to say

more than this ; in fact, I do not think that any one has a right to say more, with respect to any question about which two opinions can be held, than that his mind is perfectly open to the force of evidence.

There is another portion of the Bishop of Bedford's sermon which I think will be warmly appreciated by all honest and clear-headed men. He repudiates the views of those who say that theology and science

occupy wholly different spheres, and need in no way intermeddle with each other. They revolve, as it were, in different planes, and so never meet. Thus we may pursue scientific studies with the utmost freedom and, at the same time, may pay the most reverent regard to theology, having no fears of collision, because allowing no points of contact (p. 29).

Surely every unsophisticated mind will heartily concur with the Bishop's remark upon this convenient refuge for the descendants of Mr. Facing-both-ways. " I have never been able to understand this position, though I have often seen it assumed." Nor can any demurrer be sustained when the Bishop proceeds to point out that there are, and must be, various points of contact between theological and natural science, and therefore that it is foolish to ignore or deny the existence of as many dangers of collision.

Finally, the Bishop of Manchester freely admits the force of the objections which have been raised, on scientific grounds, to prayer, and attempts to turn them by arguing that the proper objects of prayer are not physical but spiritual. He tells us that natural accidents and moral misfortunes are not to be taken

for moral judgments of God ; he admits the propriety
of the application of scientific methods to the investigation of the origin and growth of religions ; and he
is as ready to recognise the process of evolution there,
as in the physical world.   Mark the following striking
passage :—

And how utterly all the common objections to Divine revelation vanish away when they are set in the light of this theory of
a spiritual progression.   Are we reminded that there prevailed,
in those earlier days, views of the nature of God and man, of
human life and Divine Providence, which we now find to be
untenable ?   *That*, we answer, is precisely what the theory of
development presupposes.   If early views of religion and morality had not been imperfect, where had been the development ?
If symbolical visions and mythical creations had found no place
in the early Oriental expression of Divine truth, where had been
the development ?   The sufficient answer to ninety-nine out of a
hundred of the ordinary objections to the Bible, as the record of
a divine education of our race, is asked in that one word—
development.   And to what are we indebted for that potent
word, which, as with the wand of a magician, has at the same
moment so completely transformed our knowledge and dispelled
our difficulties ?   To modern science, resolutely pursuing its
search for truth in spite of popular obloquy and—alas ! that one
should have to say it—in spite too often of theological denunciation (p. 53).

Apart from its general importance, I read this
remarkable statement with the more pleasure, since,
however imperfectly I may have endeavoured to
illustrate the evolution of theology in a paper published in this Review last year, it seems to me that in
principle, at any rate, I may hereafter claim high
theological sanction for the views there set forth.

If theologians are henceforward prepared to recognise the authority of secular science in the manner and to the extent indicated in the Manchester trilogy ; if the distinguished prelates who offer these terms are really plenipotentiaries, then, so far as I may presume to speak on such a matter, there will be no difficulty about concluding a perpetual treaty of peace, and indeed of alliance, between the high contracting powers, whose history has hitherto been little more than a record of continual warfare.   But if the great Chancellor's maxim, "Do ut des," is to form the basis of negotiation, I am afraid that secular science will be ruined ; for it seems to me that theology, under the generous impulse of a sudden conversion, has given all that she hath ; and indeed, on one point, has surrendered more than can reasonably be asked.

I suppose I must be prepared to face the reproach which attaches to those who criticise a gift, if I venture to observe that I do not think that the Bishop of Manchester need have been so much alarmed, as he evidently has been, by the objections which have often been raised to prayer, on the ground that a belief in the efficacy of prayer is inconsistent with a belief in the constancy of the order of nature.

The Bishop appears to admit that there is an antagonism between the "regular economy of nature" and the "regular economy of prayer" (p. 39), and that "prayers for the interruption of God's natural order" are of "doubtful validity" (p. 42).   It appears to me that the Bishop's difficulty simply adds another example to those which I have several times insisted upon in the pages of this Review and elsewhere, of the mischief which has been done, and is being done, by a mistaken apprehension of the real meaning of "natural order" and "law of nature."

May I, therefore, be permitted to repeat, once more, that the statements denoted by these terms have no greater value or cogency than such as may attach to generalisations from experience of the past, and to expectations for the future based upon that experience? Nobody can presume to say what the order of nature must be; all that the widest experience (even if it extended over all past time and through all space) that events had happened in a certain way could justify, would be a proportionally strong expectation that events will go on so happening, and the demand for a proportional strength of evidence in favour of any assertion that they had happened otherwise.

It is this weighty consideration, the truth of which every one who is capable of logical thought must surely admit, which knocks the bottom out of all *a priori* objections either to ordinary "miracles" or to the efficacy of prayer, in so far as the latter implies the miraculous intervention of a higher power. No one is entitled to say *a priori* that any given so-called miraculous event is impossible; and no one is entitled to say *a priori* that prayer for some change in the ordinary course of nature cannot possibly avail.

The supposition that there is any inconsistency between the acceptance of the constancy of natural order and a belief in the efficacy of prayer, is the more unaccountable as it is obviously contradicted by analogies furnished by everyday experience. The belief in the efficacy of prayer depends upon the assumption that there is somebody, somewhere, who is strong enough to deal with the earth and its contents as men deal with the 'things and events which they are strong enough to modify or control; and who is capable of being moved by appeals such as men make to one another. This belief does not even involve theism; for our earth is an insignificant particle of the solar system, while the solar system is hardly worth speaking of in relation to the All; and, for anything that can be proved to the contrary, there may be beings endowed with full powers over our system, yet, practically, as insignificant as ourselves in relation to the universe. If any one pleases, therefore, to give unrestrained liberty to his fancy, he may plead analogy in favour of the dream that there may be, somewhere, a finite being, or beings, who can play with the solar system as a child plays with a toy; and that such being may be willing to do anything which he is properly supplicated to do. For we are not justified in saying that it is impossible for beings having the nature of men, only vastly more powerful, to exist; and if they do exist, they may act as and when we ask them to do so, just as our brother men act. As a matter of fact, the great mass of the human race has believed, and still believes, in such beings, under the various names of fairies, gnomes, angels, and demons. Certainly I do not lack faith in the constancy of natural order. But I am not less convinced that if I were to ask the Bishop of Manchester to do me a kindness which lay within his power, he would do it. And I am unable to see that his action on my request involves any violation of the order of nature. On the contrary, as I have not the honour to know the Bishop personally, my action would be based upon my faith in that "law of nature," or generalisation from experience, which tells me that, as a rule, men who occupy the Bishop's position are kindly and courteous. How is the case altered if my request is preferred to some imaginary superior being, or to the Most High Being, who, by the supposition, is able to arrest disease, or make the sun stand still in the heavens, just as easily as I can stop my watch, or make it indicate any hour that pleases me ?

I repeat that it is not upon any *a priori* considerations that objections, either to the supposed efficacy of prayer in modifying the course of events, or to the supposed occurrence of miracles, can be scientifically based. The real objection, and, to my mind, the fatal objection, to both these suppositions, is the inadequacy of the evidence to prove any given case of such occurrences which has been adduced. It is a canon of common sense, to say nothing of science, that the more improbable a supposed occurrence, the more cogent ought to be the evidence in its favour. I have looked somewhat carefully into the subject, and I am unable to find in the records of any miraculous event evidence which even approximates to the fulfilment of this requirement.

But, in the case of prayer, the Bishop points out a most just and necessary distinction between its effect on the course of nature, outside ourselves, and its effect within the region of the supplicator's mind.

It is a "law of nature," verifiable by everyday experience, that our already formed convictions, our strong desires, our intent occupation with particular ideas, modify our mental operations to a most marvellous extent, and produce enduring changes in the direction and in the intensity of our intellectual and moral activities. Men can intoxicate themselves with ideas as effectually as with alcohol or with bang, and produce, by dint of intense thinking, mental conditions hardly distinguishable from monomania. Demoniac possession is mythical; but the faculty of being possessed, more or less completely, by an idea is probably the fundamental condition of what is called genius, whether it show itself in the saint, the artist, or the man of science. One calls it faith, another calls it inspiration, a third calls it insight; but the "intending of the mind," to borrow Newton's well-known phrase, the concentration of all the rays of intellectual energy on some one point, until it glows and colours the whole cast of thought with its peculiar light, is common to all.

I take it that the Bishop of Manchester has psychological science with him when he insists upon the subjective efficacy of prayer in faith, and on the

seemingly miraculous effects which such "intending of
the mind" upon religious and moral ideals may have
upon character and happiness. Scientific faith, at
present, takes it no further than the prayer which
Ajax offered ; but that petition is continually granted.

Whatever points of detail may yet remain open
for discussion, however, I repeat the opinion I have
already expressed, that the Manchester sermons
concede all that science has an indisputable right,
or any pressing need, to ask, and that not grudgingly
but generously ; and, if the three bishops of 1887
carry the Church with them, I think they will have
as good title to the permanent gratitude of posterity
as the famous seven who went to the Tower in defence
of the Church two hundred years ago.

Will their brethren follow their just and prudent
guidance ? I have no such acquaintance with the
currents of ecclesiastical opinion as would justify me
in even hazarding a guess on such a difficult topic.
But some recent omens are hardly favourable. There
seems to be an impression abroad—I do not desire to
give any countenance to it—that I am fond of reading
sermons. From time to time, unknown corre-
spondents—some apparently animated by the charitable desire to promote my conversion, and others
unmistakably anxious to spur me to the expression of
wrathful antagonism — favour me with reports or
copies of such productions.

I found one of the latter category among the
accumulated arrears to which I have already referred.

It is a full, and apparently accurate, report of a

discourse by a person of no less ecclesiastical rank than the three authors of the sermons I have hitherto been considering; but who he is, and where or when the sermon was preached, are secrets which wild horses shall not tear from me, lest I fall again under high censure for attacking a clergyman. Only if the editor of this Review thinks it his duty to have independent evidence that the sermon has a real existence, will I, in the strictest confidence, communicate it to him.

The preacher, in this case, is of a very different mind from the three bishops—and this mind is different in quality, different in spirit, and different in contents. He discourses on the *a priori* objections to miracles, apparently without being aware, in spite of all the discussions of the last seven or eight years, that he is doing battle with a shadow.

I trust I do not misrepresent the Bishop of Manchester in saying that the essence of his remarkable discourse is the insistence upon the "supreme importance of the purely spiritual in our faith," and of the relative, if not absolute, insignificance of aught else. He obviously perceives the bearing of his arguments against the alterability of the course of outward nature by prayer, on the question of miracles in general; for he is careful to say that "the possibility of miracles, of a rare and unusual transcendence of the world order is not here in question" (p. 38). It may be permitted me to suppose, however, that, if miracles were in question, the speaker who warns us "that we must look for the heart of the absolute religion in that part of it which prescribes our moral and religious relations " (p. 46) would not be disposed to advise those who had found the heart of Christianity to take much thought about its miraculous integument.

My anonymous sermon will have nothing to do with such notions as these, and its preacher is not too polite, to say nothing of charitable, towards those who entertain them.

Scientific men, therefore, are perfectly right in asserting that Christianity rests on miracles. If miracles never happened, Christianity, in any sense which is not a mockery, which does not make the term of none effect, has no reality. I dwell on this because there is now an effort making to get up a non-miraculous, invertebrate Christianity, which may escape the ban of science. And I would warn you very distinctly against this new contrivance. Christianity is essentially miraculous, and falls to the ground if miracles be impossible.

Well, warning for warning. I venture to warn this preacher and those who, with him, persist in identifying Christianity with the miraculous, that such forms of Christianity are not only doomed to fall to the ground; but that, within the last half century, they have been driving that way with continually accelerated velocity.

The so-called religious world is given to a strange delusion. It fondly imagines that it possesses the monopoly of serious and constant reflection upon the terrible problems of existence; and that those who cannot accept its shibboleths are either mere Gallios, caring for none of these things, or libertines desiring to escape from the restraints of morality. It does not appear to have entered the imaginations of these people that, outside their pale and firmly resolved never to enter it, there are thousands of men, certainly not their inferiors in character, capacity, or knowledge of the questions at issue, who estimate those purely spiritual elements of the Christian faith of which the Bishop of Manchester speaks as highly as the Bishop does; but who will have nothing to do with the Christian Churches, because in their apprehension and for them, the profession of belief in the miraculous, on the evidence offered, would be simply immoral.

So far as my experience goes, men of science are neither better nor worse than the rest of the world. Occupation with the endlessly great parts of the universe does not necessarily involve greatness of character, nor does microscopic study of the infinitely little always produce humility. We have our full share of original sin; need, greed, and vainglory beset us as they do other mortals; and our progress is, for the most part, like that of a tacking ship, the resultant of opposite divergencies from the straight path. But, for all that, there is one moral benefit which the pursuit of science unquestionably bestows. It keeps the estimate of the value of evidence up to the proper mark; and we are constantly receiving lessons, and sometimes very sharp ones, on the nature of proof. Men of science will always act up to their standard of veracity, when mankind in general leave off sinning; but that standard appears to me to be higher among them than in any other class of the community.

I do not know any body of scientific men who could be got to listen without the strongest expressions of disgusted repudiation to the exposition of a pretended scientific discovery, which had no better evidence to show for itself than the story of the devils entering a herd of swine, or of the fig-tree that was blasted for bearing no figs when "it was not the season of figs." Whether such events are possible or impossible, no man can say; but scientific ethics can and does declare that the profession of belief in them, on the evidence of documents of unknown date and of unknown authorship, is immoral. Theological apologists who insist that morality will vanish if their dogmas are exploded, would do well to consider the fact that, in the matter of intellectual veracity, science is already a long way ahead of the Churches; and that, in this particular, it is exerting an educational influence on mankind of which the Churches have shown themselves utterly incapable.

Undoubtedly that varying compound of some of the best and some of the worst elements of Paganism and Judaism, moulded in practice by the innate character of certain people of the Western world, which, since the second century, has assumed to itself the title of orthodox Christianity, "rests on miracles" and falls to the ground, not "if miracles be impossible," but if those to whom it is committed prove themselves unable to fulfil the conditions of honest belief. That this Christianity is doomed to fall is, to my mind, beyond a doubt; but its fall will be neither sudden nor speedy. The Church, with all the aid lent it by the secular arm, took many centuries to extirpate the open practice of pagan idolatry within its own fold; and those who have travelled in southern Europe will be aware that it has not extirpated the essence of such idolatry even yet. *Mutato nomine,* it is probable that there is as much sheer fetichism among the Roman populace now as there was eighteen hundred years ago; and if Marcus Antoninus could descend from his horse and ascend the steps of the Ara Cœli church about Twelfth Day, the only thing that need strike him would be the extremely contemptible character of the modern idols as works of art.

Science will certainly neither ask for, nor receive, the aid of the secular arm. It will trust to the much better and more powerful help of that education in scientific truth and in the morals of assent, which is rendered as indispensable, as it is inevitable, by the permeation of practical life with the products and ideas of science. But no one who considers the present state of even the most developed countries can doubt that the scientific light that has come into the world will, for a long time, have to shine in the midst of darkness. The urban populations, driven into contact with science by trade and manufacture, will more and more receive it, while the *pagani* will lag behind. Let us hope that no Julian may arise among them to head a forlorn hope against the inevitable. Whatever happens, science may bide her time in patience and in confidence.

But to return to my " Anonymous." I am afraid

that if he represents any great party in the Church, the spirit of justice and reasonableness which animates the three bishops has as slender a chance of being imitated, on a large scale, as their common sense and their courtesy. For, not contented with misrepresenting science on its speculative side, "Anonymous" attacks its morality.

For two whole years, investigations and conclusions which would upset the theories of Darwin on the formation of coral islands were actually suppressed, and that by the advice even of those who accepted them, *for fear of upsetting the faith and disturbing the judgment formed by the multitude on the scientific character—the infallibility—of the great master !*

So far as I know anything about the matters which are here referred to, the part of this passage which I have italicised is absolutely untrue. I believe that I am intimately acquainted with all Mr. Darwin's immediate scientific friends ; and I say that no one of them, nor any other man of science known to me, ever could, or would, have given such advice to any one—if for no other reason than that, with the example of the most candid and patient listener to objections that ever lived fresh in their memories, they could not so grossly have at once violated their highest duty and dishonoured their friend.

The charge thus brought by "Anonymous" affects the honour and the probity of men of science ; if it is true, we have forfeited all claim to the confidence of the general public. In my belief it is utterly false, and its real effect will be to discredit those who are responsible for it. As is the way with slanders, it

has grown by repetition. " Anonymous " is responsible for the peculiarly offensive form which it has
taken in his hands ; but he is not responsible for
originating it. He has evidently been inspired by an
article entitled " A Great Lesson," published in the
September number of this Review. Truly it is "a
great lesson," but not quite in the sense intended by
the giver thereof.

In the course of his doubtless well-meant admonitions, the Duke of Argyll commits himself to a greater
number of statements which are demonstrably incorrect, and which any one who ventured to write
upon the subject ought to have known to be incorrect,
than I have ever seen gathered together in so small
a space.

I submit a gathering from the rich store for the
appreciation of the public.

First :—

Mr. Murray's new explanation of the structure of coral-reefs
and islands was communicated to the Royal Society of Edinburgh in 1880, and supported with such a weight of facts and
such a close texture of reasoning, that no serious reply has ever
been attempted (p. 305).

" No serious reply has ever been attempted "! I
suppose that the Duke of Argyll may have heard of
Professor Dana, whose years of labour devoted to
corals and coral-reefs when he was naturalist of the
American expedition under Commodore Wilkes, more
than forty years ago, have ever since caused him to
be recognised as an authority of the first rank on such
subjects. Now does his Grace know, or does he not

know, that, in the year 1885, Professor Dana published
an elaborate paper " On the Origin of Coral-Reefs and
Islands," in which, after referring to a presidential
address by the Director of the Geological Survey of
Great Britain and Ireland delivered in 1883, in which
special attention is directed to Mr. Murray's views,
Professor Dana says :—

The existing state of doubt on the question has led the writer
to reconsider the earlier and later facts, and in the following
pages he gives his results.

Professor Dana then devotes many pages of his
very " serious reply " to a most admirable and weighty
criticism of the objections which have at various times
been raised to Mr. Darwin's doctrine, by Professor
Semper, by Dr. Rein, and finally by Mr. Murray, and
he states his final judgment as follows :—

With the theory of abrasion and solution incompetent, all the
hypotheses of objectors to Darwin's theory are alike weak ; for
all have made these processes their chief reliance, whether appealing to a calcareous, or a volcanic, or a mountain-peak basement
for the structure.   The subsidence which the Darwinian theory
requires has not been opposed by the mention of any fact at
variance with it, nor by setting aside Darwin's arguments in its
favour ; and it has found new support in the facts from the
Challenger's soundings off Tahiti, that had been put in array
against it, and strong corroboration in the facts from the West
Indies.

Darwin's theory, therefore, remains as the theory that accounts
for the origin of reefs and islands.[1]

Be it understood that I express no opinion on the
controverted points.   I doubt if there are ten living
men who, having a practical knowledge of what a coral-

[1] *American Journal of Science*, 1885, p. 190.

reef is, have endeavoured to master the very difficult biological and geological problems involved in their study. I happen to have spent the best part of three years among coral-reefs and to have made that attempt; and, when Mr. Murray's work appeared, I said to myself that until I had two or three months to give to the renewed study of the subject in all its bearings, I must be content to remain in a condition of suspended judgment. In the meanwhile, the man who would be voted by common acclamation as the most competent person now living to act as umpire, has delivered the verdict I have quoted; and, to go no further, has fully justified the hesitation I and others may have felt about expressing an opinion. Under these circumstances, it seems to me to require a good deal of courage to say " no serious reply has ever been attempted"; and to chide the men of science, in lofty tones, for their "reluctance to admit an error" which is not admitted; and for their "slow and sulky acquiescence" in a conclusion which they have the gravest warranty for suspecting !

Second :—

Darwin himself had lived to hear of the new solution, and, with that splendid candour which was eminent in him, his mind, though now grown old in his own early convictions, was at least ready to entertain it, and to confess that serious doubts had been awakened as to the truth of his famous theory (p. 305).

I wish that Darwin's splendid candour could be conveyed by some description of spiritual "microbe" to those who write about him. I am not aware that Mr. Darwin ever entertained "serious doubts as to

the truth of his famous theory"; and there is tolerably good evidence to the contrary. The second
edition of his work, published in 1876, proves that he
entertained no such doubts then ; a letter to Professor
Semper, whose objections, in some respects, forestalled
those of Mr. Murray, dated October 2, 1879, expresses
his continued adherence to the opinion "that the
atolls and barrier reefs in the middle of the Pacific
and Indian Oceans indicate subsidence"; and the
letter of my friend Professor Judd, printed at the end
of this article (which I had perhaps better say Professor Judd had not seen) will prove that this opinion
remained unaltered to the end of his life.

Third :—

. . . Darwin's theory is a dream. It is not only unsound,
but it is in many respects the reverse of truth. With all his
conscientiousness, with all his caution, with all his powers of observation, Darwin in this matter fell into errors as profound as
the abysses of the Pacific (p. 301).

Really? It seems to me that, under the circumstances, it is pretty clear that these lines exhibit a
lack of the qualities justly ascribed to Mr. Darwin,
which plunges their author into a much deeper abyss,
and one from which there is no hope of emergence.

Fourth :—

All the acclamations with which it was received were as the
shouts of an ignorant mob (p. 301).

But surely it should be added that the Coryphæus of
this ignorant mob, the fugleman of the shouts, was
one of the most accomplished naturalists and geologists
now living—the American Dana—who, after years of

VIII AN EPISCOPAL TRILOGY 319

independent study extending over numerous reefs in the Pacific, gave his hearty assent to Darwin's views, and, after all that had been said, deliberately reaffirmed that assent in the year 1885.

Fifth :—

> The overthrow of Darwin's speculation is only beginning to be known. It has been whispered for some time. The cherished dogma has been dropping very slowly out of sight (p. 301).

Darwin's speculation may be right or wrong, but I submit that that which has not happened cannot even begin to be known, except by those who have miraculous gifts to which we poor scientific people do not aspire. The overthrow of Darwin's views may have been whispered by those who hoped for it; and they were perhaps wise in not raising their voices above a whisper. Incorrect statements, if made too loudly, are apt to bring about unpleasant consequences.

Sixth. Mr. Murray's views, published in 1880, are said to have met with "slow and sulky acquiescence" (p. 305). I have proved that they cannot be said to have met with general acquiescence of any sort, whether quick and cheerful, or slow and sulky; and if this assertion is meant to convey the impression that Mr. Murray's views have been ignored, that there has been a conspiracy of silence against them, it is utterly contrary to notorious fact.

Professor Geikie's well-known *Textbook of Geology* was published in 1882, and at pages 457-459 of that work there is a careful exposition of Mr. Murray's views. Moreover, Professor Geikie has specially

advocated them on other occasions,[1] notably in a long
article on " The Origin of Coral-Reefs," published in
two numbers of *Nature* for 1883, and in a presidential
address delivered in the same year. If, in so short a
time after the publication of his views, Mr. Murray
could boast of a convert so distinguished and influential
as the Director of the Geological Survey, it seems to me
that this wonderful *conspiration de silence* (which
has about as much real existence as the Duke of
Argyll's other bogie, " the Reign of Terror ") must
have *ipso facto* collapsed. I wish that, when I was a
young man, my endeavours to upset some prevalent
errors had met with as speedy and effectual backing.

Seventh :—

. . . Mr. John Murray was strongly advised against the
publication of his views in derogation of Darwin's long-accepted
theory of the coral islands, and was actually induced to delay it
for two years. Yet the late Sir Wyville Thomson, who was at
the head of the naturalists of the *Challenger* expedition, was
himself convinced by Mr. Murray's reasoning (p. 307).

Clearly, then, it could not be Mr. Murray's official
chief who gave him this advice. Who was it? And
what was the exact nature of the advice given?
Until we have some precise information on this head,
I shall take leave to doubt whether this statement is
more accurate than those which I have previously
cited.

Whether such advice was wise or foolish, just or

---

[1] Professor Geikie, however, though a strong, is a fair and candid
advocate. He says of Darwin's theory, " That it may be possibly true,
in some instances, may be readily granted." For Professor Geikie,
then, it is not yet overthrown—still less a dream.

immoral, depends entirely on the motive of the person who gave it. If he meant to suggest to Mr. Murray that it might be wise for a young and comparatively unknown man to walk warily, when he proposed to attack a generalisation based on many years' labour of one undoubtedly competent person, and fortified by the independent results of the many years' labour of another undoubtedly competent person; and even, if necessary, to take two whole years in fortifying his position, I think that such advice would have been sagacious and kind. I suppose that there are few working men of science who have not kept their ideas to themselves, while gathering and sifting evidence, for a much longer period than two years.

If, on the other hand, Mr. Murray was advised to delay the publication of his criticisms, simply to save Mr. Darwin's credit and to preserve some reputation for infallibility, which no one ever heard of, then I have no hesitation in declaring that his adviser was profoundly dishonest, as well as extremely foolish, and that, if he is a man of science, he has disgraced his calling.

But, after all, this supposed scientific Achitophel has not yet made good the primary fact of his existence. Until the needful proof is forthcoming, I think I am justified in suspending my judgment as to whether he is much more than an anti-scientific myth. I leave it to the Duke of Argyll to judge of the extent of the obligation under which, for his own sake, he may lie to produce the evidence on which

Y

his aspersions of the honour of scientific men are based. I cannot pretend that we are seriously disturbed by charges which every one who is acquainted with the truth of the matter knows to be ridiculous ; but mud has a habit of staining if it lies too long, and it is as well to have it brushed off as soon as may be.

So much for the " Great Lesson." It is followed by a " Little Lesson," apparently directed against my infallibility—a doctrine about which I should be inclined to paraphrase Wilkes's remark to George the Third, when he declared that he, at any rate, was not a Wilkite. But I really should be glad to think that there are people who need the warning, because then it will be obvious that this raking up of an old story cannot have been suggested by a mere fanatical desire to damage men of science. I can but rejoice, then, that these misguided enthusiasts, whose faith in me has so far exceeded the bounds of reason, should be set right. But that " want of finish " in the matter of accuracy which so terribly mars the effect of the " Great Lesson," is no less conspicuous in the case of the " Little Lesson," and, instead of setting my too fervent disciples right, it will set them wrong.

The Duke of Argyll, in telling the story of *Bathybius,* says that my mind was " caught by this new and grand generalisation of the physical basis of life." I never have been guilty of a reclamation about anything to my credit, and I do not mean to be ; but if there is any blame going, I do not choose to be relegated to a subordinate place when I have a claim to the first. The responsibility for the first description and the naming of *Bathybius* is mine and mine only. The paper on " Some Organisms living at great depths in the Atlantic Ocean," in which I drew attention to this substance, is to be found by the curious in the 8th volume of the *Quarterly Journal of Microscopical Science*, and was published in the year 1868. Whatever errors are contained in that paper are my own peculiar property; but neither at the meeting of the British Association in 1868, nor anywhere else, have I gone beyond what is there stated; except in so far that, at a long-subsequent meeting of the Association, being importuned about the subject, I ventured to express, somewhat emphatically, the wish that the thing was at the bottom of the sea.

What is meant by my being caught by a generalisation about the physical basis of life I do not know; still less can I understand the assertion that *Bathybius* was accepted because of its supposed harmony with Darwin's speculations. That which interested me in the matter was the apparent analogy of *Bathybius* with other well-known forms of lower life, such as the plasmodia of the Myxomycetes and the Rhizopods. Speculative hopes or fears had nothing to do with the matter; and if *Bathybius* were brought up alive from the bottom of the Atlantic to-morrow, the fact would not have the slightest bearing, that I can discern, upon Mr. Darwin's speculations, or upon any of the disputed problems of biology. It would merely be one elementary organism the more added to the thousands already known.

Up to this moment I was not aware of the universal favour with which *Bathybius* was received.[1] Those simulators of an "ignorant mob" who, according to the Duke of Argyll, welcomed Darwin's theory of coral-reefs, made no demonstration in my favour, unless his Grace includes Sir Wyville Thomson, Dr. Carpenter, Dr. Bessels, and Professor Haeckel under that head. On the contrary, a sagacious friend of mine, than whom there was no more competent judge, the late Mr. George Busk, was not to be converted; while, long before the *Challenger* work, Ehrenberg wrote to me very sceptically; and I fully expected that that eminent man would favour me with pretty sharp criticism. Unfortunately he died shortly afterwards, and nothing from him, that I know of, appeared. When Sir Wyville Thomson wrote to me a brief account of the results obtained on board the *Challenger*, I sent his statement to *Nature*, in which journal it appeared the following week, without any further note or comment than was needful to explain the circumstances. In thus allowing judgment to go by default, I am afraid I showed a reckless and ungracious disregard for the feelings of the believers in my infallibility. No doubt I ought to have hedged and fenced and attenuated the effect of Sir Wyville Thomson's brief note in every possible way. Or

---

[1] I find, moreover, that I specially warned my readers against hasty judgment. After stating the facts of observation, I add, " I have, hitherto, said nothing about their meaning, as, in an inquiry so difficult and fraught with interest as this, it seems to me to be in the highest degree important to keep the questions of fact and the questions of interpretation well apart " (p. 210).

perhaps I ought to have suppressed the note altogether, on the ground that it was a mere *ex parte* statement. My excuse is that, notwithstanding a large and abiding faith in human folly, I did not know then, any more than I know now, that there was anybody foolish enough to be unaware that the only people, scientific or other, who never make mistakes are those who do nothing; or that anybody, for whose opinion I cared, would not rather see me commit ten blunders than try to hide one.

Pending the production of further evidence, I hold that the existence of people who believe in the infallibility of men of science is as purely mythical as that of the evil counsellor who advised the withholding of the truth lest it should conflict with that belief.

I venture to think, then, that the Duke of Argyll might have spared his "Little Lesson" as well as his "Great Lesson" with advantage. The paternal authority who whips the child for sins he has not committed does not strengthen his moral influence— rather excites contempt and repugnance. And if, as would seem from this and former monitory allocutions which have been addressed to us, the Duke aspires to the position of censor, or spiritual director, in relation to the men who are doing the work of physical science, he really must get up his facts better. There will be an end to all chance of our kissing the rod if his Grace goes wrong a third time. He must not say again that "no serious reply has been attempted" to a view which was discussed and repudiated, two years before, by one of the highest extant authorities on the subject; he must not say that Darwin accepted that which it can be proved he did not accept; he must not say that a doctrine has dropped into the abyss when it is quite obviously alive and kicking at the surface; he must not assimilate a man like Professor Dana to the components of an "ignorant mob"; he must not say that things are beginning to be known which are not known at all; he must not say that "slow and sulky acquiescence" has been given to that which cannot yet boast of general acquiescence of any kind; he must not suggest that a view which has been publicly advocated by the Director of the Geological Survey and no less publicly discussed by many other authoritative writers has been intentionally and systematically ignored; he must not ascribe ill motives for a course of action which is the only proper one; and finally, if any one but myself were interested, I should say that he had better not waste his time in raking up the errors of those whose lives have been occupied, not in talking about science, but in toiling, sometimes with success and sometimes with failure, to get some real work done.

The most considerable difference I note among men is not in their readiness to fall into error, but in their readiness to acknowledge these inevitable lapses. The Duke of Argyll has now a splendid opportunity for proving to the world in which of these categories it is hereafter to rank him.

DEAR PROFESSOR HUXLEY—A short time before Mr. Darwin's death, I had a conversation with him concerning the observations which had been made by Mr. Murray upon coral-reefs, and the speculations which had been founded upon those observations. I found that Mr. Darwin had very carefully considered the whole subject, and that while, on the one hand, he did not regard the actual facts recorded by Mr. Murray as absolutely inconsistent with his own theory of subsidence, on the other hand, he did not believe that they necessitated or supported the hypothesis advanced by Mr. Murray. Mr. Darwin's attitude, as I understood it, towards Mr. Murray's objections to the theory of subsidence was exactly similar to that maintained by him with respect to Professor Semper's criticism, which was of a very similar character; and his position with regard to the whole question was almost identical with that subsequently so clearly defined by Professor Dana in his well-known articles published in the *American Journal of Science* for 1885.

It is difficult to imagine how any one, acquainted with the scientific literature of the last seven years, could possibly suggest that Mr. Murray's memoir published in 1880 had failed to secure a due amount of attention. Mr. Murray, by his position in the *Challenger* office, occupied an exceptionally favourable position for making his views widely known; and he had, moreover, the singular good fortune to secure from the first the advocacy of so able and brilliant a writer as Professor Archibald Geikie, who in a special discourse and in several treatises on geology and physical geology very strongly supported the new theory. It would be an endless task to attempt to give references to the various scientific journals which have discussed the subject, but I may add that every treatise on geology which has been published, since Mr. Murray's views were made known, has dealt with his observations at considerable length. This is true of Professor A. H. Green's *Physical Geology*, published in 1882; of Professor Prestwich's *Geology, Chemical and Physical*; and of Professor James Geikie's *Outlines of Geology*, published in 1886. Similar prominence is given to the subject in De Lapparent's *Traité de Géologie*, published in 1885, and in Credner's *Elemente der Geologie*, which has appeared during the present year. If this be a "conspiracy of silence," where, alas! can the geological speculator seek for fame?—Yours very truly,

JOHN W. JUDD.

October 10, 1887.

IX

AGNOSTICISM

WITHIN the last few months the public has received much and varied information on the subject of agnostics, their tenets, and even their future. Agnosticism exercised the orators of the Church Congress at Manchester.[1] It has been furnished with a set of "articles" fewer, but not less rigid, and certainly not less consistent than the thirty-nine; its nature has been analysed, and its future severely predicted by the most eloquent of that prophetical school whose Samuel is Auguste Comte. It may still be a question, however, whether the public is as much the wiser as might be expected, considering all the trouble that has been taken to enlighten it. Not only are the three accounts of the agnostic position sadly out of harmony with one another, but I propose to show cause for my belief that all three must be seriously questioned by any one who employs the term "agnostic" in the sense in which it was originally used. The learned Principal of King's College, who brought

[1] See the *Official Report of the Church Congress held at Manchester*, October 1888, pp. 253, 254.

the topic of Agnosticism before the Church Congress, took a short and easy way of settling the business :—

But if this be so, for a man to urge, as an escape from this article of belief, that he has no means of a scientific knowledge of the unseen world, or of the future, is irrelevant. His difference from Christians lies not in the fact that he has no knowledge of these things, but that he does not believe the authority on which they are stated. He may prefer to call himself an Agnostic; but his real name is an older one—he is an infidel; that is to say, an unbeliever. The word infidel, perhaps, carries an unpleasant significance. Perhaps it is right that it should. It is, and it ought to be, an unpleasant thing for a man to have to say plainly that he does not believe in Jesus Christ.[1]

So much of Dr. Wace's address either explicitly or implicitly concerns me, that I take upon myself to deal with it; but, in so doing, it must be understood that I speak for myself alone. I am not aware that there is any sect of Agnostics; and if there be, I am not its acknowledged prophet or pope. I desire to leave to the Comtists the entire monopoly of the manufacture of imitation ecclesiasticism.

Let us calmly and dispassionately consider Dr. Wace's appreciation of agnosticism. The agnostic, according to his view, is a person who says he has no means of attaining a scientific knowledge of the unseen world or of the future; by which somewhat loose

[1] [In this place and in the eleventh essay, there are references to the late Archbishop of York which are of no importance to my main argument, and which I have expunged because I desire to obliterate the traces of a temporary misunderstanding with a man of rare ability, candour, and wit, for whom I entertained a great liking and no less respect. I rejoice to think now of the (then) Bishop's cordial hail the first time we met after our little skirmish, "Well, is it to be peace or war?" I replied, "A little of both." But there was only peace when we parted, and ever after.]

phraseology Dr. Wace presumably means the theological unseen world and future. I cannot think this description happy, either in form or substance, but for the present it may pass. Dr. Wace continues, that is not "his difference from Christians." Are there then any Christians who say that they know nothing about the unseen world and the future? I was ignorant of the fact, but I am ready to accept it on the authority of a professional theologian, and I proceed to Dr. Wace's next proposition.

The real state of the case, then, is that the agnostic "does not believe the authority" on which "these things" are stated, which authority is Jesus Christ. He is simply an old-fashioned "infidel" who is afraid to own to his right name. As "Presbyter is priest writ large," so is "agnostic" the mere Greek equivalent for the Latin "infidel." There is an attractive simplicity about this solution of the problem; and it has that advantage of being somewhat offensive to the persons attacked, which is so dear to the less refined sort of controversialist. The agnostic says, "I cannot find good evidence that so and so is true." "Ah," says his adversary, seizing his opportunity, "then you declare that Jesus Christ was untruthful, for he said so and so;" a very telling method of rousing prejudice. But suppose that the value of the evidence as to what Jesus may have said and done, and as to the exact nature and scope of his authority, is just that which the agnostic finds it most difficult to determine. If I venture to doubt that the Duke of Wellington gave the command "Up, Guards,

and at 'em!" at Waterloo, I do not think that even
Dr. Wace would accuse me of disbelieving the Duke.
Yet it would be just as reasonable to do this as to
accuse any one of denying what Jesus said before the
preliminary question as to what he did say is settled.

Now, the question as to what Jesus really said
and did is strictly a scientific problem, which is
capable of solution by no other methods than those
practised by the historian and the literary critic. It
is a problem of immense difficulty, which has occupied
some of the best heads in Europe for the last century;
and it is only of late years that their investigations
have begun to converge towards one conclusion.[1]

That kind of faith which Dr. Wace describes and
lauds is of no use here. Indeed, he himself takes
pains to destroy its evidential value.

" What made the Mahommedan world? Trust and
faith in the declarations and assurances of Mahommed.
And what made the Christian world? Trust and

---

[1] Dr. Wace tells us, " It may be asked how far we can rely on the
accounts we possess of our Lord's teaching on these subjects." And
he seems to think the question appropriately answered by the assertion
that it " ought to be regarded as settled by M. Renan's practical surrender of the adverse case." I thought I knew M. Renan's works
pretty well, but I have contrived to miss this " practical " (I wish Dr.
Wace had defined the scope of that useful adjective) surrender. However, as Dr. Wace can find no difficulty in pointing out the passage
of M. Renan's writings, by which he feels justified in making his
statement, I shall wait for further enlightenment, contenting myself,
for the present, with remarking that if M. Renan were to retract and
do penance in Notre-Dame to-morrow for any contributions to Biblical
criticism that may be specially his property, the main results of that
criticism, as they are set forth in the works of Strauss, Baur, Reuss,
and Volkmar, for example, would not be sensibly affected.

faith in the declarations and assurances of Jesus Christ and His Apostles" (*l. c.* p. 253). The triumphant tone of this imaginary catechism leads me to suspect that its author has hardly appreciated its full import. Presumably, Dr. Wace regards Mahommed as an unbeliever, or, to use the term which he prefers, infidel; and considers that his assurances have given rise to a vast delusion which has led, and is leading, millions of men straight to everlasting punishment. And this being so, the "Trust and faith" which have "made the Mahommedan world," in just the same sense as they have "made the Christian world," must be trust and faith in falsehood. No man who has studied history, or even attended to the occurrences of everyday life, can doubt the enormous practical value of trust and faith; but as little will he be inclined to deny that this practical value has not the least relation to the reality of the objects of that trust and faith. In examples of patient constancy of faith and of unswerving trust, the *Acta Martyrum* do not excel the annals of Babism.[1]

The discussion upon which we have now entered goes so thoroughly to the root of the whole matter; the question of the day is so completely, as the author of *Robert Elsmere* says, the value of testimony, that I shall offer no apology for following it out somewhat in detail; and, by way of giving substance to the

[1] [See De Gobineau, *Les Religions et les Philosophies dans l'Asie Centrale*; and the recently published work of Mr. E. G. Browne, *The Episode of the Bab.*]

argument, I shall base what I have to say upon a case, the consideration of which lies strictly within the province of natural science, and of that particular part of it known as the physiology and pathology of the nervous system.

I find, in the second Gospel (chap. v.), a statement, to all appearance intended to have the same evidential value as any other contained in that history. It is the well-known story of the devils who were cast out of a man, and ordered, or permitted, to enter into a herd of swine, to the great loss and damage of the innocent Gerasene, or Gadarene, pig owners. There can be no doubt that the narrator intends to convey to his readers his own conviction that this casting out and entering in were effected by the agency of Jesus of Nazareth ; that, by speech and action, Jesus enforced this conviction ; nor does any inkling of the legal and moral difficulties of the case manifest itself.

On the other hand, everything that I know of physiological and pathological science leads me to entertain a very strong conviction that the phenomena ascribed to possession are as purely natural as those which constitute small-pox ; everything that I know of anthropology leads me to think that the belief in demons and demoniacal possession is a mere survival of a once universal superstition, and that its persistence, at the present time, is pretty much in the inverse ratio of the general instruction, intelligence, and sound judgment of the population among whom it prevails. Everything that I know of law and

justice convinces me that the wanton destruction of
other people's property is a misdemeanour of evil
example. Again, the study of history, and especially
of that of the fifteenth, sixteenth, and seventeenth
centuries, leaves no shadow of doubt on my mind
that the belief in the reality of possession and of
witchcraft, justly based, alike by Catholics and Protestants, upon this and innumerable other passages in
both the Old and New Testaments, gave rise, through
the special influence of Christian ecclesiastics, to the
most horrible persecutions and judicial murders of
thousands upon thousands of innocent men, women,
and children. And when I reflect that the record of
a plain and simple declaration upon such an occasion
as this, that the belief in witchcraft and possession is
wicked nonsense, would have rendered the long agony
of mediæval humanity impossible, I am prompted to
reject, as dishonouring, the supposition that such declaration was withheld out of condescension to popular error.

"Come forth, thou unclean spirit, out of the
man" (Mark v. 8),[1] are the words attributed to Jesus.
If I declare, as I have no hesitation in doing, that I
utterly disbelieve in the existence of "unclean spirits,"
and, consequently, in the possibility of their "coming
forth" out of a man, I suppose that Dr. Wace will
tell me I am disregarding the testimony "of our
Lord" (l. c. p. 255). For if these words were really
used, the most resourceful of reconcilers can hardly
venture to affirm that they are compatible with a disbelief in "these things." As the learned and fair-

---

[1] Here, as always, the revised version is cited.

minded, as well as orthodox, Dr. Alexander remarks, in an editorial note to the article "Demoniacs," in the *Biblical Cyclopædia* (vol. i. p. 664, note):—

. . . On the lowest grounds on which our Lord and His Apostles can be placed they must, at least, be regarded as *honest* men. Now, though honest speech does not require that words should be used always and only in their etymological sense, it does require that they should not be used so as to affirm what the speaker knows to be false. Whilst, therefore, our Lord and His Apostles might use the word δαιμονίζεσθαι, or the phrase δαιμόνιον ἔχειν, as a popular description of certain diseases, without giving in to the belief which lay at the source of such a mode of expression, they could not speak of demons entering into a man, or being cast out of him, without pledging themselves to the belief of an actual possession of the man by the demons. (Campbell, *Prel. Diss.* vi. 1, 10.) If, consequently, they did not hold this belief, they spoke not as honest men.

The story which we are considering does not rest on the authority of the second Gospel alone. The third confirms the second, especially in the matter of commanding the unclean spirit to come out of the man (Luke viii. 29); and, although the first Gospel either gives a different version of the same story, or tells another of like kind, the essential point remains: "If thou cast us out, send us away into the herd of swine. And He said unto them: Go!" (Matt. viii. 31, 32).

If the concurrent testimony of the three synoptics, then, is really sufficient to do away with all rational doubt as to a matter of fact of the utmost practical and speculative importance — belief or disbelief in which may affect, and has affected, men's lives and their conduct towards other men in the most serious way—then I am bound to believe that Jesus implicitly affirmed himself to possess a "knowledge of the unseen world," which afforded full confirmation of the belief in demons and possession current among his contemporaries. If the story is true, the mediæval theory of the invisible world may be, and probably is, quite correct; and the witchfinders, from Sprenger to Hopkins and Mather, are much-maligned men.

On the other hand, humanity, noting the frightful consequences of this belief; common sense, observing the futility of the evidence on which it is based, in all cases that have been properly investigated; science, more and more seeing its way to enclose all the phenomena of so-called "possession" within the domain of pathology, so far as they are not to be relegated to that of the police—all these powerful influences concur in warning us, at our peril, against accepting the belief without the most careful scrutiny of the authority on which it rests.

I can discern no escape from this dilemma: either Jesus said what he is reported to have said, or he did not. In the former case, it is inevitable that his authority on matters connected with the "unseen world" should be roughly shaken; in the latter, the blow falls upon the authority of the synoptic gospels. If their report on a matter of such stupendous and far-reaching practical import as this is untrustworthy, how can we be sure of its trustworthiness in other cases? The favourite "earth," in which the hard-pressed reconciler takes refuge, that the Bible does

not profess to teach science,[1] is stopped in this instance. For the question of the existence of demons
and of possession by them, though it lies strictly
within the province of science, is also of the deepest
moral and religious significance. If physical and
mental disorders are caused by demons, Gregory of
Tours and his contemporaries rightly considered that
relics and exorcists were more useful than doctors;
the gravest questions arise as to the legal and moral
responsibilities of persons inspired by demoniacal
impulses; and our whole conception of the universe
and of our relations to it becomes totally different
from what it would be on the contrary hypothesis.

The theory of life of an average mediæval Christian
was as different from that of an average nineteenth-
century Englishman as that of a West African negro
is now, in these respects. The modern world is slowly,
but surely, shaking off these and other monstrous
survivals of savage delusions, and, whatever happens,

---

[1] Does any one really mean to say that there is any internal or
external criterion by which the reader of a biblical statement, in which
scientific matter is contained, is enabled to judge whether it is to be
taken *au sérieux* or not? Is the account of the Deluge, accepted as
true in the New Testament, less precise and specific than that of the
call of Abraham, also accepted as true therein? By what mark does
the story of the feeding with manna in the wilderness, which involves
some very curious scientific problems, show that it is meant merely
for edification, while the story of the inscription of the Law on stone
by the hand of Jahveh is literally true? If the story of the Fall is
not the true record of an historical occurrence, what becomes of
Pauline theology? Yet the story of the Fall as directly conflicts with
probability, and is as devoid of trustworthy evidence, as that of the
Creation or that of the Deluge, with which it forms an harmoniously
legendary series.

it will not return to that wallowing in the mire.
Until the contrary is proved, I venture to doubt
whether, at this present moment, any Protestant
theologian, who has a reputation to lose, will say that
he believes the Gadarene story.

The choice then lies between discrediting those
who compiled the Gospel biographies and disbelieving
the Master, whom they, simple souls, thought to
honour by preserving such traditions of the exercise
of his authority over Satan's invisible world. This
is the dilemma. No deep scholarship, nothing but a
knowledge of the revised version (on which it is to
be supposed all that mere scholarship can do has been
done), with the application thereto of the commonest
canons of common sense, is needful to enable us to
make a choice between its alternatives. It is hardly
doubtful that the story, as told in the first Gospel, is
merely a version of that told in the second and third.
Nevertheless, the discrepancies are serious and irreconcilable ; and, on this ground alone, a suspension
of judgment, at the least, is called for. But there is
a great deal more to be said. From the dawn of
scientific biblical criticism until the present day, the
evidence against the long-cherished notion that the
three synoptic Gospels are the works of three independent authors, each prompted by Divine inspiration, has steadily accumulated, until, at the present
time, there is no visible escape from the conclusion
that each of the three is a compilation consisting of
a groundwork common to all three—the threefold
tradition ; and of a superstructure, consisting, firstly,

of matter common to it with one of the others, and, secondly, of matter special to each. The use of the terms "groundwork" and "superstructure" by no means implies that the latter must be of later date than the former. On the contrary, some parts of it may be and probably are, older than some parts of the groundwork.[1]

The story of the Gadarene swine belongs to the groundwork; at least, the essential part of it, in which the belief in demoniac possession is expressed, does; and therefore the compilers of the first, second, and third Gospels, whoever they were, certainly accepted that belief (which, indeed, was universal among both Jews and pagans at that time), and attributed it to Jesus.

What, then, do we know about the originator, or originators, of this groundwork — of that threefold tradition which all three witnesses (in Paley's phrase) agree upon—that we should allow their mere statements to outweigh the counter arguments of humanity, of common sense, of exact science, and to imperil the respect which all would be glad to be able to render to their Master?

Absolutely nothing.[2] There is no proof, nothing

[1] See, for an admirable discussion of the whole subject, Dr. Abbott's article on the Gospels in the *Encyclopædia Britannica*; and the remarkable monograph by Professor Volkmar, *Jesus Nazarenus und die erste christliche Zeit* (1882). Whether we agree with the conclusions of these writers or not, the method of critical investigation which they adopt is unimpeachable.

[2] Notwithstanding the hard words shot at me from behind the hedge of anonymity by a writer in a recent number of the *Quarterly Review*, I repeat, without the slightest fear of refutation, that the four Gospels, as they have come to us, are the work of unknown writers.

more than a fair presumption, that any one of the
Gospels existed, in the state in which we find it in
the authorised version of the Bible, before the second
century, or, in other words, sixty or seventy years
after the events recorded.   And, between that time
and the date of the oldest extant manuscripts of the
Gospels, there is no telling what additions and altera-
tions and interpolations may have been made.   It
may be said that this is all·mere speculation, but it is
a good deal more.   As competent scholars and honest
men, our revisers have felt compelled to point out
that such things have happened even since the date
of the oldest known manuscripts.   The oldest two
copies of the second Gospel end with the 8th verse
of the 16th chapter; the remaining twelve verses are
spurious, and it is noteworthy that the maker of the
addition has not hesitated to introduce a speech in
which Jesus promises his disciples that " in My name
shall they cast out devils."

The other passage " rejected to the margin " is still
more instructive.   It is that touching apologue, with
its profound ethical sense, of the woman taken in
adultery—which, if internal evidence were an infallible guide, might well be affirmed to be a typical
example of the teachings of Jesus.   Yet, say the
revisers, pitilessly, " Most of the ancient authorities
omit John vii. 53–viii. 11."   Now let any reasonable
man ask himself this question.   If, after an approximate settlement of the canon of the New Testament,
and even later than the fourth and fifth centuries,
literary fabricators had the skill and the audacity to

make such additions and interpolations as these, what
may they have done when no one had thought of a
canon ; when oral tradition, still unfixed, was regarded
as more valuable than such written records as may
have existed in the latter portion of the first century ?
Or, to take the other alternative, if those who gradually settled the canon did not know of the existence
of the oldest codices which have come down to us ;
or if, knowing them, they rejected their authority,
what is to be thought of their competency as critics
of the text ?

People who object to free criticism of the Christian
Scriptures forget that they are what they are in virtue
of very free criticism ; unless the advocates of inspiration are prepared to affirm that the majority of
influential ecclesiastics during several centuries were
safeguarded against error.   For, even granting that
some books of the period were inspired, they were
certainly few amongst many ; and those who selected
the canonical books, unless they themselves were also
inspired, must be regarded in the light of mere critics,
and, from the evidence they have left of their intellectual habits, very uncritical critics.   When one thinks
that such delicate questions as those involved fell into
the hands of men like Papias (who believed in the
famous millenarian grape story) ; of Irenæus with his
" reasons " for the existence of only four Gospels ; and
of such calm and dispassionate judges as Tertullian,
with his " *Credo quia impossibile* " : the marvel is
that the selection which constitutes our New Testament is as free as it is from obviously objectionable

matter. The apocryphal Gospels certainly deserve to be apocryphal; but one may suspect that a little more critical discrimination would have enlarged the Apocrypha not inconsiderably.

At this point a very obvious objection arises and deserves full and candid consideration. It may be said that critical scepticism carried to the length suggested is historical pyrrhonism; that if we are to altogether discredit an ancient or a modern historian, because he has assumed fabulous matter to be true, it will be as well to give up paying any attention to history. It may be said, and with great justice, that Eginhard's *Life of Charlemagne* is none the less trustworthy because of the astounding revelation of credulity, of lack of judgment, and even of respect for the eighth commandment, which he has unconsciously made in the *History of the Translation of the Blessed Martyrs Marcellinus and Paul.* Or, to go no further back than the last number of this Review, surely that excellent lady, Miss Strickland, is not to be refused all credence because of the myth about the second James's remains, which she seems to have unconsciously invented.

Of course this is perfectly true. I am afraid there is no man alive whose witness could be accepted, if the condition precedent were proof that he had never invented and promulgated a myth. In the minds of all of us there are little places here and there, like the indistinguishable spots on a rock which give foothold to moss or stonecrop; on which, if the germ of a myth fall, it is certain to grow, without in the least

degree affecting our accuracy or truthfulness elsewhere. Sir Walter Scott knew that he could not
repeat a story without, as he said, "giving it a new
hat and stick." Most of us differ from Sir Walter
only in not knowing about this tendency of the
mythopœic faculty to break out unnoticed. But it is
also perfectly true that the mythopœic faculty is not
equally active in all minds, nor in all regions and
under all conditions of the same mind. David Hume
was certainly not so liable to temptation as the
Venerable Bede, or even as some recent historians
who could be mentioned; and the most imaginative
of debtors, if he owes five pounds, never makes an
obligation to pay a hundred out of it. The rule of
common sense is *primâ facie* to trust a witness in all
matters in which neither his self-interest, his passions,
his prejudices, nor that love of the marvellous, which
is inherent to a greater or less degree in all mankind,
are strongly concerned; and, when they are involved,
to require corroborative evidence in exact proportion
to the contravention of probability by the thing
testified.

Now, in the Gadarene affair, I do not think I am
unreasonably sceptical if I say that the existence of
demons who can be transferred from a man to a pig,
does thus contravene probability. Let me be perfectly candid. I admit I have no *à priori* objection
to offer. There are physical things, such as *tæniæ*
and *trichinæ*, which can be transferred from men to
pigs, and *vice versâ*, and which do undoubtedly produce most diabolical and deadly effects on both. For

anything I can absolutely prove to the contrary, there may be spiritual things capable of the same trans-migration, with like effects. Moreover I am bound to add that perfectly truthful persons, for whom I have the greatest respect, believe in stories about spirits of the present day, quite as improbable as that we are considering.

So I declare, as plainly as I can, that I am unable to show cause why these transferable devils should not exist; nor can I deny that, not merely the whole Roman Church, but many Wacean "infidels" of no mean repute, do honestly and firmly believe that the activity of such like demonic beings is in full swing in this year of grace 1889.

Nevertheless, as good Bishop Butler says, "prob-ability is the guide of. life," and it seems to me that this is just one of the cases in which the canon of credibility and testimony, which I have ventured to lay down, has full force. So that, with the most entire respect for many (by no means for all) of our witnesses for the truth of demonology, ancient and modern, I conceive their evidence on this particular matter to be ridiculously insufficient to warrant their conclusion.[1]

[1] Their arguments, in the long run, are always reducible to one form. Otherwise trustworthy witnesses affirm that such and such events took place. These events are inexplicable, except the agency of "spirits" is admitted. Therefore "spirits" were the cause of the phenomena.

And the heads of the reply are always the same. Remember Goethe's aphorism : "Alles factische ist schon Theorie." Trustworthy witnesses are constantly deceived, or deceive themselves, in their interpretation of sensible phenomena. No one can prove that the

After what has been said, I do not think that any sensible man, unless he happen to be angry, will accuse me of "contradicting the Lord and his Apostles" if I reiterate my total disbelief in the whole Gadarene story. But, if that story is discredited, all the other stories of demoniac possession fall under suspicion. And if the belief in demons and demoniac possession, which forms the sombre background of the whole picture of primitive Christianity presented to us in the New Testament, is shaken, what is to be said, in any case, of the uncorroborated testimony of the Gospels with respect to "the unseen world"?

I am not aware that I have been influenced by any more bias in regard to the Gadarene story than I have been in dealing with other cases of like kind the investigation of which has interested me. I was brought up in the strictest school of evangelical orthodoxy; and when I was old enough to think for myself, I started upon my journey of inquiry with little doubt about the general truth of what I had been taught; and with that feeling of the unpleasantness of being called an "infidel" which, we are told, is so right and proper. Near my journey's end, I find myself in a condition of something more than mere doubt about these matters.

sensible phenomena, in these cases, could be caused only by the agency of spirits : and there is abundant ground for believing that they may be produced in other ways. Therefore, the utmost that can be reasonably asked for, on the evidence as it stands, is suspension of judgment. And, on the necessity for even that suspension, reasonable men may differ, according to their views of probability.

In the course of other inquiries, I have had to do with fossil remains which looked quite plain at a distance, and became more and more indistinct as I tried to define their outline by close inspection. There was something there—something which, if I could win assurance about it, might mark a new epoch in the history of the earth; but, study as long as I might, certainty eluded my grasp. So has it been with me in my efforts to define the grand figure of Jesus as it lies in the primary strata of Christian literature. Is he the kindly, peaceful Christ depicted in the Catacombs? Or is he the stern Judge who frowns above the altar of SS. Cosmas and Damianus? Or can he be rightly represented by the bleeding ascetic, broken down by physical pain, of too many mediæval pictures? Are we to accept the Jesus of the second, or the Jesus of the fourth Gospel, as the true Jesus? What did he really say and do; and how much that is attributed to him, in speech and action, is the embroidery of the various parties into which his followers tended to split themselves within twenty years of his death, when even the threefold tradition was only nascent?

If any one will answer these questions for me with something more to the point than feeble talk about the "cowardice of agnosticism," I shall be deeply his debtor. Unless and until they are satisfactorily answered, I say of agnosticism in this matter, " *J'y suis, et j'y reste.* "

But, as we have seen, it is asserted that I have no business to call myself an agnostic; that if I am

not a Christian I am an infidel; and that I ought
to call myself by that name of "unpleasant significance." Well, I do not care much what I am called
by other people, and if I had at my side all those
who, since the Christian era, have been called infidels
by other folks, I could not desire better company.
If these are my ancestors, I prefer, with the old
Frank, to be with them wherever they are. But
there are several points in Dr. Wace's contention
which must be elucidated before I can even think
of undertaking to carry out his wishes. I must,
for instance, know what a Christian is. Now what
is a Christian? By whose authority is the signification of that term defined? Is there any doubt that
the immediate followers of Jesus, the "sect of the
Nazarenes," were strictly orthodox Jews, differing
from other Jews not more than the Sadducees, the
Pharisees, and the Essenes differed from one another;
in fact, only in the belief that the Messiah, for whom
the rest of their nation waited, had come? Was
not their chief, "James, the brother of the Lord,"
reverenced alike by Sadducee, Pharisee, and Nazarene? At the famous conference which, according
to the Acts, took place at Jerusalem, does not James
declare that "myriads" of Jews, who by that time
had become Nazarenes, were "all zealous for the
Law"? Was not the name of "Christian" first used
to denote the converts to the doctrine promulgated
by Paul and Barnabas at Antioch? Does the subsequent history of Christianity leave any doubt
that, from this time forth, the "little rift within

the lute" caused by the new teaching, developed, if not inaugurated, at Antioch, grew wider and wider, until the two types of doctrine irreconcilably diverged? Did not the primitive Nazarenism, or Ebionism, develop into the Nazarenism, and Ebionism, and Elkasaitism of later ages, and finally die out in obscurity and condemnation as damnable heresy; while the younger doctrine throve and pushed out its shoots into that endless variety of sects, of which the three strongest survivors are the Roman and Greek Churches and modern Protestantism?

Singular state of things! If I were to profess the doctrine which was held by "James, the brother of the Lord," and by every one of the "myriads" of his followers and co-religionists in Jerusalem up to twenty or thirty years after the Crucifixion (and one knows not how much later at Pella), I should be condemned with unanimity as an ebionising heretic by the Roman, Greek, and Protestant Churches! And, probably, this hearty and unanimous condemnation of the creed held by those who were in the closest personal relation with their Lord is almost the only point upon which they would be cordially of one mind. On the other hand, though I hardly dare imagine such a thing, I very much fear that the "pillars" of the primitive Hierosolymitan Church would have considered Dr. Wace an infidel. No one can read the famous second chapter of Galatians and the book of Revelations without seeing how narrow was even Paul's escape from a similar fate. And, if ecclesiastical history is to be trusted, the thirty-nine articles, be they right or wrong, diverge from the primitive doctrine of the Nazarenes vastly more than even Pauline Christianity did.

But, further than this, I have great difficulty in assuring myself that even James, "the brother of the Lord," and his "myriads" of Nazarenes, properly represented the doctrines of their Master. For it is constantly asserted by our modern "pillars" that one of the chief features of the work of Jesus was the instauration of Religion by the abolition of what our sticklers for articles and liturgies, with unconscious humour, call the narrow restrictions of the Law. Yet, if James knew this, how could the bitter controversy with Paul have arisen; and why did one or the other side not quote any of the various sayings of Jesus, recorded in the Gospels, which directly bear on the question — sometimes, apparently, in opposite directions?

So if I am asked to call myself an "infidel," I reply: To what doctrine do you ask me to be faithful? Is it that contained in the Nicene and the Athanasian Creeds? My firm belief is that the Nazarenes, say of the year 40, headed by James, would have stopped their ears and thought worthy of stoning the audacious man who propounded it to them. Is it contained in the so-called Apostles' Creed? I am pretty sure that even that would have created a recalcitrant commotion at Pella in the year 70, among the Nazarenes of Jerusalem, who had fled from the soldiers of Titus. And yet, if the unadulterated tradition of the teachings of "the Nazarene" were to be found anywhere, it surely should have been amidst those not very aged disciples who may have heard them as they were delivered.

Therefore, however sorry I may be to be unable to demonstrate that, if necessary, I should not be afraid to call myself an "infidel," I cannot do it. "Infidel" is a term of reproach, which Christians and Mahommedans, in their modesty, agree to apply to those who differ from them. If he had only thought of it, Dr. Wace might have used the term "miscreant," which, with the same etymological signification, has the advantage of being still more "unpleasant" to the persons to whom it is applied. But why should a man be expected to call himself a "miscreant" or an "infidel"? That St. Patrick "had two birthdays because he was a twin" is a reasonable and intelligible utterance beside that of the man who should declare himself to be an infidel on the ground of denying his own belief. It may be logically, if not ethically, defensible that a Christian should call a Mahommedan an infidel and *vice versâ*; but, on Dr. Wace's principles, both ought to call themselves infidels, because each applies the term to the other.

Now I am afraid that all the Mahommedan world would agree in reciprocating that appellation to Dr. Wace himself. I once visited the Hazar Mosque, the great University of Mahommedanism, in Cairo, in ignorance of the fact that I was unprovided with proper authority. A swarm of angry undergraduates, as I suppose I ought to call them, came buzzing about me and my guide ; and if I had known Arabic, I suspect that "dog of an infidel" would have been by no means the most "unpleasant" of the epithets showered upon me, before I could explain and apologise for the mistake. If I had had the pleasure of Dr. Wace's company on that occasion, the undiscriminative followers of the Prophet would, I am afraid, have made no difference between us ; not even if they had known that he was the head of an orthodox Christian seminary. And I have not the smallest doubt that even one of the learned mollahs, if his grave courtesy would have permitted him to say anything offensive to men of another mode of belief, would have told us that he wondered we did not find it "very unpleasant" to disbelieve in the Prophet of Islam.

From what precedes, I think it becomes sufficiently clear that Dr. Wace's account of the origin of the name of "Agnostic" is quite wrong. Indeed, I am bound to add that very slight effort to discover the truth would have convinced him that, as a matter of fact, the term arose otherwise. I am loath to go over an old story once more ; but more than one object which I have in view will be served by telling it a little more fully than it has yet been told.

Looking back nearly fifty years, I see myself as a boy, whose education had been interrupted, and who, intellectually, was left, for some years, altogether to his own devices. At that time, I was a voracious and omnivorous reader ; a dreamer and speculator of the first water, well endowed with that splendid courage in attacking any and every subject, which is the blessed compensation of youth and inexperience.   Among the books and essays, on all sorts of topics from metaphysics to heraldry, which I read at this time, two left indelible impressions on my mind.   One was Guizot's *History of Civilisation*, the other was Sir William Hamilton's essay *On the Philosophy of the Unconditioned*, which I came upon, by chance, in an odd volume of the *Edinburgh Review*.   The latter was certainly strange reading for a boy, and I could not possibly have understood a great deal of it ; [1] nevertheless, I devoured it with avidity, and it stamped upon my mind the strong conviction that, on even the most solemn and important of questions, men are apt to take cunning phrases for answers ; and that the limitation of our faculties, in a great number of cases, renders real answers to such questions, not merely actually impossible, but theoretically inconceivable.

Philosophy and history having laid hold of me in this eccentric fashion, have never loosened their grip.   I have no pretension to be an expert in either subject ; but the turn for philosophical and historical reading, which rendered Hamilton and Guizot attractive to me, has not only filled many

---

[1] Yet I must somehow have laid hold of the pith of the matter, for, many years afterwards, when Dean Mansell's Bampton lectures were published, it seemed to me I already knew all that this eminently agnostic thinker had to tell me.

lawful leisure hours, and still more sleepless ones, with the repose of changed mental occupation, but has not unfrequently disputed my proper work-time with my liege lady, Natural Science. In this way, I have found it possible to cover a good deal of ground in the territory of philosophy; and all the more easily that I have never cared much about A's or B's opinions, but have rather sought to know what answer he had to give to the questions I had to put to him—that of the limitation of possible knowledge being the chief. The ordinary examiner, with his "State the views of So-and-so," would have floored me at any time. If he had said what do *you* think about any given problem, I might have got on fairly well.

The reader who has had the patience to follow the enforced, but unwilling, egotism of this veritable history (especially if his studies have led him in the same direction), will now see why my mind steadily gravitated towards the conclusions of Hume and Kant, so well stated by the latter in a sentence, which I have quoted elsewhere.

"The greatest and perhaps the sole use of all philosophy of pure reason is, after all, merely negative, since it serves not as an organon for the enlargement [of knowledge], but as a discipline for its delimitation; and, instead of discovering truth, has only the modest merit of preventing error." [1]

When I reached intellectual maturity and began to ask myself whether I was an atheist, a theist, or

[1] *Kritik der reinen Vernunft.* Edit. Hartenstein, p. 256.

a pantheist; a materialist or an idealist; a Christian or a freethinker; I found that the more I learned and reflected, the less ready was the answer; until, at last, I came to the conclusion that I had neither art nor part with any of these denominations, except the last. The one thing in which most of these good people were agreed was the one thing in which I differed from them. They were quite sure they had attained a certain "gnosis,"—had, more or less success-fully, solved the problem of existence; while I was quite sure I had not, and had a pretty strong con-viction that the problem was insoluble. And, with Hume and Kant on my side, I could not think myself presumptuous in holding fast by that opinion. Like Dante,

> Nel mezzo del cammin di nostra vita
> Mi ritrovai per una selva oscura,

but, unlike Dante, I cannot add,

> Che la diritta via era smarrita.

On the contrary, I had, and have, the firmest con-viction that I never left the " verace via "—the straight road; and that this road led nowhere else but into the dark depths of a wild and tangled forest. And though I have found leopards and lions in the path; though I have made abundant acquaintance with the hungry wolf, that " with privy paw devours apace and nothing said," as another great poet says of the ravening beast; and though no friendly spectre has even yet offered his guidance, I was, and am, minded to go straight on, until I either come out on the

other side of the wood, or find there is no other side
to it, at least, none attainable by me.

This was my situation when I had the good
fortune to find a place among the members of that
remarkable confraternity of antagonists, long since
deceased, but of green and pious memory, the Metaphysical Society. Every variety of philosophical and
theological opinion was represented there, and expressed itself with entire openness; most of my
colleagues were -*ists* of one sort or another; and,
however kind and friendly they might be, I, the man
without a rag of a label to cover himself with, could
not fail to have some of the uneasy feelings which
must have beset the historical fox when, after leaving
the trap in which his tail remained, he presented
himself to his normally elongated companions. So
I took thought, and invented what I conceived to be
the appropriate title of "agnostic." It came into my
head as suggestively antithetic to the "gnostic" of
Church history, who professed to know so much
about the very things of which I was ignorant; and
I took the earliest opportunity of parading it at our
Society, to show that I, too, had a tail, like the other
foxes. To my great satisfaction, the term took; and
when the *Spectator* had stood godfather to it, any
suspicion in the minds of respectable people, that a
knowledge of its parentage might have awakened,
was, of course, completely lulled.

That is the history of the origin of the terms
"agnostic" and "agnosticism"; and it will be observed that it does not quite agree with the con-

fident assertion of the reverend Principal of King's College, that "the adoption of the term agnostic is only an attempt to shift the issue, and that it involves a mere evasion" in relation to the Church and Christianity.[1]

The last objection (I rejoice, as much as my readers must do, that it is the last) which I have to take to Dr. Wace's deliverance before the Church Congress arises, I am sorry to say, on a question of morality.

"It is, and it ought to be," authoritatively declares this official representative of Christian ethics, "an unpleasant thing for a man to have to say plainly that he does not believe in Jesus Christ" (*l. c.* p. 254).

Whether it is so depends, I imagine, a good deal on whether the man was brought up in a Christian household or not. I do not see why it should be "unpleasant" for a Mahommedan or Buddhist to say so. But that "it ought to be" unpleasant for any man to say anything which he sincerely, and after due deliberation, believes, is, to my mind, a proposition of the most profoundly immoral character. I verily believe that the great good which has been effected in the world by Christianity has been largely counteracted by the pestilent doctrine on which all the Churches have insisted, that honest disbelief in their more or less astonishing creeds is a moral offence, indeed a sin of the deepest dye, deserving

[1] *Report of the Church Congress*, Manchester, 1888, p. 252.

and involving the same future retribution as murder
and robbery.   If we could only see, in one view, the
torrents of hypocrisy and cruelty, the lies, the
slaughter, the violations of every obligation of
humanity, which have flowed from this source along
the course of the history of Christian nations, our
worst imaginations of Hell would pale beside the
vision.

A thousand times, no !   It ought *not* to be unpleasant to say that which one honestly believes or
disbelieves.   That it so constantly is painful to do
so, is quite enough obstacle to the progress of mankind in that most valuable of all qualities, honesty of
word or of deed, without erecting a sad concomitant
of human weakness into something to be admired
and cherished.   The bravest of soldiers often, and
very naturally, " feel it unpleasant " to go into action ;
but a court-martial which did its duty would make
short work of the officer who promulgated the
doctrine that his men *ought* to feel their duty unpleasant.

I am very well aware, as I suppose most thoughtful
people are in these times, that the process of breaking
away from old beliefs is extremely unpleasant ; and I
am much disposed to think that the encouragement,
the consolation, and the peace afforded to earnest
believers in even the worst forms of Christianity are
of great practical advantage to them.   What deductions must be made from this gain on the score of
the harm done to the citizen by the ascetic other-
worldliness of logical Christianity ; to the ruler, by

the hatred, malice, and all uncharitableness of
sectarian bigotry; to the legislator, by the spirit of
exclusiveness and domination of those that count
themselves pillars of orthodoxy; to the philosopher,
by the restraints on the freedom of learning and
teaching which every Church exercises, when it is
strong enough; to the conscientious soul, by the
introspective hunting after sins of the mint and
cummin type, the fear of theological error, and the
overpowering terror of possible damnation, which
have accompanied the Churches like their shadow,
I need not now consider; but they are assuredly
not small. If agnostics lose heavily on the one side,
they gain a good deal on the other. People who talk
about the comforts of belief appear to forget its dis-
comforts; they ignore the fact that the Christianity
of the Churches is something more than faith in the
ideal personality of Jesus, which they create for
themselves, *plus* so much as can be carried into
practice, without disorganising civil society, of the
maxims of the Sermon on the Mount. Trip in
morals or in doctrine (especially in doctrine), without
due repentance or retractation, or fail to get properly
baptized before you die, and a *plébiscite* of the
Christians of Europe, if they were true to their
creeds, would affirm your everlasting damnation by
an immense majority.

Preachers, orthodox and heterodox, din into our
ears that the world cannot get on without faith of
some sort. There is a sense in which that is as
eminently as obviously true; there is another, in

which, in my judgment, it is as eminently as obviously false, and it seems to me that the hortatory, or pulpit, mind is apt to oscillate between the false and the true meanings, without being aware of the fact.

It is quite true that the ground of every one of our actions, and the validity of all our reasonings, rest upon the great act of faith, which leads us to take the experience of the past as a safe guide in our dealings with the present and the future. From the nature of ratiocination it is obvious that the axioms on which it is based cannot be demonstrated by ratiocination. It is also a trite observation that, in the business of life, we constantly take the most serious action upon evidence of an utterly insufficient character. But it is surely plain that faith is not necessarily entitled to dispense with ratiocination because ratiocination cannot dispense with faith as a starting-point; and that because we are often obliged, by the pressure of events, to act on very bad evidence, it does not follow that it is proper to act on such evidence when the pressure is absent.

The writer of the epistle to the Hebrews tells us that " faith is the assurance of things hoped for, the proving of things not seen." In the authorised version " substance " stands for " assurance," and " evidence " for " proving." The question of the exact meaning of the two words, $\dot{\upsilon}\pi\acute{o}\sigma\tau\alpha\sigma\iota\varsigma$ and $\ddot{\epsilon}\lambda\epsilon\gamma\chi\sigma\varsigma$, affords a fine field of discussion for the scholar and the metaphysician. But I fancy we shall be not far from the mark if we take the writer to have had in his mind the profound psychological truth that men constantly feel certain about things for which they strongly hope, but have no evidence, in the legal or logical sense of the word; and he calls this feeling "faith." I may have the most absolute faith that a friend has not committed the crime of which he is accused. In the early days of English history, if my friend could have obtained a few more compurgators of a like robust faith, he would have been acquitted. At the present day, if I tendered myself as a witness on that score, the judge would tell me to stand down, and the youngest barrister would smile at my simplicity. Miserable indeed is the man who has not such faith in some of his fellow-men—only less miserable than the man who allows himself to forget that such faith is not, strictly speaking, evidence; and when his faith is disappointed, as will happen now and again, turns Timon and blames the universe for his own blunders. And so, if a man can find a friend, the hypostasis of all his hopes, the mirror of his ethical ideal, in the Jesus of any, or all, of the Gospels, let him live by faith in that ideal. Who shall or can forbid him? But let him not delude himself with the notion that his faith is evidence of the objective reality of that in which he trusts. Such evidence is to be obtained only by the use of the methods of science, as applied to history and to literature, and it amounts at present to very little.

It appears that Mr. Gladstone some time ago asked Mr. Laing if he could draw up a short summary

of the negative creed; a body of negative propositions, which have so far been adopted on the negative
side as to be what the Apostles' and other accepted
creeds are on the positive; and Mr. Laing at once
kindly obliged Mr. Gladstone with the desired articles
—eight of them.

If any one had preferred this request to me I
should have replied that, if he referred to agnostics,
they have no creed; and, by the nature of the case,
cannot have any. Agnosticism, in fact, is not a
creed, but a method, the essence of which lies in
the rigorous application of a single principle. That
principle is of great antiquity; it is as old as
Socrates; as old as the writer who said, "Try all
things, hold fast by that which is good;" it is the
foundation of the Reformation, which simply illustrated the axiom that every man should be able to
give a reason for the faith that is in him; it is the
great principle of Descartes; it is the fundamental
axiom of modern science. Positively the principle
may be expressed : In matters of the intellect follow
your reason as far as it will take you without regard
to any other consideration. And negatively : In
matters of the intellect do not pretend that conclusions are certain which are not demonstrated or
demonstrable. That I take to be the agnostic faith,
which if a man keep whole and undefiled, he shall
not be ashamed to look the universe in the face,
whatever the future may have in store for him.

The results of the working out of the agnostic
principle will vary according to individual knowledge

and capacity, and according to the general condition of science. That which is unproven to-day may be proven by the help of new discoveries to-morrow. The only negative fixed points will be those negations which flow from the demonstrable limitation of our faculties. And the only obligation accepted is to have the mind always open to conviction. Agnostics who never fail in carrying out their principles are, I am afraid, as rare as other people of whom the same consistency can be truthfully predicated. But, if you were to meet with such a phœnix and to tell him that you had discovered that two and two make five, he would patiently ask you to state your reasons for that conviction, and express his readiness to agree with you if he found them satisfactory. The apostolic injunction to " suffer fools gladly " should be the rule of life of a true agnostic. I am deeply conscious how far I myself fall short of this ideal, but it is my personal conception of what agnostics ought to be.

However, as I began by stating, I speak only for myself; and I do not dream of anathematizing and excommunicating Mr. Laing. But, when I consider his creed and compare it with the Athanasian, I think I have on the whole a clearer conception of the meaning of the latter. " Polarity," in Article VIII., for example, is a word about which I heard a good deal in my youth, when " Naturphilosophie " was in fashion, and greatly did I suffer from it. For many years past, whenever I have met with " polarity " anywhere but in a discussion of some purely physical topic, such as magnetism, I have shut the book. Mr.

Laing must excuse me if the force of habit was too much for me when I read his eighth article.

And now, what is to be said to Mr. Harrison's remarkable deliverance " On the future of agnosticism " ?[1]   I would that it were not my business to say anything, for I am afraid that I can say nothing which shall manifest my great personal respect for this able writer, and for the zeal and energy with which he ever and anon galvanises the weakly frame of Positivism until it looks more than ever like John Bunyan's Pope and Pagan rolled into one.   There is a story often repeated, and I am afraid none the less mythical on that account, of a valiant and loud-voiced corporal in command of two full privates who, falling in with a regiment of the enemy in the dark, orders it to surrender under pain of instant annihilation by his force ; and the enemy surrenders accordingly.   I am always reminded of this tale when I read the positivist commands to the forces of Christianity and of Science ; only the enemy show no more signs of intending to obey now than they have done any time these forty years.

The allocution under consideration has the papal flavour which is wont to hang about the utterances of the pontiffs of the Church of Comte.   Mr. Harrison speaks with authority and not as one of the common scribes of the period.   He knows not only what agnosticism is and how it has come about, but what what will become of it.   The agnostic is to content

[1] *Fortnightly Review*, Jan. 1889.

himself with being the precursor of the positivist. In
his place, as a sort of navvy levelling the ground and
cleansing it of such poor stuff as Christianity, he is a
useful creature who deserves patting on the back, on
condition that he does not venture beyond his last.
But let not these scientific Sanballats presume that
they are good enough to take part in the building of
the Temple—they are mere Samaritans, doomed to
die out in proportion as the Religion of Humanity is
accepted by mankind. Well, if that is their fate,
they have time to be cheerful. But let us hear Mr.
Harrison's pronouncement of their doom.

"Agnosticism is a stage in the evolution of religion, an entirely negative stage, the point reached
by physicists, a purely mental conclusion, with no
relation to things social at all" (p. 154). I am quite
dazed by this declaration. Are there, then, any
"conclusions" that are not "purely mental"? Is
there "no relation to things social" in "mental conclusions" which affect men's whole conception of
life? Was that prince of agnostics, David Hume,
particularly imbued with physical science? Supposing physical science to be non-existent, would not the
agnostic principle, applied by the philologist and the
historian, lead to exactly the same results? Is the
modern more or less complete suspension of judgment
as to the facts of the history of regal Rome, or the
real origin of the Homeric poems, anything but
agnosticism in history and in literature? And if so,
how can agnosticism be the "mere negation of the
physicist"?

"Agnosticism is a stage in the evolution of religion." No two people agree as to what is meant by
the term "religion"; but if it means, as I think it
ought to mean, simply the reverence and love for the
ethical ideal, and the desire to realise that ideal in
life, which every man ought to feel—then I say
agnosticism has no more to do with it than it has to
do with music or painting. If, on the other hand,
Mr. Harrison, like most people, means by "religion"
theology, then in my judgment agnosticism can be
said to be a stage in its evolution, only as death may
be said to be the final stage in the evolution of life.

When agnostic logic is simply one of the canons of thought,
agnosticism, as a distinctive faith, will have spontaneously
disappeared (p. 155).

I can but marvel that such sentences as this, and
those already quoted, should have proceeded from Mr.
Harrison's pen. Does he really mean to suggest that
agnostics have a logic peculiar to themselves? Will
he kindly help me out of my bewilderment when I
try to think of "logic" being anything else than the
canon (which, I believe, means rule) of thought? As
to agnosticism being a distinctive faith, I have already
shown that it cannot possibly be anything of the
kind, unless perfect faith in logic is distinctive of
agnostics; which, after all, it may be.

Agnosticism as a religious philosophy *per se* rests on an almost
total ignoring of history and social evolution (p. 152).

But neither *per se* nor *per aliud* has agnosticism
(if I know anything about it) the least pretension to
be a religious philosophy; so far from resting on

ignorance of history, and that social evolution of which history is the account, it is and has been the inevitable result of the strict adherence to scientific methods by historical investigators. Our forefathers were quite confident about the existence of Romulus and Remus, of King Arthur, and of Hengist and Horsa. Most of us have become agnostics in regard to the reality of these worthies. It is a matter of notoriety of which Mr. Harrison, who accuses us all so freely of ignoring history, should not be ignorant, that the critical process which has shattered the foundations of orthodox Christian doctrine owes its origin, not to the devotees of physical science, but, before all, to Richard Simon, the learned French Oratorian, just two hundred years ago. I cannot find evidence that either Simon, or any one of the great scholars and critics of the eighteenth and nineteenth centuries who have continued Simon's work, had any particular acquaintance with physical science. I have already pointed out that Hume was independent of it. And certainly one of the most potent influences in the same direction, upon history in the present century, that of Grote, did not come from the physical side. Physical science, in fact, has had nothing directly to do with the criticism of the Gospels; it is wholly incompetent to furnish demonstrative evidence that any statement made in these histories is untrue. Indeed, modern physiology can find parallels in nature for events of apparently the most eminently supernatural kind recounted in some of those histories.

It is a comfort to hear, upon Mr. Harrison's authority, that the laws of physical nature show no signs of becoming " less definite, less consistent, or less popular as time goes on " (p. 154). How a law of nature is to become indefinite, or "inconsistent," passes my poor powers of imagination. But with universal suffrage and the coach-dog theory of premiership in full view; the theory, I mean, that the whole duty of a political chief is to look sharp for the way the social coach is driving, and then run in front and bark loud—as if being the leading noise-maker and guiding were the same things—it is truly satisfactory to me to know that the laws of nature are increasing in popularity. Looking at recent developments of the policy which is said to express the great heart of the people, I have had my doubts of the fact; and my love for my fellow-countrymen has led me to reflect with dread on what will happen to them if any of the laws of nature ever become so unpopular in their eyes as to be voted down by the transcendent authority of universal suffrage. If the legion of demons, before they set out on their journey in the swine, had had time to hold a meeting and to resolve unanimously " That the law of gravitation is oppress-ive and ought to be repealed," I am afraid it would have made no sort of difference to the result, when their two thousand unwilling porters were once launched down the steep slopes of the fatal shore of Gennesaret.

The question of the place of religion as an element of human nature, as a force of human society, its origin, analysis, and functions, has never been considered at all from an agnostic point of view (p. 152).

I doubt not that Mr. Harrison knows vastly more about history than I do; in fact, he tells the public that some of my friends and I have had no opportunity of occupying ourselves with that subject. I do not like to contradict any statement which Mr. Harrison makes on his own authority; only, if I may be true to my agnostic principles, I humbly ask how he has obtained assurance on this head. I do not profess to know anything about the range of Mr. Harrison's studies; but as he has thought it fitting to start the subject, I may venture to point out that, on evidence adduced, it might be equally permissible to draw the conclusion that Mr. Harrison's absorbing labours as the *pontifex maximus* of the positivist religion have not allowed him to acquire that acquaintance with the methods and results of physical science, or with the history of philosophy, or of philological and historical criticism, which is essential to any one who desires to obtain a right understanding of agnosticism. Incompetence in philosophy, and in all branches of science except mathematics, is the well-known mental characteristic of the founder of positivism. Faithfulness in disciples is an admirable quality in itself; the pity is that it not unfrequently leads to the imitation of the weaknesses as well as of the strength of the master. It is only such over-faithfulness which can account for a "strong mind really saturated with the historical sense" (p. 153) exhibiting the extraordinary forgetfulness of the his-

2 B

torical fact of the existence of David Hume implied by the assertion that

it would be difficult to name a single known agnostic who has given to history anything like the amount of thought and study which he brings to a knowledge of the physical world (p. 153).

Whoso calls to mind what I may venture to term the bright side of Christianity—that ideal of manhood, with its strength and its patience, its justice and its pity for human frailty, its helpfulness to the extremity of self-sacrifice, its ethical purity and nobility, which apostles have pictured, in which armies · of martyrs have placed their unshakable faith, and whence obscure men and women, like Catherine of Sienna and John Knox, have derived the courage to rebuke popes and kings—is not likely to underrate the importance of the Christian faith as a factor in human history, or to doubt that if that faith should prove to be incompatible with our knowledge, or necessary want of knowledge, some other hypostasis of men's hopes, genuine enough and worthy enough to replace it, will arise. But that the incongruous mixture of bad science with eviscerated papistry, out of which Comte manufactured the positivist religion, will be the heir of the Christian ages, I have too much respect for the humanity of the future to believe. Charles the Second told his brother, "They will not kill me, James, to make you king." And if critical science is remorselessly destroying the historical foundations of the noblest ideal of humanity which mankind have yet worshipped, it is little likely to permit the pitiful reality to climb into the vacant shrine.

That a man should determine to devote himself to the service of humanity—including intellectual and moral self-culture under that name ; that this should be, in the proper sense of the word, his religion—is not only an intelligible, but, I think, a laudable resolution. And I am greatly disposed to believe that it is the only religion which will prove itself to be unassailably acceptable so long as the human race endures. But when the positivist asks me to worship "Humanity"—that is to say, to adore the generalised conception of men as they ever have been and probably ever will be—I must reply that I could just as soon bow down and worship the generalised conception of a "wilderness of apes." Surely we are not going back to the days of Paganism, when individual men were deified, and the hard good sense of a dying Vespasian could prompt the bitter jest, "Ut puto, Deus fio." No divinity doth hedge a modern man, be he even a sovereign ruler. Nor is there any one, except a municipal magistrate, who is officially declared worshipful. But if there is no spark of worship-worthy divinity in the individual twigs of humanity, whence comes that godlike splendour which the Moses of Positivism fondly imagines to pervade the whole bush.

I know no study which is so unutterably saddening as that of the evolution of humanity, as it is set forth in the annals of history. Out of the darkness of prehistoric ages man emerges with the marks of his lowly origin strong upon him. He is a brute, only more intelligent than the other brutes, a blind prey to impulses, which as often as not lead him to destruction ; a victim to endless illusions, which make his mental existence a terror and a burden, and fill his physical life with barren toil and battle. He attains a certain degree of physical comfort, and develops a more or less workable theory of life, in such favourable situations as the plains of Mesopotamia or of Egypt, and then, for thousands and thousands of years, struggles, with varying fortunes, attended by infinite wickedness, bloodshed, and misery, to maintain himself at this point against the greed and the ambition of his fellow-men. He makes a point of killing and otherwise persecuting all those who first try to get him to move on ; and when he has moved on a step, foolishly confers post-mortem deification on his victims. He exactly repeats the process with all who want to move a step yet farther. And the best men of the best epochs are simply those who make the fewest blunders and commit the fewest sins.

That one should rejoice in the good man, forgive the bad man, and pity and help all men to the best of one's ability, is surely indisputable. It is the glory of Judaism and of Christianity to have proclaimed this truth, through all their aberrations. But the worship of a God who needs forgiveness and help, and deserves pity every hour of his existence, is no better than that of any other voluntarily selected fetish. The Emperor Julian's project was hopeful in comparison with the prospects of the new Anthropolatry.

When the historian of religion in the twentieth century is writing about the nineteenth, I foresee he will say something of this kind :

The most curious and instructive events in the religious history of the preceding century are the rise and progress of two new sects called Mormons and Positivists. To the student who has carefully considered these remarkable phenomena nothing in the records of religious self-delusion can appear improbable.

The Mormons arose in the midst of the great Republic, which, though comparatively insignificant, at that time, in territory as in the number of its citizens, was (as we know from the fragments of the speeches of its orators which have come down to us) no less remarkable for the native intelligence of its population than for the wide extent of their information, owing to the activity of their publishers in diffusing all that they could invent, beg, borrow, or steal. Nor were they less noted for their perfect freedom from all restraints in thought, or speech, or deed ; except, to be sure, the beneficent and wise influence of the majority, exerted, in case of need, through an institution known as " tarring and feathering," the exact nature of which is now disputed.

There is a complete consensus of testimony that the founder of Mormonism, one Joseph Smith, was a low-minded, ignorant scamp, and that he stole the

" Scriptures " which he propounded ; not being clever
enough to forge even such contemptible stuff as they
contain.   Nevertheless he must have been a man of
some force of character, for a considerable number of
disciples soon gathered about him.   In spite of repeated outbursts of popular hatred and violence—
during one of which persecutions Smith was brutally
murdered—the Mormon body steadily increased, and
became a flourishing community.   But the Mormon
practices being objectionable to the majority, they
were, more than once, without any pretence of law,
but by force of riot, arson, and murder, driven away
from the land they had occupied.   Harried by these
persecutions, the Mormon body eventually committed
itself to the tender mercies of a desert as barren as
that of Sinai ; and after terrible sufferings and priva-
tions, reached the Oasis of Utah.   Here it grew and
flourished, sending out missionaries to, and receiving
converts from, all parts of Europe, sometimes to the
number of 10,000 in a year ; until in 1880 the rich
and flourishing community numbered 110,000 souls
in Utah alone, while there were probably 30,000 or
40,000 scattered abroad elsewhere.   In the whole
history of religions there is no more remarkable
example of the power of faith ; and, in this case, the
founder of that faith was indubitably a most despicable creature.   It is interesting to observe that
the course taken by the great Republic and its citizens
runs exactly parallel with that taken by the Roman
Empire and its citizens towards the early Christians,
except that the Romans had a certain legal excuse for

their acts of violence, inasmuch as the Christian
" sodalitia " were not licensed, and consequently were,
ipso facto, illegal assemblages.   Until, in the latter
part of the nineteenth century, the United States
legislature decreed the illegality of polygamy, the
Mormons were wholly within the law.

Nothing can present a greater contrast to all this
than the history of the Positivists.   This sect arose
much about the same time as that of the Mormons,
in the upper and most instructed stratum of the
quick-witted, sceptical population of Paris.   The
founder, Auguste Comte, was a teacher of mathe
matics, but of no eminence in that department of
knowledge, and with nothing but an amateur's acquaintance with physical, chemical, and biological
science.   His works are repulsive on account of the
dull diffuseness of their style, and a certain air, as
of a superior person, which characterises them; but
nevertheless they contain good things here and there.
It would take too much space to reproduce in detail a
system which proposes to regulate all human life by
the promulgation of a Gentile Leviticus.   Suffice it to
say, that M. Comte may be described as a syncretic,
who, like the Gnostics of early Church history, attempted to combine the substance of imperfectly
comprehended contemporary science with the form of
Roman Christianity.   It may be that this is the
reason why his disciples were so very angry with
some obscure people called Agnostics, whose views, if
we may judge by the account left in the works of a
great Positivist controversial writer, were very absurd.

To put the matter briefly, M. Comte, finding Christianity and Science at daggers drawn, seems to have said to Science, " You find Christianity rotten at the core, do you ? Well, I will scoop out the inside of it." And to Romanism : " You find Science mere dry light—cold and bare. Well, I will put your shell over it, and so, as schoolboys make a spectre out of a turnip and a tallow candle, behold the new religion of Humanity complete ! "

Unfortunately neither the Romanists nor the people who were something more than amateurs in science, could be got to worship M. Comte's new idol properly. In the native country of Positivism, one distinguished man of letters and one of science, for a time, helped to make up a roomful of the faithful, but their love soon grew cold. In England, on the other hand, there appears to be little doubt that, in the ninth decade of the century, the multitude of disciples reached the grand total of several score. They had the advantage of the advocacy of one or two most eloquent and learned apostles, and, at any rate, the sympathy of several persons of light and leading—and, if they were not seen, they were heard all over the world. On the other hand, as a sect, they laboured under the prodigious disadvantage of being refined, estimable people, living in the midst of the worn-out civilisation of the old world ; where any one who had tried to persecute them, as the Mormons were persecuted, would have been instantly hanged. But the majority never dreamed of persecuting them ; on the contrary, they were rather given to scold and otherwise try the patience of the majority.

The history of these sects in the closing years of the century is highly instructive. Mormonism . . . .

But I find I have suddenly slipped off Mr. Harrison's tripod, which I had borrowed for the occasion. The fact is, I am not equal to the prophetical business, and ought not to have undertaken it.

X

THE VALUE OF WITNESS TO THE MIRACULOUS

CHARLES, or, more properly, Karl, King of the Franks, consecrated Roman Emperor in St. Peter's on Christmas Day, A.D. 800, and known to posterity as the Great (chiefly by his agglutinative Gallicised denomination of Charlemagne), was a man great in all ways, physically and mentally. Within a couple of centuries after his death Charlemagne became the centre of innumerable legends; and the myth-making process does not seem to have been sensibly interfered with by the existence of sober and truthful histories of the Emperor and of the times which immediately preceded and followed his reign, by a contemporary writer who occupied a high and confidential position in his court, and in that of his successor. This was one Eginhard, or Einhard, who appears to have been born about A.D. 770, and spent his youth at the court, being educated along with Charles's sons. There is excellent contemporary testimony not only to Eginhard's existence, but to his abilities, and to the place which he occupied in the circle of the intimate friends of the great ruler whose life he subsequently wrote.

In fact, there is as good evidence of Eginhard's existence, of his official position, and of his being the
author of the chief works attributed to him, as can
reasonably be expected in the case of a man who
lived more than a thousand years ago, and was neither
a great king nor a great warrior. The works are—
1. *The Life of the Emperor Karl.* 2. *The Annals
of the Franks.* 3. *Letters.* 4. *The History of the
Translation of the Blessed Martyrs of Christ, SS.
Marcellinus and Petrus.*

It is to the last, as one of the most singular and
interesting records of the period during which the
Roman world passed into that of the Middle Ages,
that I wish to direct attention.[1] It was written in
the ninth century, somewhere, apparently, about the
year 830, when Eginhard, ailing in health and weary
of political life, had withdrawn to the monastery of
Seligenstadt, of which he was the founder. A manuscript copy of the work, made in the tenth century,
and once the property of the monastery of St. Bavon
on the Scheldt, of which Eginhard was Abbot, is still
extant, and there is no reason to believe that, in this
copy, the original has been in any way interpolated
or otherwise tampered with. The main features of
the strange story contained in the *Historia Translationis* are set forth in the following pages, in which,
in regard to all matters of importance, I shall adhere
as closely as possible to Eginhard's own words.

[1] My citations are made from Teulet's *Einhardi omnia quæ extant
opera*, Paris, 1840-1843, which contains a biography of the author, a
history of the text, with translations into French, and many valuable
annotations.

While I was still at Court, busied with secular affairs, I often thought of the leisure which I hoped one day to enjoy in a solitary place, far away from the crowd, with which the liberality of Prince Louis, whom I then served, had provided me. This place is situated in that part of Germany which lies between the Neckar and the Maine,[1] and is nowadays called the Odenwald by those who live in and about it. And here having built, according to my capacity and resources, not only houses and permanent dwellings, but also a basilica fitted for the performance of divine service and of no mean style of construction, I began to think to what saint or martyr I could best dedicate it. A good deal of time had passed while my thoughts fluctuated about this matter, when it happened that a certain deacon of the Roman Church, named Deusdona, arrived at the Court for the purpose of seeking the favour of the King in some affairs in which he was interested. He remained some time; and then, having transacted his business, he was about to return to Rome, when one day, moved by courtesy to a stranger, we invited him to a modest refection; and while talking of many things at table, mention was made of the translation of the body of the blessed Sebastian,[2] and of the neglected tombs of the martyrs, of which there is such a prodigious number at Rome; and the conversation having turned towards the dedication of our new basilica, I began to inquire how it might be possible for me to obtain some of the true relics of the saints which rest at Rome. He at first hesitated, and declared that he did not know how that could be done. But observing that I was both anxious and curious about the subject, he promised to give me an answer some other day.

When I returned to the question some time afterwards, he immediately drew from his bosom a paper, which he begged me to read when I was alone, and to tell him what I was disposed to think of that which was therein stated. I took the paper and, as he desired, read it alone and in secret. (Cap. i. 2, 3.)

---

[1] At present included in the Duchies of Hesse-Darmstadt and Baden.

[2] This took place in the year 826 A.D. The relics were brought from Rome and deposited in the Church of St. Medardus at Soissons.

I shall have occasion to return to Deacon Deusdona's conditions, and to what happened after Eginhard's acceptance of them. Suffice it, for the present,
to say that Eginhard's notary, Ratleicus (Ratleig),
was despatched to Rome and succeeded in securing
two bodies, supposed to be those of the holy martyrs
Marcellinus and Petrus ; and when he had got as far
on his homeward journey as the Burgundian town of
Solothurn, or Soleure,[1] notary Ratleig despatched to
his master, at St. Bavon, a letter announcing the
success of his mission.

As soon as by reading it I was assured of the arrival of the
saints, I despatched a confidential messenger to Maestricht to
gather together priests, other clerics, and also laymen, to go out
to meet the coming saints as speedily as possible. And he and
his companions, having lost no time, after a few days met those
who had charge of the saints at Solothurn. Joined with them,
and with a vast crowd of people who gathered from all parts,
singing hymns, and amidst great and universal rejoicings, they
travelled quickly to the city of Argentoratum, which is now
called Strasburg. Thence embarking on the Rhine, they came
to the place called Portus,[2] and landing on the east bank of
the river, at the fifth station thence they arrived at Michilin-
stadt,[3] accompanied by an immense multitude, praising God.
This place is in that forest of Germany which in modern times
is called the Odenwald, and about six leagues from the Maine.
And here, having found a basilica recently built by me, but not
yet consecrated, they carried the sacred remains into it and deposited them therein, as if it were to be their final resting-place.
As soon as all this was reported to me I travelled thither as
quickly as I could. (Cap. ii. 14.)

---

[1] Now included in Western Switzerland.

[2] Probably, according to Teulet, the present Sandhofer-fahrt, a
little below the embouchure of the Neckar.

[3] The present Michilstadt, thirty miles N.E. of Heidelberg.

Three days after Eginhard's arrival began the series of wonderful events which he narrates, and for which we have his personal guarantee. The first thing that he notices is the dream of a servant of Ratleig, the notary, who, being set to watch the holy relics in the church after vespers, went to sleep, and during his slumbers had a vision of two pigeons, one white and one gray and white, which came and sat upon the bier over the relics; while, at the same time, a voice ordered the man to tell his master that the holy martyrs had chosen another resting-place and desired to be transported thither without delay.

Unfortunately, the saints seem to have forgotten to mention where they wished to go; and, with the most anxious desire to gratify their smallest wishes, Eginhard was naturally greatly perplexed what to do. While in this state of mind, he was one day contemplating his "great and wonderful treasure, more precious than all the gold in the world," when it struck him that the chest in which the relics were contained was quite unworthy of its contents; and after vespers he gave orders to one of the sacristans to take the measure of the chest in order that a more fitting shrine might be constructed. The man, having lighted a wax candle and raised the pall which covered the relics, in order to carry out his master's orders, was astonished and terrified to observe that the chest was covered with a blood-like exudation (*loculum mirum in modum humore sanguineo undique distillantem*), and at once sent a message to Eginhard.

Then I and those priests who accompanied me beheld this stupendous miracle, worthy of all admiration. For just as when it is going to rain, pillars and slabs and marble images exude moisture, and, as it were sweat, so the chest which contained the most sacred relics was found moist with the blood exuding on all sides. (Cap. ii. 16.)

Three days' fast was ordained in order that the meaning of the portent might be ascertained. All that happened, however, was that at the end of that time the "blood," which had been exuding in drops all the while, dried up. Eginhard is careful to say that the liquid "had a saline taste, something like that of tears, and was thin as water, though of the colour of true blood," and he clearly thinks this satisfactory evidence that it was blood.

The same night another servant had a vision, in which still more imperative orders for the removal of the relics were given ; and, from that time forth, "not a single night passed without one, two, or even three of our companions receiving revelations in dreams that the bodies of the saints were to be transferred from that place to another." At last a priest, Hildfrid, saw, in a dream, a venerable white-haired man in a priest's vestments, who bitterly reproached Eginhard for not obeying the repeated orders of the saints, and upon this the journey was commenced. Why Eginhard delayed obedience to these repeated visions so long does not appear. He does not say so in so many words, but the general tenor of the narrative leads one to suppose that Mulinheim (after-wards Seligenstadt) is the "solitary place" in which he had built the church which awaited dedication.

In that case, all the people about him would know
that he desired that the saints should go there. If a
glimmering of secular sense led him to be a little
suspicious about the real cause of the unanimity of
the visionary beings who manifested themselves to his
entourage in favour of moving on, he does not say so.

At the end of the first day's journey the precious
relics were deposited in the church of St. Martin, in
the village of Ostheim. Hither a paralytic nun
(*sanctimonialis quædam paralytica*) of the name of
Ruodlang was brought in a car by her friends and
relatives from a monastery a league off. She spent
the night watching and praying by the bier of the
saints ; " and health returning to all her members,
on the morrow she went back to her place whence
she came, on her feet, nobody supporting her, or in
any way giving her assistance." (Cap. ii. 19.)

On the second day, the relics were carried to
Upper Mulinheim, and finally, in accordance with
the orders of the martyrs, deposited in the church of
that place, which was therefore renamed Seligenstadt.
Here, Daniel, a beggar boy of fifteen, and so bent
that " he could not look at the sky without lying on
his back," collapsed and fell down during the celebra-
tion of the Mass. " Thus he lay a long time, as if
asleep, and all his limbs straightening and his flesh
strengthening (*recepta firmitate nervorum*), he arose
before our eyes, quite well." (Cap. ii. 20.)

Some time afterwards an old man entered the
church on his hands and knees, being unable to use
his limbs properly :—

He, in presence of all of us, by the power of God and the merits of the blessed martyrs, in the same hour in which he entered was so perfectly cured that he walked without so much as a stick. And he said that, though he had been deaf for five years, his deafness had ceased along with the palsy. (Cap. iii. 33.)

Eginhard was now obliged to return to the Court at Aix-la-Chapelle, where his duties kept him through the winter; and he is careful to point out that the later miracles which he proceeds to speak of are known to him only at second hand. But, as he naturally observes, having seen such wonderful events with his own eyes, why should he doubt similar narrations when they are received from trustworthy sources?

Wonderful stories these are indeed, but as they are, for the most part, of the same general character as those already recounted, they may be passed over. There is, however, an account of a possessed maiden which is worth attention. This is set forth in a memoir, the principal contents of which are the speeches of a demon who declared himself to possess the singular appellation of " Wiggo," and revealed himself in the presence of many witnesses, before the altar, close to the relics of the blessed martyrs. It is noteworthy that the revelations appear to have been made in the shape of replies to the questions of the exorcising priest, and there is no means of judging how far the answers are, really, only the questions to which the patient replied yes or no.

The possessed girl, about sixteen years of age, was brought by her parents to the basilica of the martyrs.

When she approached the tomb containing the sacred bodies, the priest, according to custom, read the formula of exorcism over her head. When he began to ask how and when the demon had entered her, she answered, not in the tongue of the barbarians, which alone the girl knew, but in the Roman tongue. And when the priest was astonished and asked how she came to know Latin, when her parents, who stood by, were wholly ignorant of it, "Thou hast never seen my parents," was the reply. To this the priest, "Whence art thou, then, if these are not thy parents?" And the demon, by the mouth of the girl, "I am a follower and disciple of Satan, and for a long time I was gatekeeper (janitor) in hell; but, for some years, along with eleven companions, I have ravaged the kingdom of the Franks." (Cap. v. 49.)

He then goes on to tell how they blasted the crops and scattered pestilence among beasts and men, because of the prevalent wickedness of the people.[1]

The enumeration of all these iniquities, in oratorical style, takes up a whole octavo page; and at the end it is stated, "All these things the demon spoke in Latin by the mouth of the girl."

And when the priest imperatively ordered him to come out, "I shall go," said he, "not in obedience to you, but on account of the power of the saints, who do not allow me to remain any longer." And, having said this, he threw the girl down on the floor and there compelled her to lie prostrate for a time, as though she slumbered. After a little while, however, he going away, the girl, by the power of Christ and the merits of the blessed martyrs, as it were awaking from sleep, rose up quite well, to the astonishment of all present; nor after the demon had gone out was she able to speak Latin: so that it was plain enough that it was not she who had spoken in that tongue, but the demon by her mouth. (Cap. v. 51.)

---

[1] In the Middle Ages one of the most favourite accusations against witches was that they committed just these enormities.

If the *Historia Translationis* contained nothing more than has been, at present, laid before the reader, disbelief in the miracles of which it gives so precise and full a record might well be regarded as hyper-scepticism. It might fairly be said, Here you have a man, whose high character, acute intelligence, and large instruction are certified by eminent contemporaries; a man who stood high in the confidence of one of the greatest rulers of any age, and whose other works prove him to be an accurate and judicious narrator of ordinary events. This man tells you, in language which bears the stamp of sincerity, of things which happened within his own knowledge, or within that of persons in whose veracity he has entire confidence, while he appeals to his sovereign and the court as witnesses of others; what possible ground can there be for disbelieving him?

Well, it is hard upon Eginhard to say so, but it is exactly the honesty and sincerity of the man which are his undoing as a witness to the miraculous. He himself makes it quite obvious that when his profound piety comes on the stage, his good sense and even his perception of right and wrong make their exit. Let us go back to the point at which we left him, secretly perusing the letter of Deacon Deusdona. As he tells us, its contents were

that he [the deacon] had many relics of saints at home, and that he would give them to me if I would furnish him with the means of returning to Rome; he had observed that I had two mules, and if I would let him have one of them and would despatch with him a confidential servant to take charge of the relics, he would at once send them to me. This plausibly expressed proposition pleased me, and I made up my mind to test the value of the somewhat ambiguous promise at once;[1] so giving him the mule and money for his journey I ordered my notary Ratleig (who already desired to go to Rome to offer his devotions there) to go with him. Therefore, having left Aix-la-Chapelle (where the Emperor and his Court resided at the time) they came to Soissons. Here they spoke with Hildoin, abbot of the monastery of St. Medardus, because the said deacon had assured him that he had the means of placing in his possession the body of the blessed Tiburtius the Martyr. Attracted by which promises he (Hildoin) sent with them a certain priest, Hunus by name, a sharp man (*hominem callidum*), whom he ordered to receive and bring back the body of the martyr in question. And so, resuming their journey, they proceeded to Rome as fast as they could. (Cap. i. 3.)

Unfortunately, a servant of the notary, one Reginbald, fell ill of a tertian fever, and impeded the progress of the party. However, this piece of adversity had its sweet uses; for three days before they reached Rome, Reginbald had a vision. Somebody habited as a deacon appeared to him and asked why his master was in such a hurry to get to Rome; and when Reginbald explained their business, this visionary deacon, who seems to have taken the measure of his brother in the flesh with some accuracy, told him not by any means to expect that Deusdona would fulfil his promises. Moreover, taking the servant by the hand, he led him to the top of a high mountain and, showing him Rome (where the man had never been), pointed

---

[1] It is pretty clear that Eginhard had his doubts about the deacon, whose pledges he qualifies as *sponsiones incertæ*. But, to be sure, he wrote after events which fully justified scepticism.

out a church, adding "Tell Ratleig the thing he
wants is hidden there ; let him get it as quickly
as he can and go back to his master ;" and, by way
of a sign that the order was authoritative, the servant
was promised that from that time forth his fever
should disappear. And as the fever did vanish to
return no more, the faith of Eginhard's people in
Deacon Deusdona naturally vanished with it (*et
fidem diaconi promissis non haberent*). Nevertheless, they put up at the deacon's house near St. Peter
ad Vincula. But time went on and no relics made
their appearance, while the notary and the priest
were put off with all sorts of excuses—the brother
to whom the relics had been confided was gone to
Beneventum and not expected back for some time,
and so on—until Ratleig and Hunus began to despair,
and were minded to return, *infecto negotio.*

But my notary, calling to mind his servant's dream, proposed
to his companion that they should go to the cemetery which
their host had talked about without him. So, having found and
hired a guide, they went in the first place to the basilica of the
blessed Tiburtius in the Via Labicana, about three thousand paces
from the town, and cautiously and carefully inspected the tomb
of that martyr, in order to discover whether it could be opened
without any one being the wiser. Then they descended into
the adjoining crypt, in which the bodies of the blessed martyrs
of Christ, Marcellinus and Petrus were buried ; and, having made
out the nature of their tomb, they went away thinking their host
would not know what they had been about. But things fell out
differently from what they had imagined. (Cap. i. 7.)

In fact, Deacon Deusdona, who doubtless kept an
eye on his guests, knew all about their manœuvres
and made haste to offer his services, in order that

" with the help of God " (*si Deus votis eorum favere dignaretur*), they should all work together. The deacon was evidently alarmed lest they should succeed without *his* help.

So, by way of preparation for the contemplated *vol avec effraction* they fasted three days; and then, at night, without being seen, they betook themselves to the basilica of St. Tiburtius, and tried to break open the altar erected over his remains. But the marble proving too solid, they descended to the crypt, and "having evoked our Lord Jesus Christ and adored the holy martyrs," they proceeded to prise off the stone which covered the tomb, and thereby exposed the body of the most sacred martyr Marcellinus, "whose head rested on a marble tablet on which his name was inscribed." The body was taken up with the greatest veneration, wrapped in a rich covering, and given over to the keeping of the deacon and his brother, Lunison, while the stone was replaced with such care that no sign of the theft remained.

· As sacrilegious proceedings of this kind were punishable with death by the Roman law, it seems not unnatural that Deacon Deusdona should have become uneasy, and have urged Ratleig to be satisfied with what he had got and be off with his spoils. But the notary having thus cleverly captured the blessed Marcellinus, thought it a pity he should be parted from the blessed Petrus, side by side with whom he had rested for five hundred years and more in the same sepulchre (as Eginhard pathetically observes); and the pious man could neither eat, drink, nor sleep, until he had compassed his desire to re-unite the saintly colleagues. This time, apparently in consequence of Deusdona's opposition to any further resurrectionist doings, he took counsel with a Greek monk, one Basil, and, accompanied by Hunus, but saying nothing to Deusdona, they committed another sacrilegious burglary, securing this time, not only the body of the blessed Petrus, but a quantity of dust, which they agreed the priest should take, and tell his employer that it was the remains of the blessed Tiburtius. How Deusdona was "squared," and what he got for his not very valuable complicity in these transactions, does not appear. But at last the relics were sent off in charge of Lunison, the brother of Deusdona, and the priest Hunus, as far as Pavia, while Ratleig stopped behind for a week to see if the robbery was discovered, and, presumably, to act as a blind if any hue and cry was raised. But, as everything remained quiet, the notary betook himself to Pavia, where he found Lunison and Hunus awaiting his arrival. The notary's opinion of the character of his worthy colleagues, however, may be gathered from the fact that, having persuaded them to set out in advance along a road which he told them he was about to take, he immediately adopted another route, and, travelling by way of St. Maurice and the Lake of Geneva, eventually reached Soleure.

Eginhard tells all this story with the most naïve air of unconsciousness that there is anything remarkable about an abbot, and a high officer of state to

boot, being an accessory, both before and after the
fact, to a most gross and scandalous act of sacrilegious
and burglarious robbery.   And an amusing sequel to
the story proves that, where relics were concerned, his
friend Hildoin, another high ecclesiastical dignitary,
was even less scrupulous than himself.

On going to the palace early one morning, after the
saints were safely bestowed at Seligenstadt, he found
Hildoin waiting for an audience in the Emperor's
antechamber, and began to talk to him about the
miracle of the bloody exudation.   In the course of
conversation, Eginhard happened to allude to the
remarkable fineness of the garment of the blessed
Marcellinus.   Whereupon Abbot Hildoin observed
(to Eginhard's stupefaction) that his observation was
quite correct.   Much astonished at this remark from
a person who was supposed not to have seen the relics,
Eginhard asked him how he knew that ?   Upon this,
Hildoin saw that he had better make a clean breast
of it, and he told the following story, which he
had received from his priestly agent, Hunus.   While
Hunus and Lunison were at Pavia, waiting for Eginhard's notary, Hunus (according to his own account)
had robbed the robbers.   The relics were placed in a
church and a number of laymen and clerics, of whom
Hunus was one, undertook to keep watch over them.
One night, however, all the watchers, save the wide-
awake Hunus, went to sleep ; and then, according to
the story which this " sharp " ecclesiastic foisted upon
his patron,

it was borne in upon his mind that there must be some great
reason why all the people, except himself, had suddenly become
somnolent; and, determining to avail himself of the opportunity
thus offered (*oblata occasione utendum*), he rose and, having lighted
a candle, silently approached the chests. Then, having burnt
through the threads of the seals with the flame of the candle, he
quickly opened the chests, which had no locks;[1] and, taking out
portions of each of the bodies which were thus exposed, he closed
the chests and connected the burnt ends of the threads with the
seals again, so that they appeared not to have been touched;
and, no one having seen him, he returned to his place.   (Cap.
iii. 23.)

Hildoin went on to tell Eginhard that Hunus at
first declared to him that these purloined relics belonged to St. Tiburtius; but afterwards confessed, as
a great secret, how he had come by them, and he
wound up his discourse thus:

They have a place of honour beside St. Medardus, where they
are worshipped with great veneration by all the people; but
whether we may keep them or not is for your judgment.   (Cap.
iii. 23.)

Poor Eginhard was thrown into a state of great
perturbation of mind by this revelation. An acquaintance of his had recently told him of a rumour that
was spread about that Hunus had contrived to
abstract *all* the remains of SS. Marcellinus and Petrus
while Eginhard's agents were in a drunken sleep; and
that, while the real relics were in Abbot Hildoin's
hands at St. Medardus, the shrine at Seligenstadt
contained nothing but a little dust. Though greatly
annoyed by this "execrable rumour, spread every-

---

[1] The words are *scrinia sine clave*, which seems to mean " having
no key." But the circumstances forbid the idea of breaking open.

where by the subtlety of the devil," Eginhard had doubtless comforted himself by his supposed knowledge of its falsity, and he only now discovered how considerable a foundation there was for the scandal. There was nothing for it but to insist upon the return of the stolen treasures. One would have thought that the holy man, who had admitted himself to be knowingly a receiver of stolen goods, would have made instant restitution and begged only for absolution. But Eginhard intimates that he had very great difficulty in getting his brother abbot to see that even restitution was necessary.

Hildoin's proceedings were not of such a nature as to lead any one to place implicit confidence in anything he might say; still less had his agent, priest Hunus, established much claim to confidence; and it is not surprising that Eginhard should have lost no time in summoning his notary and Lunison to his presence, in order that he might hear what they had to say about the business. They, however, at once protested that priest Hunus's story was a parcel of lies, and that after the relics left Rome no one had any opportunity of meddling with them. Moreover, Lunison, throwing himself at Eginhard's feet, confessed with many tears what actually took place. It will be remembered that after the body of St. Marcellinus was abstracted from its tomb, Ratleig deposited it in the house of Deusdona, in charge of the latter's brother, Lunison. But Hunus, being very much disappointed that he could not get hold of the body of St. Tiburtius, and afraid to go back to his abbot empty-handed, bribed Lunison with four pieces of gold and five of silver to give him access to the chest. This Lunison did, and Hunus helped himself to as much as would fill a gallon measure (*vas sextarii mensuram*) of the sacred remains. Eginhard's indignation at the "rapine" of this "nequissimus nebulo" is exquisitely droll. It would appear that the adage about the receiver being as bad as the thief was not current in the ninth century.

Let us now briefly sum up the history of the acquisition of the relics. Eginhard makes a contract with Deusdona for the delivery of certain relics which the latter says he possesses. Eginhard makes no inquiry how he came by them; otherwise, the transaction is innocent enough.

Deusdona turns out to be a swindler, and has no relics. Thereupon Eginhard's agent, after due fasting and prayer, breaks open the tombs and helps himself.

Eginhard discovers by the self-betrayal of his brother abbot, Hildoin, that portions of his relics have been stolen and conveyed to the latter. With much ado he succeeds in getting them back.

Hildoin's agent, Hunus, in delivering these stolen goods to him, at first declared they were the relics of St. Tiburtius, which Hildoin desired him to obtain; but afterwards invented a story of their being the product of a theft, which the providential drowsiness of his companions enabled him to perpetrate, from the relics which Hildoin well knew were the property of his friend.

Lunison, on the contrary, swears that all this story is false, and that he himself was bribed by Hunus to allow him to steal what he pleased from the property confided to his own and his brother's care by their guest Ratleig. And the honest notary himself seems to have no hesitation about lying and stealing to any extent, where the acquisition of relics is the object in view.

For a parallel to these transactions one must read a police report of the doings of a " long firm " or of a set of horse-coupers ; yet Eginhard seems to be aware of nothing, but that he has been rather badly used by his friend Hildoin, and the " nequissimus nebulo " Hunus.

It is not easy for a modern Protestant, still less for any one who has the least tincture of scientific culture, whether physical or historical, to picture to himself the state of mind of a man of the ninth century, however cultivated, enlightened, and sincere he may have been. His deepest convictions, his most cherished hopes, were bound up with the belief in the miraculous. Life was a constant battle between saints and demons for the possession of the souls of men. The most superstitious among our modern countrymen turn to supernatural agencies only when natural causes seem insufficient ; to Eginhard and his friends the supernatural was the rule, and the sufficiency of natural causes was allowed only when there was nothing to suggest others.

Moreover, it must be recollected that the possession of miracle-working relics was greatly coveted, not only on high, but on very low grounds. To a man like Eginhard, the mere satisfaction of the religious sentiment was obviously a powerful attraction. But, more than this, the possession of such a treasure was an immense practical advantage. If the saints were duly flattered and worshipped, there was no telling what benefits might result from their interposition on your behalf. For physical evils, access to the shrine was like the grant of the use of a universal pill and ointment manufactory; and pilgrimages thereto might suffice to cleanse the performers from any amount of sin. A letter to Lupus, subsequently Abbot of Ferrara, written while Eginhard was smarting under the grief caused by the loss of his much-loved wife Imma, affords a striking insight into the current view of the relation between the glorified saints and their worshippers. The writer shows that he is anything but satisfied with the way in which he has been treated by the blessed martyrs whose remains he has taken such pains to "convey" to Seligenstadt, and to honour there as they would never have been honoured in their Roman obscurity.

> It is an aggravation of my grief and a reopening of my wound, that our vows have been of no avail, and that the faith which we placed in the merits and intervention of the martyrs has been utterly disappointed.

We may admit, then, without impeachment of Eginhard's sincerity, or of his honour under all ordinary circumstances, that when piety, self-interest, the glory of the Church in general, and that of the church at Seligenstadt in particular, all pulled one way,

even the workaday principles of morality were disregarded; and, *à fortiori*, anything like proper investigation of the reality of alleged miracles was
thrown to the winds.

And if this was the condition of mind of such a
man as Eginhard, what is it not legitimate to suppose
may have been that of Deacon Deusdona, Lunison,
Hunus, and Company, thieves and cheats by their
own confession, or of the probably hysterical nun, or
of the professional beggars, for whose incapacity to
walk and straighten themselves there is no guarantee
but their own? Who is to make sure that the
exorcist of the demon Wiggo was not just such
another priest as Hunus; and is it not at least possible, when Eginhard's servants dreamed, night after
night, in such a curiously coincident fashion, that
a careful inquirer might have found they were very
anxious to please their master?

Quite apart from deliberate and conscious fraud
(which is a rarer thing than is often supposed), people,
whose mythopœic faculty is once stirred, are capable
of saying the thing that is not, and of acting as they
should not, to an extent which is hardly imaginable
by persons who are not so easily affected by the
contagion of blind faith. There is no falsity so gross
that honest men and, still more, virtuous women,
anxious to promote a good cause, will not lend themselves to it without any clear consciousness of the
moral bearings of what they are doing.

The cases of miraculously-effected cures of which
Eginhard is ocular witness appear to belong to classes

of disease in which malingering is possible or hysteria presumable. Without modern means of diagnosis, the names given to them are quite worthless. One "miracle," however, in which the patient, a woman, was cured by the mere sight of the church in which the relics of the blessed martyrs lay, is an unmistakable case of dislocation of the lower jaw; and it is obvious that, as not unfrequently happens in such accidents in weakly subjects, the jaw slipped suddenly back into place, perhaps in consequence of a jolt, as the woman rode towards the church. (Cap. v. 53.)[1]

There is also a good deal said about a very questionable blind man—one Albricus (Alberich?)—who, having been cured, not of his blindness, but of another disease under which he laboured, took up his quarters at Seligenstadt, and came out as a prophet, inspired by the Archangel Gabriel. Eginhard intimates that his prophecies were fulfilled; but as he does not state exactly what they were or how they were accomplished, the statement must be accepted with much caution. It is obvious that he was not the man to hesitate to "ease" a prophecy until it fitted, if the credit of the shrine of his favourite saints could be increased by such a procedure. There is no impeachment of his honour in the supposition. The logic of the matter is quite simple, if somewhat sophistical. The holiness of the church of the martyrs guarantees

---

[1] Eginhard speaks with lofty contempt of the "vana ac superstitiosa præsumptio" of the poor woman's companions in trying to alleviate her sufferings with "herbs and frivolous incantations." Vain enough, no doubt, but the "mulierculæ" might have returned the epithet "superstitious" with interest.

the reality of the appearance of the Archangel Gabriel there, and what the archangel says must be true. Therefore, if anything seem to be wrong, that must be the mistake of the transmitter; and, in justice to the archangel, it must be suppressed or set right. This sort of " reconciliation " is not unknown in quite modern times, and among people who would be very much shocked to be compared with a "benighted papist" of the ninth century.

The readers of this essay are, I imagine, very largely composed of people who would be shocked to be regarded as anything but enlightened Protestants. It is not unlikely that those of them who have accompanied me thus far may be disposed to say, " Well, this is all very amusing as a story, but what is the practical interest of it? We are not likely to believe in the miracles worked by the spolia of SS. Marcellinus and Petrus, or by those of any other saints in the Roman Calendar."

The practical interest is this : if you do not believe in these miracles recounted by a witness whose character and competency are firmly established, whose sincerity cannot be doubted, and who appeals to his sovereign and other contemporaries as witnesses of the truth of what he says, in a document of which a MS. copy exists, probably dating within a century of the author's death, why do you profess to believe in stories of a like character, which are found in documents of the dates and of the authorship of which nothing is certainly determined, and no known copies of which come within two or three centuries of the

events they record. If it be true that the four
Gospels and the Acts were written by Matthew, Mark,
Luke, and John, all that we know of these persons
comes to nothing in comparison with our knowledge
of Eginhard; and not only is there no proof that the
traditional authors of these works wrote them, but
very strong reasons to the contrary may be alleged.
If, therefore, you refuse to believe that " Wiggo " was
cast out of the possessed girl on Eginhard's authority,
with what justice can you profess to believe that the
legion of devils were cast out of the man among the
tombs of the Gadarenes?   And if, on the other hand,
you accept Eginhard's evidence, why do you laugh at
the supposed efficacy of relics and the saint-worship
of the modern Romanists?   It cannot be pretended,
in the face of all evidence, that the Jews of the year
30 A.D., or thereabouts, were less imbued with the belief
in the supernatural than were the Franks of the year
800 A.D.   The same influences were at work in each
case, and it is only reasonable to suppose that the
results were the same.   If the evidence of Eginhard
is insufficient to lead reasonable men to believe in the
miracles he relates, *à fortiori* the evidence afforded
by the Gospels and the Acts must be so.[1]

But it may be said that no serious critic denies
the genuineness of the four great Pauline Epistles
— Galatians, First and Second Corinthians, and

---

[1] Of course there is nothing new in this argument; but it does
not grow weaker by age. And the case of Eginhard is far more instructive than that of Augustine, because the former has so very
frankly, though incidentally, revealed to us not only his own mental
and moral habits, but those of the people about him.

Romans—and that in three out of these four Paul lays claim to the power of working miracles.[1] Must we suppose, therefore, that the Apostle to the Gentiles has stated that which is false? But to how much does this so-called claim amount? It may mean much or little. Paul nowhere tells us what he did in this direction; and, in his sore need to justify his assumption of apostleship against the sneers of his enemies, it is hardly likely that if he had any very striking cases to bring forward he would have neglected evidence so well calculated to put them to shame. And, without the slightest impeachment of Paul's veracity, we must further remember that his strongly-marked mental characteristics, displayed in unmistakable fashion in these Epistles, are anything but those which would justify us in regarding him as a critical witness respecting matters of fact, or as a trustworthy interpreter of their significance. When a man testifies to a miracle, he not only states a fact, but he adds an interpretation of the fact. We may admit his evidence as to the former, and yet think his opinion as to the latter worthless. If Eginhard's calm and objective narrative of the historical events of his time is no guarantee for the soundness of his judgment where the supernatural is concerned, the heated rhetoric of the Apostle of the Gentiles, his absolute confidence in the "inner light," and the extraordinary conceptions of the nature and require- ments of logical proof which he betrays, in page after page of his Epistles, afford still less security.

[1] See 1 Cor. xii. 10-28 ; 2 Cor. vi. 12 ; Rom. xv. 19.

There is a comparatively modern man who shared to the full Paul's trust in the "inner light," and who, though widely different from the fiery evangelist of Tarsus in various obvious particulars, yet, if I am not mistaken, shares his deepest characteristics. I speak of George Fox, who separated himself from the current Protestantism of England, in the seventeenth century, as Paul separated himself from the Judaism of the first century, at the bidding of the "inner light"; who went through persecutions as serious as those which Paul enumerates; who was beaten, stoned, cast out for dead, imprisoned nine times, sometimes for long periods; who was in perils on land and perils at sea. George Fox was an even more widely travelled missionary; while his success in founding congregations, and his energy in visiting them, not merely in Great Britain and Ireland and the West India Islands, but on the continent of Europe and that of North America, was no less remarkable. A few years after Fox began to preach, there were reckoned to be a thousand Friends in prison in the various gaols of England; at his death, less than fifty years after the foundation of the sect, there were 70,000 Quakers in the United Kingdom. The cheerfulness with which these people—women as well as men—underwent martyrdom in this country and in the New England States is one of the most remarkable facts in the history of religion.

No one who reads the voluminous autobiography of " Honest George " can doubt the man's utter truthfulness; and though, in his multitudinous letters,

he but rarely rises far above the incoherent commonplaces of a street preacher, there can be no
question of his power as a speaker, nor any doubt
as to the dignity and attractiveness of his personality,
or of his possession of a large amount of practical
good sense and governing faculty.

But that George Fox had full faith in his own
powers as a miracle-worker, the following passage of
his autobiography (to which others might be added)
demonstrates :—

> Now after I was set at liberty from Nottingham gaol (where
> I had been kept a prisoner a pretty long time) I travelled as
> before, in the work of the Lord. And coming to Mansfield
> Woodhouse, there was a distracted woman, under a doctor's hand,
> with her hair let loose all about her ears ; and he was about
> to let her blood, she being first bound, and many people being
> about her, holding her by violence ; but he could get no blood
> from her. And I desired them to unbind her and let her alone ;
> for they could not touch the spirit in her by which she was
> tormented. So they did unbind her, and I was moved to speak
> to her, and in the name of the Lord to bid her be quiet and
> still. And she was so. And the Lord's power settled her
> mind and she mended ; and afterwards received the truth and
> continued in it to her death. And the Lord's name was honoured ;
> to whom the glory of all his works belongs. Many great and
> wonderful things were wrought by the heavenly power in those
> days. For the Lord made bare his omnipotent arm and mani-
> fested his power to the astonishment of many ; by the healing
> virtue whereof many have been delivered from great infirmities,
> and the devils were made subject through his name : of which
> particular instances might be given beyond what this unbeliev-
> ing age is able to receive or bear.[1]

It needs no long study of Fox's writings, however,

---

[1] *A Journal or Historical Account of the Life, Travels, Sufferings,
and Christian Experiences, &c., of George Fox.* Ed. 1694, pp. 27, 28.

to arrive at the conviction that the distinction between subjective and objective verities had not the
same place in his mind as it has in that of ordinary
mortals.   When an ordinary person would say "I
thought so and so," or "I made up my mind to do
so and so," George Fox says, "It was opened to me,"
or "at the command of God I did so and so."   "Then
at the command of God, on the ninth day of the
seventh month 1643 (Fox being just nineteen), I left
my relations and brake off all familiarity or friendship with young or old."   "About the beginning of
the year 1647 I was moved of the Lord to go into
Darbyshire."   Fox hears voices and he sees visions,
some of which he brings before the reader with
apocalyptic power in the simple and strong English,
alike untutored and undefiled, of which, like John
Bunyan, his contemporary, he was a master.

"And one morning, as I was sitting by the fire, a
great cloud came over me and a temptation beset
me; and I sate still.   And it was said, *All things
come by Nature.*   And the elements and stars came
over me; so that I was in a manner quite clouded
with it. . . . And as I sate still under it, and let it
alone, a living hope arose in me, and a true voice
arose in me which said, *There is a living God who
made all things.*   And immediately the cloud and
the temptation vanished away, and life rose over it
all, and my heart was glad and I praised the living
God" (p. 13).

If George Fox could speak, as he proves in this
and some other passages he could write, his astound-

ing influence on the contemporaries of Milton and of Cromwell is no mystery. But this modern reproduction of the ancient prophet, with his "Thus saith the Lord," "This is the work of the Lord," steeped in supernaturalism and glorying in blind faith, is the mental antipodes of the philosopher, founded in naturalism and a fanatic for evidence, to whom these affirmations inevitably suggest the previous question : " How do you know that the Lord saith it : " " How do you know that the Lord doeth it ? " and who is compelled to demand that rational ground for belief without which, to the man of science, assent is merely an immoral pretence.

And it is this rational ground of belief which the writers of the Gospels, no less than Paul, and Eginhard, and Fox, so little dream of offering that they would regard the demand for it as a kind of blasphemy.

XI

AGNOSTICISM: A REJOINDER

THOSE who passed from Dr. Wace's article in the last number of this Review to the anticipatory confutation of it which followed in "The New Reformation," must have enjoyed the pleasure of a dramatic surprise —just as when the fifth act of a new play proves unexpectedly bright and interesting. Mrs. Ward will, I hope, pardon the comparison, if I say that her effective clearing away of antiquated incumbrances from the lists of the controversy, reminds me of nothing so much as of the action of some neat-handed, but strong-wristed, Phyllis, who, gracefully wielding her long-handled "Turk's head," sweeps away the accumulated results of the toil of generations of spiders. I am the more indebted to this luminous sketch of the results of critical investigation, as it is carried out among those theologians who are men of science and not mere counsel for creeds, since it has relieved me from the necessity of dealing with the greater part of Dr. Wace's polemic, and enables me to devote more space to the really important issues which have been raised.[1]

[1] I may perhaps return to the question of the authorship of the

Perhaps, however, it may be well for me to observe that approbation of the manner in which a great biblical scholar, for instance, Reuss, does his work does not commit me to the adoption of all, or indeed any of his views; and, further, that the disagreements of a series of investigators do not in any way interfere with the fact that each of them has made important contributions to the body of truth ultimately established. If I cite Buffon, Linnæus, Lamarck, and Cuvier, as having each and all taken a leading share in building up modern biology, the statement that every one of these great naturalists disagreed with, and even more or less contradicted, all the rest is quite true; but the supposition that the latter assertion is in any way inconsistent with the former, would betray a strange ignorance of the manner in which all true science advances.

Dr. Wace takes a great deal of trouble to make it appear that I have desired to evade the real questions raised by his attack upon me at the Church Congress. I assure the reverend Principal that in this, as in some other respects, he has entertained a very erroneous conception of my intentions. Things would assume more accurate proportions in Dr. Wace's mind if he would kindly remember that it is just thirty years since ecclesiastical thunderbolts

Gospels. For the present I must content myself with warning my readers against any reliance upon Dr. Wace's statements as to the results arrived at by modern criticism. They are as gravely as surprisingly erroneous.

began to fly about my ears. I have had the "Lion and the Bear" to deal with, and it is long since I got quite used to the threatenings of episcopal Goliaths, whose croziers were like unto a weaver's beam. So that I almost think I might not have noticed Dr. Wace's attack, personal as it was ; and although, as he is good enough to tell us, separate copies are to be had for the modest equivalent of twopence, as a matter of fact, it did not come under my notice for a long time after it was made. May I further venture to point out that (reckoning postage) the expenditure of twopence-halfpenny, or, at the most, threepence, would have enabled Dr. Wace so far to comply with ordinary conventions, as to direct my attention to the fact that he had attacked me before a meeting at which I was not present? I really am not responsible for the five months' neglect of which Dr. Wace complains. Singularly enough, the Englishry who swarmed about the Engadine, during the three months that I was being brought back to life by the glorious air and perfect comfort of the Maloja, did not, in my hearing, say anything about the important events which had taken place at the Church Congress ; and I think I can venture to affirm that there was not a single copy of Dr. Wace's pamphlet in any of the hotel libraries which I rummaged in search of something more edifying than dull English or questionable French novels.

And now, having, as I hope, set myself right with the public as regards the sins of commission and

omission with which I have been charged, I feel free
to deal with matters to which time and type may be
more profitably devoted.

I believe that there is not a solitary argument I
have used, or that I am about to use, which is
original, or has anything to do with the fact that
I have been chiefly occupied with natural science.
They are all, facts and reasoning alike, either
identical with, or consequential upon, propositions
which are to be found in the works of scholars and
theologians of the highest repute in the only two
countries, Holland and Germany,[1] in which, at the
present time, professors of theology are to be found,
whose tenure of their posts does not depend upon the
results to which their inquiries lead them.[2] It is true
that, to the best of my ability, I have satisfied myself
of the soundness of the foundations on which my
arguments are built, and I desire to be held fully

[1] The United States ought, perhaps, to be added, but I am not
sure.

[2] Imagine that all our chairs of Astronomy had been founded in
the fourteenth century, and that their incumbents were bound to sign
Ptolemaic articles. In that case, with every respect for the efforts of
persons thus hampered to attain and expound the truth, I think men
of common sense would go elsewhere to learn astronomy. Zeller's
Vorträge und Abhandlungen were published and came into my hands
a quarter of a century ago. The writer's rank, as a theologian to
begin with, and subsequently as a historian of Greek philosophy, is of
the highest. Among these essays are two—*Das Urchristenthum* and
Die Tübinger historische Schule—which are likely to be of more use to
those who wish to know the real state of the case than all that the
official "apologists," with their one eye on truth and the other on the
tenets of their sect, have written. For the opinion of a scientific
theologian about theologians of this stamp see pp. 225 and 227 of the
Vorträge.

responsible for everything I say. But, nevertheless, my position is really no more than that of an expositor; and my justification for undertaking it is simply that conviction of the supremacy of private judgment (indeed, of the impossibility of escaping it) which is the foundation of the Protestant Reformation, and which was the doctrine accepted by the vast majority of the Anglicans of my youth, before that backsliding towards the " beggarly rudiments " of an effete and idolatrous sacerdotalism which has, even now, provided us with the saddest spectacle which has been offered to the eyes of Englishmen in this generation. A high court of ecclesiastical jurisdiction, with a host of great lawyers in battle array, is and, for Heaven knows how long, will be, occupied with these very questions of " washing of cups and pots and brazen vessels," which the Master, whose professed representatives are rending the Church over these squabbles, had in his mind when, as we are told, he uttered the scathing rebuke :—

> Well did Isaiah prophesy of you hypocrites, as it is written,
>   This people honoureth me with their lips,
>   But their heart is far from me.
>   But in vain do they worship me,
>   Teaching as their doctrines the precepts of men.
>   (Mark vii. 6-7.)

Men who can be absorbed in bickerings over miserable disputes of this kind can have but little sympathy with the old evangelical doctrine of the " open Bible," or anything but a grave misgiving of the results of diligent reading of the Bible, without the help of ecclesiastical spectacles, by the mass of the people. Greatly to the surprise of many of my friends, I have always advocated the reading of the Bible, and the diffusion of the study of that most remarkable collection of books among the people. Its teachings are so infinitely superior to those of the sects, who are just as busy now as the Pharisees were eighteen hundred years ago, in smothering them under "the precepts of men"; it is so certain, to my mind, that the Bible contains within itself the refutation of nine-tenths of the mixture of sophistical metaphysics and old-world superstition which has been piled round it by the so-called Christians of later times; it is so clear that the only immediate and ready antidote to the poison which has been mixed with Christianity, to the intoxication and delusion of mankind, lies in copious draughts from the undefiled spring, that I exercise the right and duty of free judgment on the part of every man, mainly for the purpose of inducing other laymen to follow my example. If the New Testament is translated into Zulu by Protestant missionaries, it must be assumed that a Zulu convert is competent to draw from its contents all the truths which it is necessary for him to believe. I trust that I may, without immodesty, claim to be put on the same footing as the Zulu.

The most constant reproach which is launched against persons of my way of thinking is that it is all very well for us to talk about the deductions of scientific thought, but what are the poor and the

uneducated to do ?   Has it ever occurred to those who
talk in this fashion, that their creeds and the articles
of their several confessions, their determination of the
exact nature and extent of the teachings of Jesus,
their expositions of the real meaning of that which is
written in the Epistles (to leave aside all questions
concerning the Old Testament), are nothing more than
deductions which, at any rate, profess to be the result
of strictly scientific thinking, and which are not
worth attending to unless they really possess that
character ?   If it is not historically true that such
and such things happened in Palestine eighteen
centuries ago, what becomes of Christianity ?   And
what is historical truth but that of which the evidence
bears strict scientific investigation ?   I do not call
to mind any problem of natural science which has
come under my notice which is more difficult, or
more curiously interesting as a mere problem, than
that of the origin of the Synoptic Gospels and that of
the historical value of the narratives which they contain.   The Christianity of the Churches stands or
falls by the results of the purely scientific investigation of these questions.   They were first taken up in
a purely scientific spirit just about a century ago ;
they have been studied over and over again by men
of vast knowledge and critical acumen ; but he would
be a rash man who should assert that any solution
of these problems, as yet formulated, is exhaustive.
The most that can be said is that certain prevalent
solutions are certainly false, while others are more or
less probably true.

If I am doing my best to rouse my countrymen out of their dogmatic slumbers, it is not that they may be amused by seeing who gets the best of it in a contest between a " scientist" and a theologian. The serious question is whether theological men of science, or theological special pleaders, are to have the confidence of the general public ; it is the question whether a country in which it is possible for a body of excellent clerical and lay gentlemen to discuss, in public meeting assembled, how much it is desirable to let the congregations of the faithful know of the results of biblical criticism, is likely to wake up with anything short of the grasp of a rough lay hand upon its shoulder ; it is the question whether the New Testament books, being, as I believe they were, written and compiled by people who, according to their lights, were perfectly sincere, will not, when properly studied as ordinary historical documents, afford us the means of self-criticism. And it must be remembered that the New Testament books are not responsible for the doctrine invented by the Churches that they are anything but ordinary historical documents. The author of the third gospel tells us, as straightforwardly as a man can, that he has no claim to any other character than that of an ordinary compiler and editor, who had before him the works of many and variously qualified predecessors.

In my former papers, according to Dr. Wace, I have evaded giving an answer to his main proposition, which he states as follows—

Apart from all disputed points of criticism, no one practically doubts that our Lord lived, and that He died on the cross, in the most intense sense of filial relation to His Father in Heaven, and that He bore testimony to that Father's providence, love, and grace towards mankind. The Lord's Prayer affords a sufficient evidence on these points. If the Sermon on the Mount alone be added, the whole unseen world, of which the Agnostic refuses to know anything, stands unveiled before us. . . . If Jesus Christ preached that Sermon, made those promises, and taught that prayer, then any one who says that we know nothing of God, or of a future life, or of an unseen world, says that he does not believe Jesus Christ (pp. 354-355).

Again—

The main question at issue, in a word, is one which Professor Huxley has chosen to leave entirely on one side—whether, namely, allowing for the utmost uncertainty on other points of the criticism to which he appeals, there is any reasonable doubt that the Lord's Prayer and the Sermon on the Mount afford a true account of our Lord's essential belief and cardinal teaching (p. 355).

I certainly was not aware that I had evaded the questions here stated ; indeed I should say that I have indicated my reply to them pretty clearly ; but, as Dr. Wace wants a plainer answer, he shall certainly be gratified. If, as Dr. Wace declares it is, his "whole case is involved in" the argument as stated in the latter of these two extracts, so much the worse for his whole case. For I am of opinion that there is the gravest reason for doubting whether the "Sermon on the Mount" was ever preached, and whether the so-called "Lord's Prayer" was ever prayed, by Jesus of Nazareth. My reasons for this opinion are, among others, these :—There is now no doubt that the three Synoptic Gospels, so far from being the work of three independent writers, are closely interdependent,[1] and that in one of two ways. Either all three contain, as their foundation, versions, to a large extent verbally identical, of one and the same tradition; or two of them are thus closely dependent on the third; and the opinion of the majority of the best critics has of late years more and more converged towards the conviction that our canonical second gospel (the so-called " Mark's " Gospel) is that which most closely represents the primitive groundwork of the three.[2] That I take to be one of the most valid results of New Testament criticism, of immeasurably greater importance than the discussion about dates and authorship.

But if, as I believe to be the case, beyond any rational doubt or dispute, the second gospel is the nearest extant representative of the oldest tradition,

[1] I suppose this is what Dr. Wace is thinking about when he says that I allege that there " is no visible escape " from the supposition of an *Ur-Marcus* (p. 367). That a " theologian of repute " should confound an indisputable fact with one of the modes of explaining that fact is not so singular as those who are unaccustomed to the ways of theologians might imagine.

[2] Any examiner whose duty it has been to examine into a case of " copying " will be particularly well prepared to appreciate the force of the case stated in that most excellent little book, *The Common Tradition of the Synoptic Gospels,* by Dr. Abbott and Mr. Rushbrooke (Macmillan, 1884). To those who have not passed through such painful experiences I may recommend the brief discussion of the genuineness of the " Casket Letters " in my friend Mr. Skelton's interesting book, *Maitland of Lethington.* The second edition of Holtzmann's *Lehrbuch,* published in 1886, gives a remarkably fair and full account of the present results of criticism. At p. 366 he writes that the present burning question is whether the " relatively primitive narrative and the root of the other synoptic texts is contained in Matthew or in Mark. It is only on this point that properly-informed (*sachkundige*) critics differ," and he decides in favour of Mark.

whether written or oral, how comes it that it contains neither the " Sermon on the Mount " nor the " Lord's Prayer," those typical embodiments, according to Dr. Wace, of the " essential belief and cardinal teaching " of Jesus ? Not only does " Mark's " gospel fail to contain the " Sermon on the Mount," or anything but a very few of the sayings contained in that collection ; but, at the point of the history of Jesus where the " Sermon " occurs in " Matthew," there is in " Mark " an apparently unbroken narrative from the calling of James and John to the healing of Simon's wife's mother. Thus the oldest tradition not only ignores the " Sermon on the Mount," but, by implication, raises a probability against its being delivered when and where the later " Matthew " inserts it in his compilation.

And still more weighty is the fact that the third gospel, the author of which tells us that he wrote after " many " others had " taken in hand " the same enterprise ; who should therefore have known the first gospel (if it existed), and was bound to pay to it the deference due to the work of an apostolic eye-witness (if he had any reason for thinking it was so)—this writer, who exhibits far more literary competence than the other two, ignores any " Sermon on the Mount," such as that reported by " Matthew," just as much as the oldest authority does. Yet " Luke " has a great many passages identical, or parallel, with those in " Matthew's " " Sermon on the Mount," which are, for the most part, scattered about in a totally different connection.

Interposed, however, between the nomination of the Apostles and a visit to Capernaum; occupying, therefore, a place which answers to that of the "Sermon on the Mount" in the first gospel, there is, in the third gospel, a discourse which is as closely similar to the "Sermon on the Mount," in some particulars, as it is widely unlike it in others.

This discourse is said to have been delivered in a "plain" or "level place" (Luke vi. 17), and by way of distinction we may call it the "Sermon on the Plain."

I see no reason to doubt that the two Evangelists are dealing, to a considerable extent, with the same traditional material; and a comparison of the two "Sermons" suggests very strongly that "Luke's" version is the earlier. The correspondences between the two forbid the notion that they are independent. They both begin with a series of blessings, some of which are almost verbally identical. In the middle of each (Luke vi. 27-38, Matt. v. 43-48) there is a striking exposition of the ethical spirit of the command given in Leviticus xix. 18. And each ends with a passage containing the declaration that a tree is to be known by its fruit, and the parable of the house built on the sand. But while there are only 29 verses in the "Sermon on the Plain" there are 107 in the "Sermon on the Mount;" the excess in length of the latter being chiefly due to the long interpolations, one of 30 verses before and one of 34 verses after, the middlemost parallelism with Luke. Under these circumstances it is quite impossible to admit that there is more probability that "Matthew's" version of the Sermon is historically accurate than there is that Luke's version is so; and they cannot both be accurate.

"Luke" either knew the collection of loosely-connected and aphoristic utterances which appear under the name of the "Sermon on the Mount" in "Matthew;" or he did not. If he did not, he must have been ignorant of the existence of such a document as our canonical "Matthew," a fact which does not make for the genuineness, or the authority, of that book. If he did, he has shown that he does not care for its authority on a matter of fact of no small importance; and that does not permit us to conceive that he believed the first gospel to be the work of an authority to whom he ought to defer, let alone that of an apostolic eye-witness.

The tradition of the Church about the second gospel, which I believe to be quite worthless, but which is all the evidence there is for "Mark's" authorship, would have us believe that "Mark" was little more than the mouthpiece of the apostle Peter. Consequently, we are to suppose that Peter either did not know, or did not care very much for, that account of the "essential belief and cardinal teaching" of Jesus which is contained in the Sermon on the Mount; and, certainly, he could not have shared Dr. Wace's view of its importance.[1]

---

[1] Holtzmann (*Die synoptischen Evangelien*, 1863, p. 75), following Ewald, argues that the "Source A" (= the threefold tradition, more or less) contained something that answered to the "Sermon on the

I thought that all fairly attentive and intelligent students of the gospels, to say nothing of theologians of reputation, knew these things. But how can any one who does know them have the conscience to ask whether there is " any reasonable doubt" that the Sermon on the Mount was preached by Jesus of Nazareth ? If conjecture is permissible, where nothing else is possible, the most probable conjecture seems to be that " Matthew," having a *cento* of sayings attributed—rightly or wrongly it is impossible to say—to Jesus, among his materials, thought they were, or might be, records of a continuous discourse, and put them in at the place he thought likeliest. Ancient historians of the highest character saw no harm in composing long speeches which never were spoken, and putting them into the mouths of statesmen and warriors ; and I presume that whoever is represented by " Matthew " would have been grievously astonished to find that any one objected to his following the example of the best models accessible to him.

So with the " Lord's Prayer." Absent in our representative of the oldest tradition, it appears in both " Matthew " and " Luke." There is reason to believe that every pious Jew, at the commencement of our era, prayed three times a day, according to a formula which is embodied in the present *Schmone-Esre* [1] of

Plain " immediately after the words of our present Mark, " And he cometh into a house " (iii. 19). But what conceivable motive could " Mark " have for omitting it ? Holtzmann has no doubt, however, that the " Sermon on the Mount " is a compilation, or, as he calls it in his recently-published *Lehrbuch* (p. 372), " an artificial mosaic work."

[1] See Schürer, *Geschichte des jüdischen Volkes*, Zweiter Theil, p. 384.

the Jewish prayer-book. Jesus, who was assuredly, in all respects, a pious Jew, whatever else he may have been, doubtless did the same. Whether he modified the current formula, or whether the so-called "Lord's Prayer" is the prayer substituted for the *Schmone-Esre* in the congregations of the Gentiles, is a question which can hardly be answered.

In a subsequent passage of Dr. Wace's article (p. 356) he adds to the list of the verities which he imagines to be unassailable, "The Story of the Passion." I am not quite sure what he means by this. I am not aware that any one (with the exception of certain ancient heretics) has propounded doubts as to the reality of the crucifixion; and certainly I have no inclination to argue about the precise accuracy of every detail of that pathetic story of suffering and wrong. But, if Dr. Wace means, as I suppose he does, that that which, according to the orthodox view, happened after the crucifixion, and which is, in a dogmatic sense, the most important part of the story, is founded on solid historical proofs, I must beg leave to express a diametrically opposite conviction.

What do we find when the accounts of the events in question, contained in the three Synoptic gospels, are compared together? In the oldest, there is a simple, straightforward statement which, for anything that I have to urge to the contrary, may be exactly true. In the other two, there is, round this possible and probable nucleus, a mass of accretions of the most questionable character.

The cruelty of death by crucifixion depended very much upon its lingering character. If there were a support for the weight of the body, as not unfrequently was the practice, the pain during the first hours of the infliction was not, necessarily, extreme; nor need any serious physical symptoms, at once, arise from the wounds made by the nails in the hands and feet, supposing they were nailed, which was not invariably the case. When exhaustion set in, and hunger, thirst, and nervous irritation had done their work, the agony of the sufferer must have been terrible; and the more terrible that, in the absence of any effectual disturbance of the machinery of physical life, it might be prolonged for many hours, or even days. Temperate, strong men, such as were the ordinary Galilean peasants, might live for several days on the cross. It is necessary to bear these facts in mind when we read the account contained in the fifteenth chapter of the second gospel.

Jesus was crucified at the third hour (xv. 25), and the narrative seems to imply that he died immediately after the ninth hour (v. 34). In this case, he would have been crucified only six hours; and the time spent on the cross cannot have been much longer, because Joseph of Arimathæa must have gone to Pilate, made his preparations, and deposited the body in the rock-cut tomb before sunset, which, at that time of the year, was about the twelfth hour. That any one should die after only six hours' crucifixion could not have been at all in accordance with Pilate's large experience of the effects of that method of punishment. It, therefore, quite agrees with what might be expected if Pilate "marvelled if he were already dead" and required to be satisfied on this point by the testimony of the Roman officer who was in command of the execution party. Those who have paid attention to the extraordinarily difficult question, What are the indisputable signs of death ?—will be able to estimate the value of the opinion of a rough soldier on such a subject ; even if his report to the Procurator were in no wise affected by the fact that the friend of Jesus, who anxiously awaited his answer, was a man of influence and of wealth.

The inanimate body, wrapped in linen, was deposited in a spacious,[1] cool rock chamber, the entrance of which was closed, not by a well-fitting door, but by a stone rolled against the opening, which would of course allow free passage of air. A little more than thirty-six hours afterwards (Friday 6 P.M., to Sunday 6 A.M., or a little after) three women visit the tomb and find it empty. And they are told by a young man " arrayed in a white robe " that Jesus is gone to his native country of Galilee, and that the disciples and Peter will find him there.

Thus it stands, plainly recorded, in the oldest tradition that, for any evidence to the contrary, the sepulchre may have been vacated at any time during the Friday or Saturday nights. If it is said that no Jew would have violated the Sabbath by taking the former course, it is to be recollected that Joseph of

---

[1] Spacious, because a young man could sit in it " on the right side " (xv. 5), and therefore with plenty of room to spare.

Arimathæa might well be familiar with that wise and liberal interpretation of the fourth commandment, which permitted works of mercy to men—nay even the drawing of an ox or an ass out of a pit—on the Sabbath. At any rate, the Saturday night was free to the most scrupulous of observers of the Law.

These are the facts of the case as stated by the oldest extant narrative of them. I do not see why any one should have a word to say against the inherent probability of that narrative; and, for my part, I am quite ready to accept it as an historical fact, that so much and no more is positively known of the end of Jesus of Nazareth. On what grounds can a reasonable man be asked to believe any more? So far as the narrative in the first gospel, on the one hand, and those in the third gospel and the Acts, on the other, go beyond what is stated in the second gospel, they are hopelessly discrepant with one another. And this is the more significant because the pregnant phrase "some doubted," in the first gospel, is ignored in the third.

But it is said that we have the witness Paul speaking to us directly in the Epistles. There is little doubt that we have, and a very singular witness he is. According to his own showing, Paul, in the vigour of his manhood, with every means of becoming acquainted, at first hand, with the evidence of eye-witnesses, not merely refused to credit them, but "persecuted the church of God and made havoc of it." The reasoning of Stephen fell dead upon the acute intellect of this zealot for the traditions of

his fathers : his eyes were blind to the ecstatic
illumination of the martyr's countenance " as it had
been the face of an angel ; " and when, at the words
" Behold, I see the heavens opened and the Son of
Man standing on the right hand of God," the mur-
derous mob rushed upon and stoned the rapt disciple
of Jesus, Paul ostentatiously made himself their
official accomplice.

Yet this strange man, because he has a vision one
day, at once, and with equally headlong zeal, flies to
the opposite pole of opinion. And he is most careful
to tell us that he abstained from any re-examination
of the facts.

> Immediately I conferred not with flesh and blood ; neither
> went I up to Jerusalem to them which were Apostles before me ;
> but I went away into Arabia. (Galatians i. 16, 17.)

I do not presume to quarrel with Paul's procedure.
If it satisfied him, that was his affair ; and, if it satisfies any else, I am not called upon to dispute the
right of that person to be satisfied. But I certainly
have the right to say that it would not satisfy me, in
like case ; that I should be very much ashamed to
pretend that it could, or ought to, satisfy me ; and
that I can entertain but a very low estimate of the
value of the evidence of people who are to be satisfied
in this fashion, when questions of objective fact, in
which their faith is interested, are concerned. So that
when I am called upon to believe a great deal more
than the oldest gospel tells me about the final events
of the history of Jesus on the authority of Paul (1
Corinthians xv. 5-8) I must pause. Did he think it,

at any subsequent time, worth while "to confer with flesh and blood," or, in modern phrase, to re-examine the facts for himself? or was he ready to accept anything that fitted in with his preconceived ideas? Does he mean, when he speaks of all the appearances of Jesus after the crucifixion as if they were of the same kind, that they were all visions, like the manifestation to himself? And, finally, how is this account to be reconciled with those in the first and third gospels—which, as we have seen, disagree with one another?

Until these questions are satisfactorily answered, I am afraid that, so far as I am concerned, Paul's testimony cannot be seriously regarded, except as it may afford evidence of the state of traditional opinion at the time at which he wrote, say between 55 and 60 A.D.; that is, more than twenty years after the event; a period much more than sufficient for the development of any amount of mythology about matters of which nothing was really known. A few years later, among the contemporaries and neighbours of the Jews, and, if the most probable interpretation of the Apocalypse can be trusted, among the followers of Jesus also, it was fully believed, in spite of all the evidence to the contrary, that the Emperor Nero was not really dead, but that he was hidden away somewhere in the East, and would speedily come again at the head of a great army, to be revenged upon his enemies.[1]

Thus, I conceive that I have shown cause for the

[1] King Herod had not the least difficulty in supposing the resurrection of John the Baptist—"John, whom I beheaded, he is risen" (Mark vi. 16).

opinion that Dr. Wace's challenge touching the Sermon
on the Mount, the Lord's Prayer, and the Passion was
more valorous than discreet. After all this discussion, I am still at the agnostic point. Tell me, first,
what Jesus can be proved to have been, said, and
done, and I will say whether I believe him, or
in him,[1] or not. As Dr. Wace admits that I have
dissipated his lingering shade of unbelief about the
bedevilment of the Gadarene pigs, he might have
done something to help mine. Instead of that, he
manifests a total want of conception of the nature
of the obstacles which impede the conversion of his
" infidels."

The truth I believe to be, that the difficulties in
the way of arriving at a sure conclusion as to these
matters, from the Sermon on the Mount, the Lord's
Prayer, or any other data offered by the Synoptic
gospels (and *à fortiori* from the fourth gospel), are
insuperable. Every one of these records is coloured
by the prepossessions of those among whom the
primitive traditions arose, and of those by whom they
were collected and edited; and the difficulty of
making allowance for these prepossessions is enhanced
by our ignorance of the exact dates at which the
documents were first put together; of the extent to

---

[1] I am very sorry for the interpolated " in," because citation ought
to be accurate in small things as in great. But what difference it
makes whether one " believes Jesus " or " believes in Jesus " much
thought has not enabled me to discover. If you " believe him " you
must believe him to be what he professed to be—that is, " believe in
him ; " and if you " believe in him " you must necessarily " believe
him."

which they have been subsequently worked over and interpolated; and of the historical sense, or want of sense, and the dogmatic tendencies of their compilers and editors. Let us see if there is any other road which will take us into something better than negation.

There is a widespread notion that the "primitive Church," while under the guidance of the Apostles and their immediate successors, was a sort of dogmatic dovecot, pervaded by the most loving unity and doctrinal harmony. Protestants, especially, are fond of attributing to themselves the merit of being nearer "the Church of the Apostles" than their neighbours; and they are the less to be excused for their strange delusion because they are great readers of the documents which prove the exact contrary. The fact is that, in the course of the first three centuries of its existence, the Church rapidly underwent a process of evolution of the most remarkable character, the final stage of which is far more different from the first than Anglicanism is from Quakerism. The key to the comprehension of the problem of the origin of that which is now called "Christianity," and its relation to Jesus of Nazareth, lies here. Nor can we arrive at any sound conclusion as to what it is probable that Jesus actually said and did without being clear on this head. By far the most important and subsequently influential steps in the evolution of Christianity took place in the course of the century, more or less, which followed upon the crucifixion. It is almost the darkest period of Church history, but, most fortunately, the beginning and the end of the period are brightly illuminated by the contemporary evidence of two writers of whose historical existence there is no doubt,[1] and against the genuineness of whose most important works there is no widely-admitted objection. These are Justin, the philosopher and martyr, and Paul, the Apostle to the Gentiles. I shall call upon these witnesses only to testify to the condition of opinion among those who called themselves disciples of Jesus in their time.

Justin, in his Dialogue with Trypho the Jew, which was written somewhere about the middle of the second century, enumerates certain categories of persons who, in his opinion, will, or will not, be saved.[2] These are :—

1. Orthodox Jews who refuse to believe that Jesus is the Christ. *Not saved.*

2. Jews who observe the Law ; believe Jesus to be the Christ ; but who insist on the observance of the Law by Gentile converts. *Not saved.*

3. Jews who observe the Law ; believe Jesus to be the Christ, and hold that Gentile converts need not observe the Law. *Saved* (in Justin's opinion ; but some of his fellow-Christians think the contrary).

4. Gentile converts to the belief in Jesus as the Christ, who observe the Law. *Saved* (possibly).

5. Gentile believers in Jesus as the Christ, who do

---

[1] True for Justin : but there is a school of theological critics, who more or less question the historical reality of Paul and the genuineness of even the four cardinal epistles.

[2] See *Dial. cum Tryphone*, § 47 and § 35. It is to be understood that Justin does not arrange these categories in order, as I have done.

not observe the Law themselves (except so far as the refusal of idol sacrifices), but do not consider those who do observe it heretics. *Saved* (this is Justin's own view).

6. Gentile believers who do not observe the Law, except in refusing idol sacrifices, and hold those who do observe it to be heretics. *Saved.*

7. Gentiles who believe Jesus to be the Christ and call themselves Christians, but who eat meats sacrificed to idols. *Not saved.*

8. Gentiles who disbelieve in Jesus as the Christ. *Not saved.*

Justin does not consider Christians who believe in the natural birth of Jesus, of whom he implies that there is a respectable minority, to be heretics, though he himself strongly holds the preternatural birth of Jesus and his pre-existence as the "Logos" or "Word." He conceives the Logos to be a second God, inferior to the first, unknowable, God, with respect to whom Justin, like Philo, is a complete agnostic. The Holy Spirit is not regarded by Justin as a separate personality, and is often mixed up with the "Logos." The doctrine of the natural immortality of the soul is, for Justin, a heresy; and he is as firm a believer in the resurrection of the body, as in the speedy Second Coming and establishment of the millennium.

This pillar of the Church in the middle of the second century—a much-travelled native of Samaria —was certainly well acquainted with Rome, probably with Alexandria, and it is likely that he knew the state of opinion throughout the length and breadth of the Christian world as well as any man of his time. If the various categories above enumerated are arranged in a series thus :—

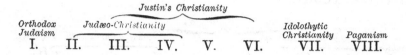

Justin's Christianity

Orthodox Judaism | Judæo-Christianity | Idolothytic Christianity | Paganism

I.   II.   III.   IV.   V.   VI.   VII.   VIII.

it is obvious that they form a gradational series from orthodox Judaism, on the extreme left, to Paganism, whether philosophic or popular, on the extreme right; and it will further be observed that, while Justin's conception of Christianity is very broad, he rigorously excludes two classes of persons who, in his time, called themselves Christians; namely, those who insist on circumcision and other observances of the Law on the part of Gentile converts; that is to say, the strict Judæo-Christians (II.); and, on the other hand, those who assert the lawfulness of eating meat offered to idols—whether they are Gnostic or not (VII.) These last I have called " idolothytic " Christians, because I cannot devise a better name, not because it is strictly defensible etymologically.

At the present moment, I do not suppose there is an English missionary in any heathen land who would trouble himself whether the materials of his dinner had been previously offered to idols or not. On the other hand, I suppose there is no Protestant sect within the pale of orthodoxy, to say nothing of the Roman and Greek Churches, which would hesitate to declare the practice of circumcision and the observance of the Jewish Sabbath and dietary rules, shockingly heretical.

Modern Christianity has, in fact, not only shifted far to the right of Justin's position, but it is of much narrower compass.

For, though it includes VII., and even, in saint and relic worship, cuts a "monstrous cantle" out of paganism, it excludes, not only all Judæo-Christians, but all who doubt that such are heretics. Ever since the thirteenth century, the Inquisition would have cheerfully burned, and in Spain did abundantly burn, all persons who came under the categories II., III., IV., V. And the wolf would play the same havoc now, if it could only get its blood-stained jaws free from the muzzle imposed by the secular arm.

Further, there is not a Protestant body except the Unitarian, which would not declare Justin himself a heretic, on account of his doctrine of the inferior godship of the Logos; while I am very much afraid that, in strict logic, Dr. Wace would be under the necessity, so painful to him, of calling him an "infidel," on the same and on other grounds.

Now let us turn to our other authority. If there is any result of critical investigations of the sources of Christianity which is certain,[1] it is that Paul of

---

[1] I guard myself against being supposed to affirm that even the four cardinal epistles of Paul may not have been seriously tampered with. See note 1, p. 429 above.

Tarsus wrote the Epistle to the Galatians somewhere between the years 55 and 60 A.D., that is to say, roughly, twenty, or five-and-twenty years after the crucifixion. If this is so, the Epistle to the Galatians is one of the oldest, if not the very oldest, of extant documentary evidences of the state of the primitive Church. And, be it observed, if it is Paul's writing, it unquestionably furnishes us with the evidence of a participator in the transactions narrated. With the exception of two or three of the other Pauline epistles, there is not one solitary book in the New Testament of the authorship and authority of which we have such good evidence.

And what is the state of things we find disclosed ? A bitter quarrel, in his account of which Paul by no means minces matters, or hesitates to hurl defiant sarcasms against those who were "reputed to be pillars : " James, " the brother of the Lord," Peter, the rock on whom Jesus is said to have built his Church, and John, " the beloved disciple." And no deference toward " the rock " withholds Paul from charging Peter to his face with " dissimulation."

The subject of the hot dispute was simply this. Were Gentile converts bound to obey the Law or not ? Paul answered in the negative ; and, acting upon his opinion, had created at Antioch (and else-where) a specifically " Christian " community, the sole qualifications for admission into which were the confession of the belief that Jesus was the Messiah, and baptism upon that confession. In the epistle in question, Paul puts this—his " gospel," as he calls it—in its most extreme form. Not only does he deny the necessity of conformity with the Law, but he declares such conformity to have a negative value. " Behold, I, Paul, say unto you, that if ye receive circumcision, Christ will profit you nothing " (Galatians v. 2). He calls the legal observances " beggarly rudiments," and anathematises every one who preaches to the Galatians any other gospel than his own. That is to say, by direct consequence, he anathematises the Nazarenes of Jerusalem, whose zeal for the Law is testified by James in a passage of the Acts cited further on. In the first Epistle to the Corinthians, dealing with the question of eating meat offered to idols, it is clear that Paul himself thinks it a matter of indifference; but he advises that it should not be done, for the sake of the weaker brethren. On the other hand, the Nazarenes of Jerusalem most strenuously opposed Paul's " gospel," insisting on every convert becoming a regular Jewish proselyte, and consequently on his observance of the whole Law; and this party was led by James and Peter and John (Galatians ii. 9). Paul does not suggest that the question of principle was settled by the discussion referred to in Galatians. All he says is that it ended in the practical agreement that he and Barnabas should do as they had been doing, in respect to the Gentiles; while James and Peter and John should deal in their own fashion with Jewish converts. Afterwards, he complains bitterly of Peter, because, when on a visit to Antioch, he at first inclined to Paul's view, and ate with the Gentile converts; but when " certain came from James,"

" drew back, and separated himself, fearing them that were of the circumcision. And the rest of the Jews dissembled likewise with him; insomuch that even Barnabas was carried away with their dissimulation" (Galatians ii. 12-13).

There is but one conclusion to be drawn from Paul's account of this famous dispute, the settlement of which determined the fortunes of the nascent religion. It is that the disciples at Jerusalem, headed by " James, the Lord's brother," and by the leading apostles, Peter and John, were strict Jews, who had objected to admit any converts into their body, unless these, either by birth, or by becoming proselytes, were also strict Jews. In fact, the sole difference between James and Peter and John, with the body of the disciples whom they led, and the Jews by whom they were surrounded, and with whom they for many years shared the religious observances of the Temple, was that they believed that the Messiah, whom the leaders of the nation yet looked for, had already come in the person of Jesus of Nazareth.

The Acts of the Apostles is hardly a very trust-worthy history ; it is certainly of later date than the Pauline Epistles, supposing them to be genuine. And the writer's version of the conference of which Paul gives so graphic a description, if that is correct, is unmistakably coloured with all the art of a recon-ciler, anxious to cover up a scandal. But it is none the less instructive on this account. The judgment of the " council" delivered by James is that the Gentile converts shall merely " abstain from things sacrificed to idols, and from blood and from things strangled, and from fornication." But notwithstanding the accommodation in which the writer of the Acts would have us believe, the Jerusalem Church held to its endeavour to retain the observance of the Law. Long after the conference, some time after the writing of the Epistles to the Galatians and Corinthians, and immediately after the despatch of that to the Romans, Paul makes his last visit to Jerusalem, and presents himself to James·and all the elders. And this is what the Acts tells us of the interview :—

> And they said unto him, Thou seest, brother, how many thousands [or myriads] there are among the Jews of them which have believed ; and they are all zealous for the law ; and they have been informed concerning thee, that thou teachest all the Jews which are among the Gentiles to forsake Moses, telling them not to circumcise their children, neither to walk after the customs. (Acts xxi. 20, 21.)

They therefore request that he should perform a certain public religious act in the Temple, in order that

> all shall know that there is no truth in the things whereof they have been informed concerning thee ; but that thou thyself walkest orderly, keeping the law (*ibid.* 24).[1]

How far Paul could do what he is here requested to do, and which the writer of the Acts goes on to say he did, with a clear conscience, if he wrote the Epistles to the Galatians and Corinthians, I may leave any candid reader of these epistles to decide. The

---

[1] [Paul, in fact, is required to commit in Jerusalem, an act of the same character as that which he brands as "dissimulation" on the part of Peter in Antioch.]

point to which I wish to direct attention is the
declaration that the Jerusalem Church, led by the
brother of Jesus and by his personal disciples and
friends, twenty years and more after his death, consisted of strict and zealous Jews.

Tertullus, the orator, caring very little about the
internal dissensions of the followers of Jesus, speaks
of Paul as a " ringleader of the sect of the Nazarenes "
(Acts xxiv. 5), which must have affected James
much in the same way as it would have moved the
Archbishop of Canterbury, in George Fox's day,
to hear the latter called a "ringleader of the sect
of Anglicans." In fact, " Nazarene " was, as is well
known, the distinctive appellation applied to Jesus ;
his immediate followers were known as Nazarenes ;
while the congregation of the disciples, and, later, of
converts at Jerusalem—the Jerusalem Church—was
emphatically the "sect of the Nazarenes," no more in
itself to be regarded as anything outside Judaism
than the sect of the Sadducees or of the Essenes.[1]
In fact, the tenets of both the Sadducees and the
Essenes diverged much more widely from the Pharisaic standard of orthodoxy than Nazarenism did.

Let us consider the position of affairs now (A.D.
50-60) in relation to that which obtained in Justin's
time, a century later. It is plain that the Nazarenes
—presided over by James, " the brother of the Lord,"
and comprising within their body all the twelve apostles

---

[1] All this was quite clearly pointed out by Ritschl nearly forty
years ago. See *Die Entstehung der alt-katholischen Kirche* (1850), p.
108.

—belonged to Justin's second category of " Jews who observe the Law, believe Jesus to be the Christ, but who insist on the observance of the Law by Gentile converts," up till the time at which the controversy reported by Paul arose. They then, according to Paul, simply allowed him to form his congregations of non-legal Gentile converts at Antioch and elsewhere ; and it would seem that it was to these converts, who would come under Justin's fifth category, that the title of "Christian" was first applied. If any of these Christians had acted upon the more than half-permission given by Paul, and had eaten meats offered to idols, they would have belonged to Justin's seventh category.

Hence, it appears that, if Justin's opinion, which was probably that of the Church generally in the middle of the second century, was correct, James and Peter and John and their followers could not be saved ; neither could Paul, if he carried into practice his views as to the indifference of eating meats offered to idols. Or, to put the matter another way, the centre of gravity of orthodoxy, which is at the extreme right of the series in the nineteenth century, was at the extreme left, just before the middle of the first century, when the "sect of the Nazarenes" constituted the whole church founded by Jesus and the apostles ; while, in the time of Justin, it lay midway between the two. It is therefore a profound mistake to imagine that the Judæo-Christians (Nazarenes and Ebionites) of later times were heretical outgrowths from a primitive universalist "Christianity." On the

contrary, the universalist " Christianity " is an out-
growth from the primitive, purely Jewish, Nazarenism;
which, gradually eliminating all the ceremonial and
dietary parts of the Jewish law, has thrust aside its
parent, and all the intermediate stages of its development, into the position of damnable heresies.

Such being the case, we are in a position to form
a safe judgment of the limits within which the
teaching of Jesus of Nazareth must have been confined. Ecclesiastical authority would have us believe
that the words which are given at the end of the
first Gospel, " Go ye, therefore, and make disciples of
all the nations, baptizing them in the name of the
Father and of the Son and of the Holy Ghost," are
part of the last commands of Jesus, issued at the
moment of his parting with the eleven. If so, Peter
and John must have heard these words; they are too
plain to be misunderstood; and the occasion is too
solemn for them ever to be forgotten. Yet the
" Acts " tell us that Peter needed a vision to enable
him so much as to baptize Cornelius; and Paul, in
the Galatians, knows nothing of words which would
have completely borne him out as against those who,
though they heard, must be supposed to have either
forgotten or ignored them. On the other hand, Peter
and John, who are supposed to have heard the
" Sermon on the Mount," know nothing of the saying
that Jesus had not come to destroy the Law, but that
every jot and tittle of the Law must be fulfilled,
which surely would have been pretty good evidence
for their view of the question.

We are sometimes told that the personal friends and daily companions of Jesus remained zealous Jews and opposed Paul's innovations, because they were hard of heart and dull of comprehension. This hypothesis is hardly in accordance with the concomitant faith of those who adopt it, in the miraculous insight and superhuman sagacity of their Master; nor do I see any way of getting it to harmonise with the orthodox postulate; namely, that Matthew was the author of the first gospel and John of the fourth. If that is so, then, most assuredly, Matthew was no dullard; and as for the fourth gospel—a theosophic romance of the first order—it could have been written by none but a man of remarkable literary capacity, who had drunk deep of Alexandrian philosophy. Moreover, the doctrine of the writer of the fourth gospel is more remote from that of the " sect of the Nazarenes " than is that of Paul himself. I am quite aware that orthodox critics have been capable of maintaining that John, the Nazarene, who was probably well past fifty years of age when he is supposed to have written the most thoroughly Judaising book in the New Testament— the Apocalypse—in the roughest of Greek, underwent an astounding metamorphosis of both doctrine and style by the time he reached the ripe age of ninety or so, and provided the world with a history in which the acutest critic cannot [always] make out where the speeches of Jesus end and the text of the narrative begins; while that narrative is utterly irreconcilable, in regard to matters of fact, with that of his fellow-apostle, Matthew.

The end of the whole matter is this :—The " sect of the Nazarenes," the brother and the immediate followers of Jesus, commissioned by him as apostles, and those who were taught by them up to the year 50 A.D., were not " Christians " in the sense in which that term has been understood ever since its asserted origin at Antioch, but Jews—strict orthodox Jews— whose belief in the Messiahship of Jesus never led to their exclusion from the Temple services, nor would have shut them out from the wide embrace of Judaism.[1] The open proclamation of their special view about the Messiah was doubtless offensive to the Pharisees, just as rampant Low Churchism is offensive to bigoted High Churchism in our own country ; or as any kind of dissent is offensive to fervid religionists of all creeds. To the Sadducees, no doubt, the political danger of any Messianic movement was serious ; and they would have been glad to put down Nazarenism, lest it should end in useless rebellion against their Roman masters, like that other Galilean movement headed by Judas, a generation earlier. Galilee was always a hotbed of seditious enthusiasm against the rule of Rome ; and high priest and procurator alike had need to keep a sharp eye upon natives of that district. On the whole, however, the Nazarenes were but little troubled for the first twenty years of their existence ; and the undying hatred of the Jews against those later con-

---

[1] " If every one was baptized as soon as he acknowledged Jesus to be the Messiah, the first Christians can have been aware of no other essential differences from the Jews."—Zeller, *Vorträge* (1865), p. 26.

verts, whom they regarded as apostates and fautors of a sham Judaism, was awakened by Paul. From their point of view, he was a mere renegade Jew, opposed alike to orthodox Judaism and to orthodox Nazarenism, and whose teachings threatened Judaism with destruction. And, from their point of view, they were quite right. In the course of a century, Pauline influences had a large share in driving primitive Nazarenism from being the very heart of the new faith into the position of scouted error; and the spirit of Paul's doctrine continued its work of driving Christianity farther and farther away from Judaism, until "meats offered to idols" might be eaten without scruple, while the Nazarene methods of observing even the Sabbath, or the Passover, were branded with the mark of Judaising heresy.

But if the primitive Nazarenes of whom the Acts speaks were orthodox Jews, what sort of probability can there be that Jesus was anything else? How can he have founded the universal religion which was not heard of till twenty years after his death?[1] That Jesus possessed in a rare degree the gift of attaching men to his person and to his fortunes; that he was the author of many a striking saying, and the advocate of equity, of love, and of humility; that he may have disregarded the subtleties of the bigots for legal observance, and appealed rather to those noble

[1] Dr. Harnack, in the lately-published second edition of his *Dogmengeschichte*, says (p. 39), "Jesus Christ brought forward no new doctrine;" and again (p. 65), "It is not difficult to set against every portion of the utterances of Jesus an observation which deprives him of originality." See also Zusatz 4, on the same page.

conceptions of religion which constituted the pith and
kernel of the teaching of the great prophets of his
nation seven hundred years earlier; and that, in the
last scenes of his career, he may have embodied the
ideal sufferer of Isaiah, may be, as I think it is, extremely probable. But all this involves not a step
beyond the borders of orthodox Judaism. Again,
who is to say whether Jesus proclaimed himself the
veritable Messiah, expected by his nation since the
appearance of the pseudo-prophetic work of Daniel, a
century and a half before his time ; or whether the
enthusiasm of his followers gradually forced him to
assume that position ?

But one thing is quite certain : if that belief in
the speedy second coming of the Messiah which was
shared by all parties in the primitive Church, whether
Nazarene or Pauline; which Jesus is made to prophesy, over and over again, in the Synoptic gospels ;
and which dominated the life of Christians during the
first century after the crucifixion ;—if he believed and
taught that, then assuredly he was under an illusion,
and he is responsible for that which the mere effluxion
of time has demonstrated to be a prodigious error.

When I ventured to doubt "whether any Protestant theologian who has a reputation to lose will
say that he believes the Gadarene story," it appears
that I reckoned without Dr. Wace, who, referring to
this passage in my paper, says :—

He will judge whether I fall under his description ; but I
repeat that I believe it, and that he has removed the only objection
to my believing it (p. 363).

Far be it from me to set myself up as a judge of any such delicate question as that put before me; but I think I may venture to express the conviction that, in the matter of courage, Dr. Wace has raised for himself a monument *ære perennius.*. For really, in my poor judgment, a certain splendid intrepidity, such as one admires in the leader of a forlorn hope, is manifested by Dr. Wace when he solemnly affirms that he believes the Gadarene story on the evidence offered. I feel less complimented perhaps than I ought to do, when I am told that I have been an accomplice in extinguishing in Dr. Wace's mind the last glimmer of doubt which common sense may have suggested. In fact, I must disclaim all responsibility for the use to which the information I supplied has been put. I formally decline to admit that the expression of my ignorance whether devils, in the existence of which I do not believe, if they did exist, might or might not be made to go out of men into pigs, can, as a matter of logic, have been of any use whatever to a person who already believed in devils and in the historical accuracy of the gospels.

Of the Gadarene story, Dr. Wace, with all solemnity and twice over, affirms that he " believes it." I am sorry to trouble him further, but what does he mean by "it"? Because there are two stories, one in " Mark" and " Luke," and the other in " Matthew." In the former, which I quoted in my previous paper, there is one possessed man; in the latter there are two. The story is told fully, with the vigorous homely diction and the picturesque details of a piece of folklore, in the second gospel. The immediately antecedent event is the storm on the Lake of Gennesaret. The immediately consequent events are the message from the ruler of the synagogue and the healing of the woman with an issue of blood. In the third gospel, the order of events is exactly the same, and there is an extremely close general and verbal correspondence between the narratives of the miracle. Both agree in stating that there was only one possessed man, and that he was the residence of many devils, whose name was " Legion."

In the first gospel, the event which immediately precedes the Gadarene affair is, as before, the storm ; the message from the ruler and the healing of the issue are separated from it by the accounts of the healing of a paralytic, of the calling of Matthew, and of a discussion with some Pharisees. Again, while the second gospel speaks of the country of the " Gerasenes " as the locality of the event, the third gospel has " Gerasenes," " Gergesenes," and " Gadarenes " in different ancient MSS. ; while the first has " Gadarenes."

The really important points to be noticed, however, in the narrative of the first gospel, are these—that there are two possessed men instead of one ; and that while the story is abbreviated by omissions, what there is of it is often verbally identical with the corresponding passages in the other two gospels. The most unabashed of reconcilers cannot well say that one man is the same as two, or two as one ; and, though the suggestion really has been made, that two different miracles, agreeing in all essential particulars, except the number of the possessed, were effected immediately after the storm on the lake, I should be sorry to accuse any one of seriously adopting it. Nor will it be pretended that the allegory refuge is accessible in this particular case.

So, when Dr. Wace says that he believes in the synoptic evangelists' account of the miraculous bedevilment of swine, I may fairly ask which of them does he believe? Does he hold by the one evangelist's story, or by that of the two evangelists? And having made his election, what reasons has he to give for his choice? If it is suggested that the witness of two is to be taken against that of one, not only is the testimony dealt with in that common-sense fashion against which the theologians of his school protest so warmly; not only is all question of inspiration at an end, but the further inquiry arises, After all, is it the testimony of two against one? Are the authors of the versions in the second and the third gospels really independent witnesses? In order to answer this question, it is only needful to place the English versions of the two side by side, and compare them carefully. It will then be seen that the coincidences between them, not merely in substance, but in arrangement, and in the use of identical words in the same order, are such, that only two alternatives are conceivable : either one evangelist freely copied from the other, or both based themselves upon a common source, which may either have been a written docu-

ment, or a definite oral tradition learned by heart.
Assuredly, these two testimonies are not those of
independent witnesses. Further, when the narrative
in the first gospel is compared with that in the
other two, the same fact comes out.

Supposing, then, that Dr. Wace is right in his
assumption that Matthew, Mark, and Luke wrote
the works which we find attributed to them by
tradition, what is the value of their agreement, even
that something more or less like this particular
miracle occurred, since it is demonstrable, either
that all depend on some antecedent statement, of
the authorship of which nothing is known, or that
two are dependent upon the third?

Dr. Wace says he believes the Gadarene story;
whichever version of it he accepts, therefore, he
believes that Jesus said what he is stated in all the
versions to have said, and thereby virtually declared
that the theory of the nature of the spiritual world
involved in the story is true. Now I hold that this
theory is false, that it is a monstrous and mischievous
fiction; and I unhesitatingly express my disbelief in
any assertion that it is true, by whomsoever made.
So that, if Dr. Wace is right in his belief, he is also
quite right in classing me among the people he calls
"infidels"; and although I cannot fulfil the eccentric
expectation that I shall glory in a title which, from
my point of view, it would be simply silly to adopt,
I certainly shall rejoice not to be reckoned among
"Christians" so long as the profession of belief in
such stories as the Gadarene pig affair, on the

strength of a tradition of unknown origin, of which two discrepant reports, also of unknown origin, alone remain, forms any part of the Christian faith.   And, although I have, more than once, repudiated the gift of prophecy, yet I think I may venture to express the anticipation, that if " Christians " generally are going to follow the line taken by Dr. Wace, it will not be long before all men of common sense qualify for a place among the " infidels."

XII

AGNOSTICISM AND CHRISTIANITY

Nemo ergo ex me scire quærat, quod me nescire scio, nisi forte ut nescire discat.—AUGUSTINUS, *De Civ. Dei*, xii. 7.

[1] THE present discussion has arisen out of the use, which has become general in the last few years, of the terms " Agnostic " and " Agnosticism."

The people who call themselves " Agnostics " have been charged with doing so because they have not the courage to declare themselves " Infidels." It has been insinuated that they have adopted a new name in order to escape the unpleasantness which attaches to their proper denomination. To this wholly erroneous imputation, I have replied by showing that the term " Agnostic " did, as a matter of fact, arise in a manner which negatives it; and my statement has not been, and cannot be, refuted. Moreover, speaking for myself, and without impugning the right of any other person to use the term in another sense, I further say that Agnosticism

---

[1] The substance of a paragraph which precedes this has been transferred to the Prologue.

is not properly described as a "negative" creed, nor indeed as a creed of any kind, except in so far as it expresses absolute faith in the validity of a principle, which is as much ethical as intellectual. This principle may be stated in various ways, but they all amount to this: that it is wrong for a man to say that he is certain of the objective truth of any proposition unless he can produce evidence which logically justifies that certainty. This is what Agnosticism asserts; and, in my opinion, it is all that is essential to Agnosticism. That which Agnostics deny and repudiate, as immoral, is the contrary doctrine, that there are propositions which men ought to believe, without logically satisfactory evidence; and that reprobation ought to attach to the profession of disbelief in such inadequately supported propositions. The justification of the Agnostic principle lies in the success which follows upon its application, whether in the field of natural, or in that of civil, history; and in the fact that, so far as these topics are concerned, no sane man thinks of denying its validity.

Still speaking for myself, I add, that though Agnosticism is not, and cannot be, a creed, except in so far as its general principle is concerned; yet that the application of that principle results in the denial of, or the suspension of judgment concerning, a number of propositions respecting which our contemporary ecclesiastical "gnostics" profess entire certainty. And, in so far as these ecclesiastical persons can be justified in their old-established custom (which many nowadays think more honoured in the breach than the observance) of using opprobrious names to those who differ from them, I fully admit their right to call me and those who think with me " Infidels : " all I have ventured to urge is that they must not expect us to speak of ourselves by that title.

The extent of the region of the uncertain, the number of the problems the investigation of which ends in a verdict of not proven, will vary according to the knowledge and the intellectual habits of the individual Agnostic. I do not very much care to speak of anything as " unknowable." What I am sure about is that there are many topics about which I know nothing ; and which, so far as I can see, are out of reach of my faculties. But whether these things are knowable by any one else is exactly one of those matters which is beyond my knowledge, though I may have a tolerably strong opinion as to the probabilities of the case. Relatively to myself, I am quite sure that the region of uncertainty—the nebulous country in which words play the part of realities—is far more extensive than I could wish. Materialism and Idealism ; Theism and Atheism ; the doctrine of the soul and its mortality or immortality—appear in the history of philosophy like the shades of Scandinavian heroes, eternally slaying one another and eternally coming to life again in a metaphysical " Nifelheim." It is getting on for twenty-five centuries, at least, since mankind began seriously to give their minds to these topics.

Generation after generation, philosophy has been doomed to roll the stone uphill; and, just as all the world swore it was at the top, down it has rolled to the bottom again. All this is written in innumerable books; and he who will toil through them will discover that the stone is just where it was when the work began. Hume saw this; Kant saw it; since their time, more and more eyes have been cleansed of the films which prevented them from seeing it; until now the weight and number of those who refuse to be the prey of verbal mystifications has begun to tell in practical life.

It was inevitable that a conflict should arise between Agnosticism and Theology; or rather, I ought to say, between Agnosticism and Ecclesiasticism. For Theology, the science, is one thing; and Ecclesiasticism, the championship of a foregone conclusion [1] as to the truth of a particular form of Theology, is another. With scientific Theology, Agnosticism has no quarrel. On the contrary, the Agnostic, knowing too well the influence of prejudice and idiosyncrasy, even on those who desire most earnestly to be impartial, can wish for nothing more urgently than that the scientific theologian should not only be at perfect liberty to thresh out the matter in his own fashion; but that he should, if he can, find flaws in the Agnostic position; and, even if demonstration is not to be had, that he should put, in their full force, the grounds of the conclusions he thinks probable. The

---

[1] " Let us maintain, before we have proved. This seeming paradox is the secret of happiness " (Dr. Newman : Tract 85, p. 85).

scientific theologian admits the Agnostic principle, however widely his results may differ from those reached by the majority of Agnostics.

But, as between Agnosticism and Ecclesiasticism, or, as our neighbours across the Channel call it, Clericalism, there can be neither peace nor truce. The Cleric asserts that it is morally wrong not to believe certain propositions, whatever the results of a strict scientific investigation of the evidence of these propositions. He tells us "that religious error is, in itself, of an immoral nature."[1] He declares that he has prejudged certain conclusions, and looks upon those who show cause for arrest of judgment as emissaries of Satan. It necessarily follows that, for him, the attainment of faith, not the ascertainment of truth, is the highest aim of mental life. And, on careful analysis of the nature of this faith, it will too often be found to be, not the mystic process of unity with the Divine, understood by the religious enthusiast—but that which the candid simplicity of a Sunday scholar once defined it to be. "Faith," said this unconscious plagiarist of Tertullian, "is the power of saying you believe things which are incredible."

Now I, and many other Agnostics, believe that faith, in this sense, is an abomination; and though we do not indulge in the luxury of self-righteousness so far as to call those who are not of our way of thinking hard names, we do feel that the disagreement between ourselves and those who hold this

---

[1] Dr. Newman, *Essay on Development*, p 357.

doctrine is even more moral than intellectual. It is desirable there should be an end of any mistakes on this topic. If our clerical opponents were clearly aware of the real state of the case, there would be an end of the curious delusion, which often appears between the lines of their writings, that those whom they are so fond of calling " Infidels " are people who not only ought to be, but in their hearts are, ashamed of themselves. It would be discourteous to do more than hint the antipodal opposition of this pleasant dream of theirs to facts.

The clerics and their lay allies commonly tell us, that if we refuse to admit that there is good ground for expressing definite convictions about certain topics, the bonds of human society will dissolve and mankind lapse into savagery. There are several answers to this assertion. One is that the bonds of human society were formed without the aid of their theology ; and, in the opinion of not a few competent judges, have been weakened rather than strengthened by a good deal of it. Greek science, Greek art, the ethics of old Israel, the social organisation of old Rome, contrived to come into being without the help of any one who believed in a single distinctive article of the simplest of the Christian creeds. The science, the art, the jurisprudence, the chief political and social theories, of the modern world have grown out of those of Greece and Rome—not by favour of, but in the teeth of, the fundamental teachings of early Christianity, to which science, art, and any serious occupation with the things of this world, were alike despicable.

Again, all that is best in the ethics of the modern world, in so far as it has not grown out of Greek thought, or Barbarian manhood, is the direct development of the ethics of old Israel. There is no code of legislation, ancient or modern, at once so just and so merciful, so tender to the weak and poor, as the Jewish law; and, if the Gospels are to be trusted, Jesus of Nazareth himself declared that he taught nothing but that which lay implicitly, or explicitly, in the religious and ethical system of his people.

And the scribe said unto him, Of a truth, Teacher, thou hast well said that He is one; and there is none other but He: and to love Him with all the heart, and with all the understanding, and with all the strength, and to love his neighbour as himself, is much more than all whole burnt offerings and sacrifices. (Mark xii. 32, 33).

Here is the briefest of summaries of the teaching of the prophets of Israel of the eighth century; does the Teacher, whose doctrine is thus set forth in his presence, repudiate the exposition? Nay; we are told, on the contrary, that Jesus saw that he "answered discreetly," and replied, "Thou art not far from the Kingdom of God."

So that I think that even if the creeds, from the so-called "Apostles'" to the so-called "Athanasian," were swept into oblivion; and even if the human race should arrive at the conclusion that, whether a bishop washes a cup or leaves it unwashed, is not a matter of the least consequence, it will get on very well. The causes which have led to the development of morality in mankind, which have guided or impelled us all the way from the savage to the civilised state, will not cease to operate because a number of ecclesiastical hypotheses turn out to be baseless. And, even if the absurd notion that morality is more the child of speculation than of practical necessity and inherited instinct, had any foundation; if all the world is going to thieve, murder, and otherwise misconduct itself as soon as it discovers that certain portions of ancient history are mythical; what is the relevance of such arguments to any one who holds by the Agnostic principle?

Surely, the attempt to cast out Beelzebub by the aid of Beelzebub is a hopeful procedure as compared to that of preserving morality by the aid of immorality. For I suppose it is admitted that an Agnostic may be perfectly sincere, may be competent, and may have studied the question at issue with as much care as his clerical opponents. But, if the Agnostic really believes what he says, the " dreadful consequence" argufier (consistently, I admit, with his own principles) virtually asks him to abstain from telling the truth, or to say what he believes to be untrue, because of the supposed injurious consequences to morality. " Beloved brethren, that we may be spotlessly moral, before all things let us lie," is the sum total of many an exhortation addressed to the " Infidel." Now, as I have already pointed out, we cannot oblige our exhorters. We leave the practical application of the convenient doctrines of " Reserve" and " Non-natural interpretation" to those who invented them.

I trust that I have now made amends for any ambiguity, or want of fulness, in my previous exposition of that which I hold to be the essence of the Agnostic doctrine. Henceforward, I might hope to hear no more of the assertion that we are necessarily Materialists, Idealists, Atheists, Theists, or any other *ists*, if experience had led me to think that the proved falsity of a statement was any guarantee against its repetition. And those who appreciate the nature of our position will see, at once, that when Ecclesiasticism declares that we ought to believe this, that, and the other, and are very wicked if we don't, it is impossible for us to give any answer but this : We have not the slightest objection to believe anything you like, if you will give us good grounds for belief; but, if you cannot, we must respectfully refuse, even if that refusal should wreck morality and insure our own damnation several times over. We are quite content to leave that to the decision of the future. The course of the past has impressed us with the firm conviction that no good ever comes of falsehood, and we feel warranted in refusing even to experiment in that direction.

In the course of the present discussion it has been asserted that the "Sermon on the Mount" and the "Lord's Prayer" furnish a summary and condensed view of the essentials of the teaching of Jesus of Nazareth, set forth by himself. Now this supposed *Summa* of Nazarene theology distinctly affirms the existence of a spiritual world, of a Heaven, and of a

Hell of fire; it teaches the Fatherhood of God and the malignity of the Devil; it declares the superintending providence of the former and our need of deliverance from the machinations of the latter; it affirms the fact of demoniac possession and the power of casting out devils by the faithful. And, from these premises, the conclusion is drawn, that those Agnostics who deny that there is any evidence of such a character as to justify certainty, respecting the existence and the nature of the spiritual world, contradict the express declarations of Jesus. I have replied to this argumentation by showing that there is strong reason to doubt the historical accuracy of the attribution to Jesus of either the " Sermon on the Mount" or the " Lord's Prayer"; and, therefore, that the conclusion in question is not warranted, at any rate on the grounds set forth.

But, whether the Gospels contain trustworthy statements about this and other alleged historical facts or not, it is quite certain that from them, taken together with the other books of the New Testament, we may collect a pretty complete exposition of that theory of the spiritual world which was held by both Nazarenes and Christians; and which was undoubtedly supposed by them to be fully sanctioned by Jesus, though it is just as clear that they did not imagine it contained any revelation by him of something heretofore unknown. If the pneumatological doctrine which pervades the whole New Testament is nowhere systematically stated, it is everywhere assumed. The writers of the Gospels and of the Acts take it for granted, as a matter of common knowledge; and it is easy to gather from these sources a series of propositions, which only need arrangement to form a complete system.

In this system, Man is considered to be a duality formed of a spiritual element, the soul; and a corporeal[1] element, the body.   And this duality is repeated in the Universe, which consists of a corporeal world embraced and interpenetrated by a spiritual world.   The former consists of the earth, as its principal and central constituent, with the subsidiary sun, planets, and stars.   Above the earth is the air, and below it the watery abyss.   Whether the heaven, which is conceived to be above the air, and the hell in, or below, the subterranean deeps, are to be taken as corporeal or incorporeal is not clear.   However this may be, the heaven and the air, the earth and the abyss, are peopled by innumerable beings analogous in nature to the spiritual element in man, and these spirits are of two kinds, good and bad. The chief of the good spirits, infinitely superior to all the others, and their creator, as well as the creator of the corporeal world and of the bad spirits, is God.   His residence is heaven, where he is sur-

---

[1] It is by no means to be assumed that "spiritual" and "corporeal" are exact equivalents of "immaterial" and "material" in the minds of ancient speculators on these topics. The "spiritual body" of the risen dead (1 Cor. xv.) is not the "natural" "flesh and blood" body. Paul does not teach the resurrection of the body in the ordinary sense of the word "body"; a fact, often overlooked, but pregnant with many consequences.

rounded by the ordered hosts of good spirits; his
angels, or messengers, and the executors of his will
throughout the universe.

On the other hand, the chief of the bad spirits is
Satan, *the* devil *par excellence*. He and his company
of demons are free to roam through all parts of the
universe, except the heaven. These bad spirits are far
superior to man in power and subtlety, and their
whole energies are devoted to bringing physical and
moral evils upon him, and to thwarting, so far as their
power goes, the benevolent intentions of the Supreme
Being. In fact, the souls and bodies of men form
both the theatre and the prize of an incessant warfare
between the good and the evil spirits—the powers of
light and the powers of darkness. By leading Eve
astray, Satan brought sin and death upon mankind.
As the gods of the heathen, the demons are the
founders and maintainers of idolatry; as the " powers
of the air " they afflict mankind with pestilence and
famine; as " unclean spirits " they cause disease of
mind and body.

The significance of the appearance of Jesus, in the
capacity of the Messiah or Christ, is the reversal
of the satanic work by putting an end to both sin
and death. He announces that the kingdom of
God is at hand, when the " Prince of this world " shall
be finally " cast out " (John xii. 31) from the cosmos,
as Jesus, during his earthly career, cast him out from
individuals. Then will Satan and all his devilry,
along with the wicked whom they have seduced to
their destruction, be hurled into the abyss of un-

quenchable fire—there to endure continual torture, without a hope of winning pardon from the merciful God, their Father; or of moving the glorified Messiah to one more act of pitiful intercession; or even of interrupting, by a momentary sympathy with their wretchedness, the harmonious psalmody of their brother angels and men, eternally lapped in bliss unspeakable.

The straitest Protestant, who refuses to admit the existence of any source of Divine truth, except the Bible, will not deny that every point of the pneumatological theory here set forth has ample scriptural warranty. The Gospels, the Acts, the Epistles, and the Apocalypse assert the existence of the devil, of his demons and of Hell, as plainly as they do that of God and his angels and Heaven. It is plain that the Messianic and the Satanic conceptions of the writers of these books are the obverse and the reverse of the same intellectual coinage. If we turn from Scripture to the traditions of the Fathers and the confessions of the Churches, it will appear that, in this one particular, at any rate, time has brought about no important deviation from primitive belief. From Justin onwards, it may often be a fair question whether God, or the devil, occupies a larger share of the attention of the Fathers. It is the devil who instigates the Roman authorities to persecute; the gods and goddesses of paganism are devils, and idolatry itself is an invention of Satan; if a saint falls away from grace, it is by the seduction of the demon; if heresy arises, the devil has suggested it;

and some of the Fathers[1] go so far as to challenge the pagans to a sort of exorcising match, by way of testing the truth of Christianity. Mediæval Christianity is at one with patristic on this head. The masses, the clergy, the theologians, and the philosophers alike, live and move and have their being in a world full of demons, in which sorcery and possession are everyday occurrences. Nor did the Reformation make any difference. Whatever else Luther assailed, he left the traditional demonology untouched; nor could any one have entertained a more hearty and uncompromising belief in the devil, than he and, at a later period, the Calvinistic fanatics of New England did. Finally, in these last years of the nineteenth century, the demonological hypotheses of the first century are, explicitly or implicitly, held and occasionally acted upon by the immense majority of Christians of all confessions.

Only here and there has the progress of scientific thought, outside the ecclesiastical world, so far affected Christians, that they and their teachers fight shy of the demonology of their creed. They are fain to conceal their real disbelief in one half of Christian doctrine by judicious silence about it; or by flight to those refuges for the logically destitute, accommodation or allegory. But the faithful who fly to allegory in order to escape absurdity resemble nothing so

[1] Tertullian (*Apolog. adv. Gentes,* cap. xxiii.) thus challenges the Roman authorities: let them bring a possessed person into the presence of a Christian before their tribunal; and, if the demon does not confess himself to be such, on the order of the Christian, let the Christian be executed out of hand.

much as the sheep in the fable who—to save their
lives—jumped into the pit. The allegory pit is too
commodious, is ready to swallow up so much more
than one wants to put into it. If the story of the
temptation is an allegory ; if the early recognition of
Jesus as the Son of God by the demons is an allegory ;
if the plain declaration of the writer of the first
Epistle of John (iii. 8), " To this end was the Son of
God manifested, that He might destroy the works of
the devil," is allegorical, then the Pauline version of
the Fall may be allegorical, and still more the words
of consecration of the Eucharist, or the promise of the
second coming ; in fact, there is not a dogma of
ecclesiastical Christianity the scriptural basis of which
may not be whittled away by a similar process.

As to accommodation, let any honest man who
can read the New Testament ask himself whether
Jesus and his immediate friends and disciples can be
dishonoured more grossly than by the supposition
that they said and did that which is attributed to
them ; while, in reality, they disbelieved in Satan
and his demons, in possession and in exorcism ? [1]

An eminent theologian has justly observed that
we have no right to look at the propositions of the
Christian faith with one eye open and the other shut.
(Tract 85, p. 29.) It really is not permissible to see,
with one eye, that Jesus is affirmed to declare the
personality and the Fatherhood of God, His loving
providence and His accessibility to prayer ; and to

[1] See the expression of orthodox opinion upon the " accommodation " subterfuge already cited above, p. 336.

shut the other to the no less definite teaching ascribed to Jesus in regard to the personality and the misanthropy of the devil, his malignant watchfulness, and his subjection to exorcistic formulæ and rites. Jesus is made to say that the devil " was a murderer from the beginning " (John viii. 44) by the same authority as that upon which we depend for his asserted declaration that " God is a spirit " (John iv. 24).

To those who admit the authority of the famous Vincentian dictum that the doctrine which has been held " always, everywhere, and by all " is to be received as authoritative, the demonology must possess a higher sanction than any other Christian dogma, except, perhaps, those of the Resurrection and of the Messiahship of Jesus; for it would be difficult to name any other points of doctrine on which the Nazarene does not differ from the Christian, and the different historical stages and contemporary subdivisions of Christianity from one another. And, if the demonology is accepted, there can be no reason for rejecting all those miracles in which demons play a part. The Gadarene story fits into the general scheme of Christianity; and the evidence for " Legion " and their doings is just as good as any other in the New Testament for the doctrine which the story illustrates.

It was with the purpose of bringing this great fact into prominence; of getting people to open both their eyes when they look at Ecclesiasticism; that I devoted so much space to that miraculous story which happens to be one of the best types of its class. And

I could not wish for a better justification of the course I have adopted, than the fact that my heroically consistent adversary has declared his implicit belief in the Gadarene story and (by necessary consequence) in the Christian demonology as a whole. It must be obvious, by this time, that, if the account of the spiritual world given in the New Testament, professedly on the authority of Jesus, is true, then the demonological half of that account must be just as true as the other half. And, therefore, those who question the demonology, or try to explain it away, deny the truth of what Jesus said, and are, in ecclesiastical terminology, " Infidels " just as much as those who deny the spirituality of God. This is as plain as anything can well be, and the dilemma for my opponent was either to assert that the Gadarene pig-bedevilment actually occurred, or to write himself down an " Infidel." As was to be expected, he chose the former alternative ; and I may express my great satisfaction at finding that there is one spot of common ground on which both he and I stand. So far as I can judge, we are agreed to state one of the broad issues between the consequences of agnostic principles (as I draw them), and the consequences of ecclesiastical dogmatism (as he accepts it), as follows.

Ecclesiasticism says : The demonology of the Gospels is an essential part of that account of that spiritual world, the truth of which it declares to be certified by Jesus.

Agnosticism (*me judice*) says : There is no good

evidence of the existence of a demonic spiritual world,
and much reason for doubting it.

Hereupon the ecclesiastic may observe: Your
doubt means that you disbelieve Jesus; therefore
you are an "Infidel" instead of an "Agnostic."
To which the agnostic may reply: No; for two
reasons: first, because your evidence that Jesus said
what you say he said is worth very little; and
secondly, because a man may be an agnostic, in the
sense of admitting he has no positive knowledge, and
yet consider that he has more or less probable ground
for accepting any given hypothesis about the spiritual
world. Just as a man may frankly declare that he
has no means of knowing whether the planets
generally are inhabited or not, and yet may think
one of the two possible hypotheses more likely than
the other, so he may admit that he has no means of
knowing anything about the spiritual world, and yet
may think one or other of the current views on the
subject, to some extent, probable.

The second answer is so obviously valid that it
needs no discussion. I draw attention to it simply
in justice to those agnostics who may attach greater
value than I do to any sort of pneumatological speculations, and not because I wish to escape the responsibility of declaring that, whether Jesus sanctioned the
demonological part of Christianity or not, I unhesitatingly reject it. The first answer, on the other
hand, opens up the whole question of the claim of the
biblical and other sources, from which hypotheses
concerning the spiritual world are derived, to be re-

garded as unimpeachable historical evidence as to matters of fact.

Now, in respect of the trustworthiness of the Gospel narratives, I was anxious to get rid of the common assumption that the determination of the authorship and of the dates of these works is a matter of fundamental importance. That assumption is based upon the notion that what contemporary witnesses say must be true, or, at least, has always a *primâ facie* claim to be so regarded; so that if the writers of any of the Gospels were contemporaries of the events (and still more if they were in the position of eye-witnesses) the miracles they narrate must be historically true, and, consequently, the demonology which they involve must be accepted. But the story of the *Translation of the blessed martyrs Marcellinus and Petrus*, and the other considerations (to which endless additions might have been made from the Fathers and the mediæval writers) set forth in a preceding essay, yield, in my judgment, satisfactory proof that, where the miraculous is concerned, neither considerable intellectual ability, nor undoubted honesty, nor knowledge of the world, nor proved faithfulness as civil historians, nor profound piety, on the part of eye-witnesses and contemporaries, affords any guarantee of the objective truth of their statements, when we know that a firm belief in the miraculous was ingrained in their minds, and was the pre-supposition of their observations and reasonings.

Therefore, although it be, as I believe, demonstrable that we have no real knowledge of the

authorship, or of the date of composition of the
Gospels, as they have come down to us, and
that nothing better than more or less probable
guesses can be arrived at on that subject, I have
not cared to expend any space on the question. It
will be admitted, I suppose, that the authors of the
works attributed to Matthew, Mark, Luke, and John,
whoever they may be, are personages whose capacity
and judgment in the narration of ordinary events
are not quite so well certified as those of Eginhard;
and we have seen what the value of Eginhard's
evidence is when the miraculous is in question.

I have been careful to explain that the arguments
which I have used in the course of this discussion
are not new; that they are historical and have
nothing to do with what is commonly called science;
and that they are all, to the best of my belief, to be
found in the works of theologians of repute.
The position which I have taken up, that the evidence in favour of such miracles as those recorded by
Eginhard, and consequently of mediæval demonology,
is quite as good as that in favour of such miracles
as the Gadarene, and consequently of Nazarene
demonology, is none of my discovery. Its strength
was, wittingly or unwittingly, suggested, a century
and a half ago, by a theological scholar of eminence;
and it has been, if not exactly occupied, yet so
fortified with bastions and redoubts by a living
ecclesiastical Vauban, that, in my judgment, it has
been rendered impregnable. In the early part of

the last century, the ecclesiastical mind in this country was much exercised by the question, not exactly of miracles, the occurrence of which in biblical times was axiomatic, but by the problem : When did miracles cease ?   Anglican divines were quite sure that no miracles had happened in their day, nor for some time past; they were equally sure that they happened sixteen or seventeen centuries earlier.   And it was a vital question for them to determine at what point of time, between this *terminus a quo* and that *terminus ad quem*, miracles came to an end.

The Anglicans and the Romanists agreed in the assumption that the possession of the gift of miracle-working was *primâ facie* evidence of the soundness of the faith of the miracle-workers.   The supposition that miraculous powers might be wielded by heretics (though it might be supported by high authority) led to consequences too frightful to be entertained by people who were busied in building their dogmatic house on the sands of early Church history. If, as the Romanists maintained, an unbroken series of genuine miracles adorned the records of their Church, throughout the whole of its existence, no Anglican could lightly venture to accuse them of doctrinal corruption.   Hence, the Anglicans, who indulged in such accusations, were bound to prove the modern, the mediæval Roman, and the later Patristic, miracles false ; and to shut off the wonder-working power from the Church at the exact point of time when Anglican doctrine ceased and Roman doctrine

began. With a little adjustment—a squeeze here
and a pull there—the Christianity of the first three
or four centuries might be made to fit, or seem to
fit, pretty well into the Anglican scheme. So the
miracles, from Justin say to Jerome, might be recognised; while, in later times, the Church having
become "corrupt"—that is to say, having pursued
one and the same line of development further than
was pleasing to Anglicans—its alleged miracles must
needs be shams and impostures.

Under these circumstances, it may be imagined
that the establishment of a scientific frontier between
the earlier realm of supposed fact and the later of
asserted delusion, had its difficulties; and torrents
of theological special pleading about the subject
flowed from clerical pens; until that learned and
acute Anglican divine, Conyers Middleton, in his
Free Inquiry, tore the sophistical web they had
laboriously woven to pieces, and demonstrated that
the miracles of the patristric age, early and late,
must stand or fall together, inasmuch as the evidence
for the later is just as good as the evidence for the
earlier wonders. If the one set are certified by
contemporaneous witnesses of high repute, so are
the other; and, in point of probability, there is not
a pin to choose between the two. That is the solid
and irrefragable result of Middleton's contribution
to the subject. But the Free Inquirer's freedom had
its limits; and he draws a sharp line of demarcation
between the patristic and the New Testament miracles
—on the professed ground that the accounts of the

latter, being inspired, are out of the reach of criticism.

A century later, the question was taken up by another divine, Middleton's equal in learning and acuteness, and far his superior in subtlety and dialectic skill; who, though an Anglican, scorned the name of Protestant; and, while yet a Churchman, made it his business to parade, with infinite skill, the utter hollowness of the arguments of those of his brother Churchmen who dreamed that they could be both Anglicans and Protestants. The argument of the *Essay on the Miracles recorded in the Ecclesiastical History of the Early Ages*,[1] by the present Roman Cardinal, but then Anglican Doctor, John Henry Newman, is compendiously stated by himself in the following passage :—

If the miracles of Church history cannot be defended by the arguments of Leslie, Lyttleton, Paley, or Douglas, how many of the Scripture miracles satisfy their conditions ? (p. cvii).

And, although the answer is not given in so many words, little doubt is left on the mind of the reader, that, in the mind of the writer, it is : None. In fact, this conclusion is one which cannot be resisted, if the argument in favour of the Scripture miracles is based upon that which laymen, whether lawyers, or men of science, or historians, or ordinary men of affairs, call evidence. But there is something really impressive

---

[1] I quote the first edition (1843). A second edition appeared in 1870. Tract 85 of the *Tracts for the Times* should be read with this *Essay*. If I were called upon to compile a Primer of " Infidelity," I think I should save myself trouble by making a selection from these works, and from the *Essay on Development* by the same author.

in the magnificent contempt with which, at times, Dr. Newman sweeps aside alike those who offer and those who demand such evidence.

> Some infidel authors advise us to accept no miracles which would not have a verdict in their favour in a court of justice; that is, they employ against Scripture a weapon which Protestants would confine to attacks upon the Church; as if moral and religious questions required legal proof, and evidence were the test of truth [1] (p. cvii).

" As if evidence were the test of truth " !—although the truth in question is the occurrence, or the non-occurrence of certain phenomena at a certain time and in a certain place. This sudden revelation of the great gulf fixed between the ecclesiastical and the scientific mind is enough to take away the breath of any one unfamiliar with the clerical organon. As if, one may retort, the assumption that miracles may, or have, served a moral or a religious end, in any way alters the fact that they profess to be historical events, things that actually happened; and, as such, must needs be exactly those subjects about which evidence is appropriate and legal proofs (which are such merely because they afford adequate evidence) may be justly demanded. The Gadarene miracle either happened, or it did not. Whether the Gadarene " question " is moral or religious, or not, has nothing to do with the fact that it is a purely

---

[1] Yet, when it suits his purpose, as in the Introduction to the *Essay on Development*, Dr. Newman can demand strict evidence in religious questions as sharply as any "infidel author"; and he can even profess to yield to its force (*Essay on Miracles*, 1870, note, p. 391).

historical question whether the demons said what they are declared to have said, and the devil-possessed pigs did, or did not, rush over the cliffs bounding the Lake of Gennesaret on a certain day of a certain year, after A.D. 26 and before A.D. 36 : for vague and uncertain as New Testament chronology is, I suppose it may be assumed that the event in question, if it happened at all, took place during the procuratorship of Pilate. If that is not a matter about which evidence ought to be required, and not only legal, but strict scientific proof demanded by sane men who are asked to believe the story—what is ? Is a reasonable being to be seriously asked to credit statements, which, to put the case gently, are not exactly probable, and on the acceptance or rejection of which his whole view of life may depend, without asking for as much " legal " proof as would send an alleged pickpocket to gaol, or as would suffice to prove the validity of a disputed will ?

" Infidel authors " (if, as I am assured, I may answer for them) will decline to waste time on mere darkenings of counsel of this sort ; but to those Anglicans who accept his premises, Dr. Newman is a truly formidable antagonist. What, indeed, are they to reply when he puts the very pertinent question :—

whether persons who not merely question, but prejudge the Ecclesiastical miracles on the ground of their want of resemblance, whatever that be, to those contained in Scripture—as if the Almighty could not do in the Christian Church what He had not already done at the time of its foundation, or under the Mosaic Covenant—whether such reasoners are not siding with the sceptic, and

whether it is not a happy inconsistency by which they continue to believe the Scriptures while they reject the Church [1]
(p. liii).

Again, I invite Anglican orthodoxy to consider this
passage :—

the narrative of the combats of St. Antony with evil spirits, is a
development rather than a contradiction of revelation, viz. of
such texts as speak of Satan being cast out by prayer and fasting.   To be shocked, then, at the miracles of Ecclesiastical
history, or to ridicule them for their strangeness, is no part of a
scriptural philosophy (pp. liii-liv).

Further on, Dr. Newman declares that it has been
admitted

that a distinct line can be drawn in point of character and circumstance between the miracles of Scripture and of Church history;
but this is by no means the case (p. lv). . . . specimens are not
wanting in the history of the Church, of miracles as awful in
their character and as momentous in their effects as those which
are recorded in Scripture.   The fire interrupting the rebuilding
of the Jewish temple, and the death of Arius, are instances, in
Ecclesiastical history, of such solemn events.   On the other
hand, difficult instances in the Scripture history are such as
these : the serpent in Eden, the Ark, Jacob's vision for the
multiplication of his cattle, the speaking of Balaam's ass, the
axe swimming at Elisha's word, the miracle on the swine, and
various instances of prayers or prophecies, in which, as in that
of Noah's blessing and curse, words which seem the result of
private feeling are expressly or virtually ascribed to a Divine
suggestion (p. lvi).

Who is to gainsay our ecclesiastical authority

---

[1] Compare Tract 85, p. 110 : " I am persuaded that were men but
consistent who oppose the Church doctrines as being unscriptural,
they would vindicate the Jews for rejecting the Gospel."

here ? "Infidel authors" might be accused of a wish to ridicule the Scripture miracles by putting them on a level with the remarkable story about the fire which stopped the rebuilding of the Temple, or that about the death of Arius—but Dr. Newman is above suspicion. The pity is that his list of what he delicately terms "difficult" instances is so short. Why omit the manufacture of Eve out of Adam's rib, on the strict historical accuracy of which the chief argument of the defenders of an iniquitous portion of our present marriage law depends ? Why leave out the account of the "Bene Elohim" and their gallantries, on which a large part of the worst practices of the mediæval inquisitors into witchcraft was based ? Why forget the angel who wrestled with Jacob, and, as the account suggests, somewhat over-stepped the bounds of fair play, at the end of the struggle ? Surely, we must agree with Dr. Newman that, if all these camels have gone down, it savours of affectation to strain at such gnats as the sudden ailment of Arius in the midst of his deadly, if prayerful,[1] enemies; and the fiery explosion which

[1] According to Dr. Newman, "This prayer [that of Bishop Alexander, who begged God to 'take Arius away'] is said to have been offered about 3 P.M. on the Saturday ; that same evening Arius was in the great square of Constantine, when he was suddenly seized with indisposition" (p. clxx). The "infidel" Gibbon seems to have dared to suggest that "an option between poison and miracle" is pre-sented by this case ; and, it must be admitted, that, if the Bishop had been within the reach of a modern police magistrate, things might have gone hardly with him. Modern "Infidels," possessed of a slight knowledge of chemistry, are not unlikely, with no less audacity, to suggest an "option between fire-damp and miracle" in seeking for the cause of the fiery outburst at Jerusalem.

stopped the Julian building operations. Though the *words* of the "Conclusion" of the *Essay on Miracles* may, perhaps, be quoted against me, I may express my satisfaction at finding myself in substantial accordance with a theologian above all suspicion of heterodoxy. With all my heart, I can declare my belief that there is just as good reason for believing in the miraculous slaying of the man who fell short of the Athanasian power of affirming contradictories, with respect to the nature of the Godhead, as there is for believing in the stories of the serpent and the ark told in Genesis, the speaking of Balaam's ass in Numbers, or the floating of the axe, at Elisha's order, in the second book of Kings.

It is one of the peculiarities of a really sound argument that it is susceptible of the fullest development; and that it sometimes leads to conclusions unexpected by those who employ it. To my mind, it is impossible to refuse to follow Dr. Newman when he extends his reasoning from the miracles of the patristic and mediæval ages backward in time as far as miracles are recorded. But, if the rules of logic are valid, I feel compelled to extend the argument forward to the alleged Roman miracles of the present day, which Dr. Newman might not have admitted, but which Cardinal Newman may hardly reject. Beyond question, there is as good, or perhaps better, evidence for the miracles worked by our Lady of Lourdes, as there is for the floating of Elisha's axe, or the speaking of Balaam's ass. But we must go still further; there is a modern system of thaumaturgy and demonology which is just as well certified as the ancient.[1] Veracious, excellent, sometimes learned and acute persons, even philosophers of no mean pretensions, testify to the "levitation" of bodies much heavier than Elisha's axe; to the existence of "spirits" who, to the mere tactile sense, have been indistinguishable from flesh and blood, and, occasionally, have wrestled with all the vigour of Jacob's opponent; yet, further, to the speech, in the language of raps, of spiritual beings, whose discourses, in point of coherence and value, are far inferior to that of Balaam's humble but sagacious steed. I have not the smallest doubt that, if these were persecuting times, there is many

---

[1] A writer in a spiritualist journal takes me roundly to task for venturing to doubt the historical and literal truth of the Gadarene story. The following passage in his letter is worth quotation : " Now to the materialistic and scientific mind, to the uninitiated in spiritual verities, certainly this story of the Gadarene or Gergesene swine presents insurmountable difficulties ; it seems grotesque and nonsensical. To the experienced, trained, and cultivated Spiritualist this miracle is, as I am prepared to show, one of the most instructive, the most profoundly useful, and the most beneficent which Jesus ever wrought in the whole course of His pilgrimage of redemption on earth." Just so. And the first page of this same journal presents the following advertisement, among others of the same kidney :—

"To WEALTHY SPIRITUALISTS.—A Lady Medium of tried power wishes to meet with an elderly gentleman who would be willing to give her a comfortable home and maintenance in Exchange for her Spiritualistic services, as her guides consider her health is too delicate for public sittings : London preferred.—Address "Mary," Office of Light."

Are we going back to the days of the Judges, when wealthy Micah set up his private ephod, teraphim, and Levite ?

a worthy "spiritualist" who would cheerfully go to the stake in support of his pneumatological faith, and furnish evidence, after Paley's own heart, in proof of the truth of his doctrines.   Not a few modern divines, doubtless struck by the impossibility of refusing, the spiritualist evidence, if the ecclesiastical evidence is accepted, and deprived of any *à priori* objection by their implicit belief in Christian Demonology, show themselves ready to take poor Sludge seriously, and to believe that he is possessed by other devils than those of need, greed, and vainglory.

Under these circumstances, it was to be expected, though it is none the less interesting to note the fact, that the arguments of the latest school of "spiritualists" present a wonderful family likeness to those which adorn the subtle disquisitions of the advocate of ecclesiastical miracles of forty years ago. It is unfortunate for the "spiritualists" that, over and over again, celebrated and trusted media, who really, in some respects, call to mind the Montanist[1] and gnostic seers of the second century, are either proved in courts of law to be fraudulent impostors;

---

[1] Consider Tertullian's "sister" ("hodie apud nos"), who conversed with angels, saw and heard mysteries, knew men's thoughts, and prescribed medicine for their bodies (*De Anima*, cap. 9).   Tertullian tells us that this woman saw the soul as corporeal, and described its colour and shape.   The "infidel" will probably be unable to refrain from insulting the memory of the ecstatic saint by the remark, that Tertullian's known views about the corporeality of the soul may have had something to do with the remarkable perceptive powers of the Montanist medium, in whose revelations of the spiritual world he took such profound interest.

or, in sheer weariness, as it would seem, of the honest
dupes who swear by them, spontaneously confess
their long-continued iniquities, as the Fox women
did the other day in New York.[1] But, whenever a
catastrophe of this kind takes place, the believers
are no wise dismayed by it. They freely admit
that not only the media, but the spirits whom they
summon, are sadly apt to lose sight of the elementary
principles of right and wrong; and they triumphantly
ask: How does the occurrence of occasional impostures disprove the genuine manifestations (that
is to say, all those which have not yet been proved
to be impostures or delusions)? And, in this, they
unconsciously plagiarise from the churchman, who
just as freely admits that many ecclesiastical miracles
may have been forged; and asks, with calm contempt, not only of legal proofs, but of common-sense
probability, Why does it follow that none are to be
supposed genuine? I must say, however, that the
spiritualists, so far as I know, do not venture to
outrage right reason so boldly as the ecclesiastics.
They do not sneer at "evidence"; nor repudiate
the requirement of legal proofs. In fact, there can
be no doubt that the spiritualists produce better
evidence for their manifestations than can be shown
either for the miraculous death of Arius, or for the
Invention of the Cross.[2]

[1] See the New York *World* for Sunday, 21st October 1888; and
the *Report of the Seybert Commission*, Philadelphia, 1887.

[2] Dr. Newman's observation that the miraculous multiplication of
the pieces of the true cross (with which "the whole world is filled,"
according to Cyril of Jerusalem; and of which some say there are

From the "levitation" of the axe at one end of a period of near three thousand years to the "levitation" of Sludge & Co. at the other end, there is a complete continuity of the miraculous, with every gradation from the childish to the stupendous, from the gratification of a caprice to the illustration of sublime truth. There is no drawing a line in the series that might be set out of plausibly attested cases of spiritual intervention. If one is true, all may be true; if one is false, all may be false.

This is, to my mind, the inevitable result of that method of reasoning which is applied to the confutation of Protestantism, with so much success, by one of the acutest and subtlest disputants who have ever championed Ecclesiasticism—and one cannot put his claims to acuteness and subtlety higher.

. . . the Christianity of history is not Protestantism. If ever there were a safe truth it is this. . . . "To be deep in history is to cease to be a Protestant." [1]

I have not a shadow of doubt that these anti-Protestant epigrams are profoundly true. But I have as little that, in the same sense, the "Christianity of history is not" Romanism; and that to be deeper in history is to cease to be a Romanist. The reasons which compel my doubts about the compatibility of the Roman doctrine, or any other form

enough extant to build a man-of-war) is no more wonderful than that of the loaves and fishes is one that I do not see my way to contradict. See *Essay on Miracles*, 2d ed. p. 163.

[1] *An Essay on the Development of Christian Doctrine*, by J. H. Newman, D.D., pp. 7 and 8. (1878.)

of Catholicism, with history, arise out of exactly the same line of argument as that adopted by Dr. Newman in the famous essay which I have just cited. If, with one hand, Dr. Newman has destroyed Protestantism, he has annihilated Romanism with the other; and the total result of his ambidextral efforts is to shake Christianity to its foundations. Nor was any one better aware that this must be the inevitable result of his arguments—if the world should refuse to accept Roman doctrines and Roman miracles—than the writer of Tract 85.

Dr. Newman made his choice and passed over to the Roman Church half a century ago. Some of those who were essentially in harmony with his views preceded, and many followed him. But many remained; and, as the quondam Puseyite and present Ritualistic party, they are continuing that work of sapping and mining the Protestantism of the Anglican Church which he and his friends so ably commenced. At the present time, they have no little claim to be considered victorious all along the line. I am old enough to recollect the small beginnings of the Tractarian party; and I am amazed when I consider the present position of their heirs. Their little leaven has leavened, if not the whole, yet a very large lump of the Anglican Church; which is now pretty much of a preparatory school for Papistry. So that it really behoves Englishmen (who, as I have been informed by high authority, are all, legally, members of the State Church, if they profess to belong to no other sect) to wake up to what that

powerful organisation is about, and whither it is
tending.   On this point, the writings of Dr. Newman, while he still remained within the Anglican
fold, are a vast store of the best and the most
authoritative information.   His doctrines on Ecclesiastical miracles and on Development are the
corner-stones of the Tractarian fabric.   He believed
that his arguments led either Romeward, or to what
ecclesiastics call " Infidelity," and I call Agnosticism.
I believe that he was quite right in this conviction ;
but while he chooses the one alternative, I choose the
other ; as he rejects Protestantism on the ground of
its incompatibility with history, so, *à fortiori*, I conceive that Romanism ought to be rejected, and that
an impartial consideration of the evidence must refuse
the authority of Jesus to anything more than the
Nazarenism of James and Peter and John.   And
let it not be supposed that this is a mere " infidel "
perversion of the facts.   No one has more openly
and clearly admitted the possibility that they may
be fairly interpreted in this way than Dr. Newman.
If, he says, there are texts which seem to show that
Jesus contemplated the evangelisation of the heathen :

. . . Did not the Apostles hear our Lord ? and what was
their impression from what they heard ?   Is it not certain that
the Apostles did not gather this truth from His teaching ?
(Tract 85, p. 63).

He said, "Preach the Gospel to every creature."   These
words *need* have only meant "Bring all men to Christianity
through Judaism."   Make them Jews, that they may enjoy
Christ's privileges, which are lodged in Judaism ; teach them
those rites and ceremonies, circumcision and the like, which

hitherto have been dead ordinances, and now are living : and so the Apostles seem to have understood them (*ibid.* p. 65).

So far as Nazarenism differentiated itself from contemporary orthodox Judaism, it seems to have tended towards a revival of the ethical and religious spirit of the prophetic age, accompanied by the belief in Jesus as the Messiah, and by various accretions which had grown round Judaism subsequently to the exile. To these belong the doctrines of the Resurrection, of the Last Judgment, of Heaven and Hell; of the hierarchy of good angels; of Satan and the hierarchy of evil spirits. And there is very strong ground for believing that all these doctrines, at least in the shapes in which they were held by the post-exilic Jews, were derived from Persian and Babylonian [1] sources, and are essentially of heathen origin.

How far Jesus positively sanctioned all these indrainings of circumjacent Paganism into Judaism; how far any one has a right to declare, that the refusal to accept one or other of these doctrines, as ascertained verities, comes to the same thing as contradicting Jesus, it appears to me not easy to say. But it is hardly less difficult to conceive that he could have distinctly negatived any of them; and, more especially, that demonology which has been accepted by

---

[1] Dr. Newman faces this question with his customary ability. "Now, I own, I am not at all solicitous to deny that this doctrine of an apostate Angel and his hosts was gained from Babylon : it might still be Divine nevertheless. God who made the prophet's ass speak, and thereby instructed the prophet, might instruct His Church by means of heathen Babylon" (Tract 85, p. 83). There seems to be no end to the apologetic burden that Balaam's ass can carry.

the Christian Churches in every age and under all their mutual antagonisms. But, I repeat my conviction that, whether Jesus sanctioned the demonology of his time and nation or not, it is doomed. The future of Christianity, as a dogmatic system and apart from the old Israelitish ethics which it has appropriated and developed, lies in the answer which mankind will eventually give to the question whether they are prepared to believe such stories as the Gadarene and the pneumatological hypotheses which go with it, or not. My belief is they will decline to do anything of the sort, whenever and wherever their minds have been disciplined by science. And that discipline must, and will, at once follow and lead the footsteps of advancing civilisation.

The preceding pages were written before I became acquainted with the contents of the May number of the *Nineteenth Century*, wherein I discover many things which are decidedly not to my advantage. It would appear that " evasion " is my chief resource, " incapacity for strict argument " and " rottenness of ratiocination " my main mental characteristics, and that it is " barely credible " that a statement which I profess to make of my own knowledge is true. All which things I notice, merely to illustrate the great truth, forced on me by long experience, that it is only from those who enjoy the blessing of a firm hold of the Christian faith that such manifestations of meekness, patience, and charity are to be expected.

I had imagined that no one who had read my preceding papers, could entertain a doubt as to my position in respect of the main issue as it has been stated and restated by my opponent:

an Agnosticism which knows nothing of the relation of man to God must not only refuse belief to our Lord's most undoubted teaching, but must deny the reality of the spiritual convictions in which He lived.[1]

That is said to be "the simple question which is at issue between us," and the three testimonies to that teaching and those convictions selected are the Sermon on the Mount, the Lord's Prayer, and the Story of the Passion.

My answer, reduced to its briefest form, has been: In the first place, the evidence is such that the exact nature of the teachings and the convictions of Jesus is extremely uncertain, so that what ecclesiastics are pleased to call a denial of them may be nothing of the kind. And, in the second place, if Jesus taught the demonological system involved in the Gadarene story—if a belief in that system formed a part of the spiritual convictions in which he lived and died—then I, for my part, unhesitatingly refuse belief in that teaching, and deny the reality of those spiritual convictions. And I go further and add, that, exactly in so far as it can be proved that Jesus sanctioned the essentially pagan demonological theories current among the Jews of his age, exactly in so far, for me, will his authority in any matter touching the spiritual world be weakened.

---

[1] *Nineteenth Century*, May 1889 (p. 701).

With respect to the first half of my answer, I have pointed out that the Sermon on the Mount, as given in the first Gospel, is, in the opinion of the best critics, a "mosaic work" of materials derived from different sources, and I do not understand that this statement is challenged. The only other Gospel, the third, which contains something like it, makes, not only the discourse, but the circumstances under which it was delivered, very different. Now, it is one thing to say that there was something real at the bottom of the two discourses—which is quite possible ; and another to affirm that we have any right to say what that something was, or to fix upon any particular phrase and declare it to be a genuine utterance. Those who pursue theology as a science, and bring to the study an adequate knowledge of the ways of ancient historians, will find no difficulty in providing illustrations of my meaning. I may supply one which has come within range of my own limited vision.

In Josephus's *History of the Wars of the Jews* (chap. xix.), that writer reports a speech which he says Herod made at the opening of a war with the Arabians. It is in the first person, and would naturally be supposed by the reader to be intended for a true version of what Herod said. In the *Antiquities*, written some seventeen years later, the same writer gives another report, also in the first person, of Herod's speech on the same occasion. This second oration is twice as long as the first, and though the general tenour of the two speeches is pretty much the same, there is hardly any verbal identity, and a good deal of matter is introduced into the one, which is absent from the other. Josephus prides himself on his accuracy; people whose fathers might have heard Herod's oration were his contemporaries; and yet his historical sense is so curiously undeveloped that he can, quite innocently, perpetrate an obvious literary fabrication; for one of the two accounts must be incorrect. Now, if I am asked whether I believe that Herod made some particular statement on this occasion; whether, for example, he uttered the pious aphorism, "Where God is, there is both multitude and courage," which is given in the *Antiquities*, but not in the *Wars*, I am compelled to say I do not know. One of the two reports must be erroneous, possibly both are: at any rate, I cannot tell how much of either is true. And, if some fervent admirer of the Idumean should build up a theory of Herod's piety upon Josephus's evidence that he propounded the aphorism, is it a "mere evasion" to say, in reply, that the evidence that he did utter it is worthless?

It appears again that, adopting the tactics of Conachar when brought face to face with Hal o' the Wynd, I have been trying to get my simple-minded adversary to follow me on a wild-goose chase through the early history of Christianity, in the hope of escaping · impending defeat on the main issue. But I may be permitted to point out that there is an alternative hypothesis which equally fits the facts; and that, after all, there

may have been method in the madness of my
supposed panic.

For suppose it to be established that Gentile
Christianity was a totally different thing from the
Nazarenism of Jesus and his immediate disciples;
suppose it to be demonstrable that, as early as the
sixth decade of our era at least, there were violent
divergencies of opinion among the followers of Jesus;
suppose it to be hardly doubtful that the Gospels and
the Acts took their present shapes under the influence
of these divergencies; suppose that their authors, and
those through whose hands they passed, had notions
of historical veracity not more eccentric than those
which Josephus occasionally displays: surely the
chances that the Gospels are altogether trustworthy
records of the teachings of Jesus become very slender.
And since the whole of the case of the other side is
based on the supposition that they are accurate
records (especially of speeches, about which ancient
historians are so curiously loose), I really do venture
to submit that this part of my argument bears very
seriously on the main issue; and, as ratiocination, is
sound to the core.

Again, when I passed by the topic of the speeches
of Jesus on the Cross, it appears that I could have
had no other motive than the dictates of my native
evasiveness. An ecclesiastical dignitary may have
respectable reasons for declining a fencing match "in
sight of Gethsemane and Calvary"; but an ecclesiastical "Infidel"! Never. It is obviously impossible
that, in the belief that " the greater includes the less,"

I, having declared the Gospel evidence in general, as to the sayings of Jesus, to be of questionable value, thought it needless to select for illustration of my views, those particular instances which were likely to be most offensive to persons of another way of thinking. But any supposition that may have been entertained that the old familiar tones of the ecclesiastical war-drum will tempt me to engage in such needless discussion had better be renounced. I shall do nothing of the kind. Let it suffice that I ask my readers to turn to the twenty-third chapter of Luke (revised version), verse thirty-four, and he will find in the margin

> Some ancient authorities omit: And Jesus said "Father forgive them, for they know not what they do."

So that, even as late as the fourth century, there were ancient authorities, indeed some of the most ancient and weightiest, who either did not know of this utterance, so often quoted as characteristic of Jesus, or did not believe it had been uttered.

Many years ago, I received an anonymous letter, which abused me heartily for my want of moral courage in not speaking out. I thought that one of the oddest charges an anonymous letter-writer could bring. But I am not sure that the plentiful sowing of the pages of the article with which I am dealing with accusations of evasion, may not seem odder to those who consider that the main strength of the answers with which I have been favoured (in this review and elsewhere) is devoted, not to anything in the text of my first paper, but to a note which occurs at p. 171.   In this I say :

> Dr. Wace tells us : " It may be asked how far we can rely on the accounts we possess of our Lord's teaching on these subjects." And he seems to think the question appropriately answered by the assertion that it "ought to be regarded as settled by M. Renan's practical surrender of the adverse case."

I requested Dr. Wace to point out the passages of M. Renan's works in which, as he affirms, this "practical surrender" (not merely as to the age and authorship of the Gospels, be it observed, but as to their historical value) is made, and he has been so good as to do so.   Now let us consider the parts of Dr. Wace's citation from Renan which are relevant to the issue :—

> The author of this Gospel [Luke] is certainly the same as the author of the Acts of the Apostles.   Now the author of the Acts seems to be a companion of St. Paul—a character which accords completely with St. Luke.   I know that more than one objection may be opposed to this reasoning ; but one thing, at all events, is beyond doubt, namely, that the author of the third Gospel and of the Acts is a man who belonged to the second apostolic generation ; and this suffices for our purpose.

This is a curious "practical surrender of the adverse case."   M. Renan thinks that there is no doubt that the author of the third Gospel is the author of the Acts—a conclusion in which I suppose critics generally agree.   He goes on to remark that this person *seems* to be a companion of St. Paul, and adds that Luke was a companion of St. Paul.   Then, somewhat needlessly, M. Renan points out that there is more than one objection to jumping, from such data as these, to

the conclusion that " Luke " is the writer of the third
Gospel. And, finally, M. Renan is content to reduce
that which is " beyond doubt " to the fact that the
author of the two books is a man of the second
apostolic generation. Well, it seems to me that I
could agree with all that M. Renan considers " beyond
doubt " here, without surrendering anything, either
" practically " or theoretically.

Dr. Wace (*Nineteenth Century*, March, p. 363)
states that he derives the above citation from the
preface to the 15th edition of the *Vie de Jésus*. My
copy of *Les Évangiles*, dated 1877, contains a list of
Renan's *Œuvres Complètes*, at the head of which I
find *Vie de Jésus*, 15ᵉ édition. It is, therefore, a later
work than the edition of the *Vie de Jésus* which Dr.
Wace quotes. Now *Les Évangiles*, as its name im
plies, treats fully of the questions respecting the date
and authorship of the Gospels ; and any one who
desired, not merely to use M. Renan's expressions for
controversial purposes, but to give a fair account of
his views in their full significance, would, I think,
refer to the later source.

If this course had been taken, Dr. Wace might have
found some as decided expressions of opinion in favour
of Luke's authorship of the third Gospel as he has discovered in *The Apostles*. I mention this circumstance
because I desire to point out that, taking even the
strongest of Renan's statements, I am still at a loss
to see how it justifies that large-sounding phrase,
" practical surrender of the adverse case." For, on
p. 438 of *Les Évangiles*, Renan speaks of the way in

which Luke's " excellent intentions " have led him to
torture history in the Acts ; he declares Luke to be
the founder of that " eternal fiction which is called
ecclesiastical history " ; and, on the preceding page,
he talks of the " myth " of the Ascension—with its
" *mise en scène voulue.*"　At p. 435, I find " Luc, ou
l'auteur quel qu'il soit du troisième Évangile " ; at p.
280, the accounts of the Passion, the death and the
resurrection of Jesus, are said to be " peu historiques " ;
at p. 283, " La valeur historique du troisième Évangile
est sûrement moindre que celles des deux premiers."
A Pyrrhic sort of victory for orthodoxy this " surrender !"　And, all the while, the scientific student
of theology knows that the more reason there may be
to believe that Luke was the companion of Paul, the
more doubtful becomes his credibility, if he really
wrote the Acts.　For, in that case, he could not
fail to have been acquainted with Paul's account of
the Jerusalem conference, and he must have consciously misrepresented it.

We may next turn to the essential part of Dr.
Wace's citation (*Nineteenth Century*, p. 365) touching the first Gospel :—

St. Matthew evidently deserves peculiar confidence for the
discourses.　Here are the " oracles "—the very notes taken while
the memory of the instruction of Jesus was living and definite.

M. Renan here expresses the very general opinion
as to the existence of a collection of " logia," having
a different origin from the text in which they are
embedded, in Matthew.　" Notes " are somewhat suggestive of a shorthand writer, but the suggestion is

unintentional, for M. Renan assumes that these " notes " were taken, not at the time of the delivery of the " logia " but subsequently, while (as he assumes) the memory of them was living and definite ; so that, in this very citation, M. Renan leaves open the question of the general historical value of the first Gospel, while it is obvious that the accuracy of " notes " taken, not at the time of delivery, but from memory, is a matter about which more than one opinion may be fairly held. Moreover, Renan expressly calls attention to the difficulty of distinguishing the authentic " logia " from later additions of the same kind (*Les Évangiles*, p. 201). The fact is, there is no contradiction here to that opinion about the first Gospel which is expressed in *Les Évangiles* (p. 175).

The text of the so-called Matthew supposes the pre-existence of that of Mark, and does little more than complete it. He completes it in two fashions—first, by the insertion of those long discourses which gave their chief value to the Hebrew Gospels ; then by adding traditions of a more modern formation, results of successive developments of the legend, and to which the Christian consciousness already attached infinite value.

M. Renan goes on to suggest that besides " Mark," " pseudo-Matthew " used an Aramaic version of the Gospel originally set forth in that dialect. Finally, as to the second Gospel (*Nineteenth Century*, p. 365):—

He [Mark] is full of minute observations, proceeding, beyond doubt, from an eye-witness. There is nothing to conflict with the supposition that this eye-witness . . . was the Apostle Peter himself, as Papias has it.

Let us consider this citation by the light of *Les Évangiles* :—

This work, although composed after the death of Peter, was, in a sense, the work of Peter; it represents the way in which Peter was accustomed to relate the life of Jesus (p. 116).

M. Renan goes on to say that, as an historical document, the Gospel of Mark has a great superiority (p. 116); but Mark has a motive for omitting the discourses, and he attaches a "puerile importance" to miracles (p. 117). The Gospel of Mark is less a legend than a biography written with credulity (p. 118). It would be rash to say that Mark has not been interpolated and retouched (p. 120).

If any one thinks that I have not been warranted in drawing a sharp distinction between "scientific theologians" and "counsels for creeds"; or that my warning against the too ready acceptance of certain declarations as to the state of biblical criticism was needless; or that my anxiety as to the sense of the word "practical" was superfluous; let him compare the statement that M. Renan has made a "practical surrender of the adverse case" with the facts just set forth. For what is the adverse case? The question, as Dr. Wace puts it, is, "It may be asked how far can we rely on the accounts we possess of our Lord's teaching on these subjects." It will be obvious that M. Renan's statements amount to an adverse answer —to a "practical" denial that any great reliance can be placed on these accounts. He does not believe that Matthew, the apostle, wrote the first Gospel; he does not profess to know who is responsible for the collection of "logia," or how many of them are authentic; though he calls the second Gospel the

most historical, he points out that it is written
with credulity, and may have been interpolated
and retouched; and, as to the author, "quel
qu'il soit," of the third Gospel, who is to "rely
on the accounts" of a writer who deserves the
cavalier treatment which "Luke" meets with at M.
Renan's hands?

I repeat what I have already more than once said,
that the question of the age and the authorship of the
Gospels has not, in my judgment, the importance
which is so commonly assigned to it; for the simple
reason that the reports, even of eye-witnesses, would
not suffice to justify belief in a large and essential
part of their contents; on the contrary, these reports
would discredit the witnesses. The Gadarene miracle,
for example, is so extremely improbable, that the fact
of its being reported by three, even independent,
authorities could not justify belief in it unless we had
the clearest evidence as to their capacity as observers
and as interpreters of their observations. But it
is evident that the three authorities are not independent; that they have simply adopted a legend,
of which there were two versions; and instead of
their proving its truth, it suggests their superstitious
credulity: so that if "Matthew," "Mark," and
"Luke" are really responsible for the Gospels, it
is not the better for the Gadarene story, but the
worse for them.

A wonderful amount of controversial capital has
been made out of my assertion in the note to which I
have referred, as an *obiter dictum* of no consequence

to my argument, that if Renan's work [1] were non-extant, the main results of biblical criticism, as set forth in the works of Strauss, Baur, Reuss, and Volkmar, for example, would not be sensibly affected. I thought I had explained it satisfactorily already, but it seems that my explanation has only exhibited still more of my native perversity, so I ask for one more chance.

In the course of the historical development of any branch of science, what is universally observed is this : that the men who make epochs, and are the real architects of the fabric of exact knowledge, are those who introduce fruitful ideas or methods. As a rule, the man who does this pushes his idea, or his method, too far ; or, if he does not, his school is sure to do so, and those who follow have to reduce his work to its proper value, and assign it its place in the whole. Not unfrequently they, in their turn, overdo the critical process, and, in trying to eliminate error, throw away truth.

Thus, as I said, Linnæus, Buffon, Cuvier, Lamarck, really "set forth the results" of a developing science, although they often heartily contradict one another. Notwithstanding this circumstance, modern classificatory method and nomenclature have largely grown out of the work of Linnæus ; the modern conception of biology, as a science, and of its relation to climatology, geography, and geology, are as largely rooted in the results of the labours of Buffon ; comparative

---

[1] I trust it may not be supposed that I undervalue M. Renan's labours, or intended to speak slightingly of them.

anatomy and palæontology owe a vast debt to Cuvier's results; while invertebrate zoology and the revival of the idea of evolution are intimately dependent on the results of the work of Lamarck. In other words, the main results of biology up to the early years of this century are to be found in, or spring out of, the works of these men.

So, if I mistake not, Strauss, if he did not originate the idea of taking the mythopœic faculty into account in the development of the Gospel narratives, and though he may have exaggerated the influence of that faculty, obliged scientific theology hereafter to take that element into serious consideration; so Baur, in giving prominence to the cardinal fact of the divergence of the Nazarene and Pauline tendencies in the primitive Church; so Reuss, in setting a marvellous example of the cool and dispassionate application of the principles of scientific criticism over the whole field of Scripture; so Volkmar, in his clear and forcible statement of the Nazarene limitations of Jesus, contributed results of permanent value in scientific theology. I took these names as they occurred to me. Undoubtedly, I might have advantageously added to them; perhaps I might have made a better selection. But it really is absurd to try to make out that I did not know that these writers widely disagree; and I believe that no scientific theologian will deny that, in principle, what I have said is perfectly correct. Ecclesiastical advocates, of course, cannot be expected to take this view of the matter. To them, these mere seekers after truth, in so far as their results are unfavourable to the creed the clerics have to support, are more or less "infidels," or favourers of "infidelity"; and the only thing they care to see, or probably can see, is the fact that, in a great many matters, the truth-seekers differ from one another, and therefore can easily be exhibited to the public, as if they did nothing else; as if any one who referred to their having, each and all, contributed his share to the results of theological science, was merely showing his ignorance; and as if a charge of inconsistency could be based on the fact that he himself often disagrees with what they say. I have never lent a shadow of foundation to the assumption that I am a follower of either Strauss, or Baur, or Reuss, or Volkmar, or Renan; my debt to these eminent men —so far my superiors in theological knowledge—is, indeed, great; yet it is not for their opinions, but for those I have been able to form for myself, by their help.

In *Agnosticism: a Rejoinder* (p. 410), I have referred to the difficulties under which those professors of the science of theology, whose tenure of their posts depends on the results of their investigations, must labour; and, in a note, I add—

Imagine that all our chairs of Astronomy had been founded in the fourteenth century, and that their incumbents were bound to sign Ptolemaic articles. In that case, with every respect for the efforts of persons thus hampered to attain and expound the truth, I think men of common sense would go elsewhere to learn astronomy.

I did not write this paragraph without a know-

ledge that its sense would be open to the kind of
perversion which it has suffered; but, if that was
clear, the necessity for the statement was still
clearer. It is my deliberate opinion : I reiterate it ;
and I say that, in my judgment, it is extremely
inexpedient that any subject which calls itself a
science should be entrusted to teachers who are
debarred from freely following out scientific methods
to their legitimate conclusions, whatever those conclusions may be. If I may borrow a phrase paraded
at the Church Congress, I think it " ought to be
unpleasant " for any man of science to find himself
in the position of such a teacher.

Human nature is not altered by seating it in a
professorial chair, even of theology. I have very
little doubt that if, in the year 1859, the tenure of
my office had depended upon my adherence to the
doctrines of Cuvier, the objections to those set forth
in the *Origin of Species* would have had a halo of
gravity about them that, being free to teach what I
pleased, I failed to discover. And, in making that
statement, it does not appear to me that I am con-
fessing that I should have been debarred by " selfish
interests " from making candid inquiry, or that I
should have been biassed by " sordid motives." I
hope that even such a fragment of moral sense as
may remain in an ecclesiastical " infidel " might have
got me through the difficulty ; but it would be
unworthy to deny or disguise the fact that a very
serious difficulty must have been created for me by
the nature of my tenure. And let it be observed that

the temptation, in my case, would have been far slighter than in that of a professor of theology ; whatever biological doctrine I had repudiated, nobody I cared for would have thought the worse of me for so doing. No scientific journals would have howled me down, as the religious newspapers howled down my too honest friend, the late Bishop of Natal ; nor would my colleagues of the Royal Society have turned their backs upon me, as his episcopal colleagues boycotted him.

I say these facts are obvious, and that it is wholesome and needful that they should be stated. It is in the interests of theology, if it be a science, and it is in the interests of those teachers of theology who desire to be something better than counsel for creeds, that it should be taken to heart. The seeker after theological truth and that only, will no more suppose that I have insulted him, than the prisoner who works in fetters will try to pick a quarrel with me, if I suggest that he would get on better if the fetters were knocked off ; unless indeed, as it is said does happen in the course of long captivities, that the victim at length ceases to feel the weight of his chains, or even takes to hugging them, as if they were honourable ornaments.[1]

[1] To-day's *Times* contains a report of a remarkable speech by Prince Bismarck, in which he tells the Reichstag that he has long given up investing in foreign stock, lest so doing should mislead his judgment in his transactions with foreign states. Does this declaration prove that the Chancellor accuses himself of being "sordid" and "selfish," or does it not rather show that, even in dealing with himself, he remains the man of realities ?

XIII

THE LIGHTS OF THE CHURCH AND THE LIGHT OF SCIENCE

THERE are three ways of regarding any account of past occurrences, whether delivered to us orally or recorded in writing.

The narrative may be exactly true. That is to say, the words, taken in their natural sense, and interpreted according to the rules of grammar, may convey to the mind of the hearer, or of the reader, an idea precisely correspondent with one which would have remained in the mind of a witness. For example, the statement that King Charles the First was beheaded at Whitehall on the 30th day of January 1649, is as exactly true as any proposition in mathematics or physics; no one doubts that any person of sound faculties, properly placed, who was present at Whitehall throughout that day, and who used his eyes, would have seen the King's head cut off; and that there would have remained in his mind an idea of that occurrence which he would have put into words of the same value as those which we use to express it.

Or the narrative may be partly true and partly false. Thus, some histories of the time tell us what the King said, and what Bishop Juxon said; or report royalist conspiracies to effect a rescue; or detail the motives which induced the chiefs of the Commonwealth to resolve that the King should die. One account declares that the King knelt at a high block, another that he lay down with his neck on a mere plank. And there are contemporary pictorial representations of both these modes of procedure. Such narratives, while veracious as to the main event, may and do exhibit various degrees of unconscious and conscious misrepresentation, suppression, and invention, till they become hardly distinguishable from pure fictions. Thus, they present a transition to narratives of a third class, in which the fictitious element predominates. Here, again, there are all imaginable gradations, from such works as Defoe's quasi-historical account of the Plague year, which probably gives a truer conception of that dreadful time than any authentic history, through the historical novel, drama, and epic, to the purely phantasmal creations of imaginative genius, such as the old *Arabian Nights*, or the modern *Shaving of Shagpat*. It is not strictly needful for my present purpose that I should say anything about narratives which are professedly fictitious. Yet it may be well, perhaps, if I disclaim any intention of derogating from their value, when I insist upon the paramount necessity of recollecting that there is no sort of relation between the ethical, or the æsthetic, or even the scientific importance of such works, and their worth as historical documents. Unquestionably, to the poetic artist, or even to the student of psychology, *Hamlet* and *Macbeth* may be better instructors than all the books of a wilderness of professors of æsthetics or of moral philosophy. But, as evidence of occurrences in Denmark, or in Scotland, at the times and places indicated, they are out of court; the profoundest admiration for them, the deepest gratitude for their influence, are consistent with the knowledge that, historically speaking, they are worthless fables, in which any foundation of reality that may exist is submerged beneath the imaginative superstructure.

At present, however, I am not concerned to dwell upon the importance of fictitious literature and the immensity of the work which it has effected in the education of the human race. I propose to deal with the much more limited inquiry: Are there two other classes of consecutive narratives (as distinct from statements of individual facts), or only one? Is there any known historical work which is throughout exactly true, or is there not? In the case of the great majority of histories the answer is not doubtful: they are all only partially true. Even those venerable works which bear the names of some of the greatest of ancient Greek and Roman writers, and which have been accepted by generation after generation, down to modern times, as stores of unquestionable truth, have been compelled by scientific criticism, after a long battle, to descend to the common level, and to confess to a large admixture of error. I might

fairly take this for granted; but it may be well that
I should entrench myself behind the very apposite
words of a historical authority who is certainly not
obnoxious to even a suspicion of sceptical tendencies.

> Time was—and that not very long ago—when all the rela-
> tions of ancient authors concerning the old world were received
> with a ready belief; and an unreasoning and uncritical faith
> accepted with equal satisfaction the narrative of the campaigns
> of Cæsar and of the doings of Romulus, the account of Alex-
> ander's marches and of the conquests of Semiramis. We can
> most of us remember when, in this country, the whole story of
> regal Rome, and even the legend of the Trojan settlement in
> Latium, were seriously placed before boys as history, and dis-
> coursed of as unhesitatingly and in as dogmatic a tone as the
> tale of the Catiline Conspiracy or the Conquest of Britain. . . .
>   But all this is now changed. The last century has seen the
> birth and growth of a new science—the Science of Historical
> Criticism. . . . The whole world of profane history has been
> revolutionised. . . .[1]

If these utterances were true when they fell from
the lips of a Bampton lecturer in 1859, with how
much greater force do they appeal to us now, when
the immense labours of the generation now passing
away constitute one vast illustration of the power
and fruitfulness of scientific methods of investigation in history, no less than in all other departments
of knowledge.

At the present time, I suppose, there is no one
who doubts that histories which appertain to any

---

[1] *Bampton Lectures* (1859), on "The Historical Evidences of the
Truth of the Scripture Records stated anew, with Special Reference to
the Doubts and Discoveries of Modern Times," by the Rev. G. Rawlinson, M.A., pp. 5-6.

other people than the Jews, and their spiritual
progeny in the first century, fall within the second
class of the three enumerated. Like Goethe's
Autobiography, they might all be entitled " Wahrheit
und Dichtung "—"Truth and Fiction." The proportion of the two constituents changes indefinitely ;
and the quality of the fiction varies through the
whole gamut of unveracity. But " Dichtung " is
always there. For the most acute and learned of
historians cannot remedy the imperfections of his
sources of information ; nor can the most impartial wholly escape the influence of the " personal
equation " generated by his temperament and by
his education. Therefore, from the narratives of
Herodotus to those set forth in yesterday's *Times*,
all history is to be read subject to the warning that
fiction has its share therein. The modern vast
development of fugitive literature cannot be the
unmitigated evil that some do vainly say it is,
since it has put an end to the popular delusion of
less press-ridden times, that what appears in print
must be true. We should rather hope that some
beneficent influence may create among the erudite
a like healthy suspicion of manuscripts and inscriptions, however ancient ; for a bulletin may
lie, even though it be written in cuneiform characters. Hotspur's starling, that was to be taught
to speak nothing but " Mortimer " into the ears of
King Henry the Fourth, might be a useful inmate
of every historian's library, if " Fiction " were substituted for the name of Harry Percy's friend.

But it was the chief object of the lecturer to the congregation gathered in St. Mary's, Oxford, thirty-one years ago, to prove to them, by evidence gathered with no little labour and marshalled with much skill, that one group of historical works was exempt from the general rule; and that the narratives contained in the canonical Scriptures are free from any admixture of error. With justice and candour, the lecturer impresses upon his hearers that the special distinction of Christianity, among the religions of the world, lies in its claim to be historical; to be surely founded upon events which have happened, exactly as they are declared to have happened in its sacred books; which are true, that is, in the sense that the statement about the execution of Charles the First is true. Further, it is affirmed that the New Testament presupposes the historical exactness of the Old Testament; that the points of contact of " sacred " and " profane " history are innumerable; and that the demonstration of the falsity of the Hebrew records, especially in regard to those narratives which are assumed to be true in the New Testament, would be fatal to Christian theology.

My utmost ingenuity does not enable me to discover a flaw in the argument thus briefly summarised. I am fairly at a loss to comprehend how any one, for a moment, can doubt that Christian theology must stand or fall with the historical trustworthiness of the Jewish Scriptures. The very conception of the Messiah, or Christ, is inextricably interwoven with Jewish history; the identification

XIII LIGHTS OF THE CHURCH AND LIGHT OF SCIENCE 507

of Jesus of Nazareth with that Messiah rests upon the interpretation of passages of the Hebrew Scriptures which have no evidential value unless they possess the historical character assigned to them. If the covenant with Abraham was not made ; if circumcision and sacrifices were not ordained by Jahveh ; if the " ten words " were not written by God's hand on the stone tables ; if Abraham is more or less a mythical hero, such as Theseus ; the story of the Deluge a fiction ; that of the Fall a legend.; and that of the Creation the dream of a seer ; if all these definite and detailed narratives of apparently real events have no more value as history than have the stories of the regal period of Rome—what is to be said about the Messianic doctrine, which is so much less clearly enunciated ?   And what about the authority of the writers of the books of the New Testament, who, on this theory, have not merely accepted flimsy fictions for solid truths, but have built the very foundations of Christian dogma upon legendary quicksands ?

But these may be said to be merely the carpings of that carnal reason which the profane call common sense ; I hasten, therefore, to bring up the forces of unimpeachable ecclesiastical authority in support of my position.   In a sermon preached last December, in St. Paul's Cathedral,[1] Canon Liddon declares :—

---

[1] *The Worth of the Old Testament*, a Sermon preached in St. Paul's Cathedral on the Second Sunday in Advent, 8th Dec. 1889, by H. P. Liddon, D.D., D.C.L., Canon and Chancellor of St. Paul's.   Second edition, revised and with a new preface, 1890.

For Christians it will be enough to know that our Lord Jesus Christ set the seal of His infallible sanction on the whole of the Old Testament.   He found the Hebrew Canon as we have it in our hands to-day, and he treated it as an authority which was above discussion.   Nay more: He went out of His way—if we may reverently speak thus—to sanction not a few portions of it which modern scepticism rejects.   When he would warn His hearers against the dangers of spiritual relapse, He bids them remember "Lot's wife."[1]   When He would point out how worldly engagements may blind the soul to a coming judgment, He reminds them how men ate, and drank, and married, and were given in marriage, until the day that Noah entered into the ark, and the Flood came and destroyed them all.[2]   If He would put His finger on a fact in past Jewish history which, by its admitted reality, would warrant belief in His own coming Resurrection, He points to Jonah's being three days and three nights in the whale's belly (p. 23).[3]

The preacher proceeds to brush aside the common —I had almost said vulgar—apologetic pretext that Jesus was using *ad hominem* arguments, or " accommodating " his better knowledge to popular ignorance, as well as to point out the inadmissibility of the other alternative, that he shared the popular ignorance. And to those who hold the latter view sarcasm is dealt out with no niggard hand.

But they will find it difficult to persuade mankind that, if He could be mistaken on a matter of such strictly religious importance as the value of the sacred literature of His countrymen, He can be safely trusted about anything else.   The trustworthiness of the Old Testament is, in fact, inseparable from the trustworthiness of our Lord Jesus Christ; and if we believe that He is the true Light of the world, we shall close our ears against suggestions impairing the credit of those Jewish Scriptures which have received the stamp of His Divine authority (p. 25).

---

[1] St. Luke xvii. 32.   [2] *Ibid.* 27.   [3] St. Matt. xii. 40.

Moreover, I learn from the public journals that a brilliant and sharply-cut view of orthodoxy, of like hue and pattern, was only the other day exhibited in that great theological kaleidoscope, the pulpit of St. Mary's, recalling the time so long past by, when a Bampton lecturer, in the same place, performed the unusual feat of leaving the faith of old-fashioned Christians undisturbed.

Yet many things have happened in the intervening thirty-one years. The Bampton lecturer of 1859 had to grapple only with the infant Hercules of historical criticism ; and he is now a full-grown athlete, bearing on his shoulders the spoils of all the lions that have stood in his path. Surely a martyr's courage, as well as a martyr's faith, is needed by any one who, at this time, is prepared to stand by the following plea for the veracity of the Pentateuch :—

Adam, according to the Hebrew original, was for 243 years contemporary with Methuselah, who conversed for a hundred years with Shem. Shem was for fifty years contemporary with Jacob, who probably saw Jochebed, Moses's mother. Thus, Moses might by oral tradition have obtained the history of Abraham, and even of the Deluge, at third hand ; and that of the Temptation and the Fall at fifth hand. . . .

If it be granted—as it seems to be—that the great and stirring events in a nation's life will, under ordinary circumstances, be remembered (apart from all written memorials) for the space of 150 years, being handed down through five generations, it must be allowed (even on mere human grounds) that the account which Moses gives of the Temptation and the Fall is to be depended upon, if it passed through no more than four hands between him and Adam.[1]

---

[1] *Bampton Lectures*, 1859, pp. 50-51.

If "the trustworthiness of our Lord Jesus Christ" is to stand or fall with the belief in the sudden transmutation of the chemical components of a woman's body into sodium chloride, or on the "admitted reality" of Jonah's ejection, safe and sound, on the shores of the Levant, after three day's sea-journey in the stomach of a gigantic marine animal, what possible pretext can there be for even hinting a doubt as to the precise truth of the longevity attributed to the Patriarchs? Who that has swallowed the camel of Jonah's journey will be guilty of the affectation of straining at such a historical gnat—nay midge—as the supposition that the mother of Moses was told the story of the Flood by Jacob; who had it straight from Shem; who was on friendly terms with Methuselah; who knew Adam quite well?

Yet, by the strange irony of things, the illustrious brother of the divine who propounded this remarkable theory, has been the guide and foremost worker of that band of investigators of the records of Assyria and of Babylonia, who have opened to our view, not merely a new chapter, but a new volume of primeval history, relating to the very people who have the most numerous points of contact with the life of the ancient Hebrews. Now, whatever imperfections may yet obscure the full value of the Mesopotamian records, everything that has been clearly ascertained tends to the conclusion that the assignment of no more than 4000 years to the period between the time of the origin of mankind and that of Augustus Cæsar, is wholly inadmissible. Therefore the Biblical chronology, which Canon Rawlinson trusted so implicitly in 1859, is relegated by all serious critics to the domain of fable.

But if scientific method, operating in the region of history, of philology, of archæology, in the course of the last thirty or forty years, has become thus formidable to the theological dogmatist, what may not be said about scientific method working in the province of physical science? For, if it be true that the Canonical Scriptures have innumerable points of contact with civil history, it is no less true that they have almost as many with natural history; and their accuracy is put to the test as severely by the latter as by the former. The origin of the present state of the heavens and the earth is a problem which lies strictly within the province of physical science; so is that of the origin of man among living things; so is that of the physical changes which the earth has undergone since the origin of man; so is that of the origin of the various races and nations of men, with all their varieties of language and physical conformation. Whether the earth moves round the sun or the contrary; whether the bodily and mental diseases of men and animals are caused by evil spirits or not; whether there is such an agency as witchcraft or not—all these are purely scientific questions; and to all of them the canonical Scriptures profess to give true answers. And though nothing is more common than the assumption that these books come into conflict only with the speculative part of modern physical science, no assumption can have less foundation.

The antagonism between natural knowledge and the Pentateuch would be as great if the speculations of our time had never been heard of. It arises out of contradiction upon matters of fact. The books of ecclesiastical authority declare that certain events happened in a certain fashion; the books of scientific authority say they did not. As it seems that this unquestionable truth has not yet penetrated among many of those who speak and write on these subjects, it may be useful to give a full illustration of it. And for that purpose I propose to deal, at some length, with the narrative of the Noachian Deluge given in Genesis.

The Bampton lecturer in 1859, and the Canon of St. Paul's in 1890, are in full agreement that this history is true, in the sense in which I have defined historical truth. The former is of opinion that the account attributed to Berosus records a tradition—

not drawn from the Hebrew record, much less the foundation of that record; yet coinciding with it in the most remarkable way. The Babylonian version is tricked out with a few extravagances, as the monstrous size of the vessel and the translation of Xisuthros; but otherwise it is the Hebrew history *down to its minutiæ* (p. 64).

Moreover, correcting Niebuhr, the Bampton lecturer points out that the narrative of Berosus implies the universality of the Flood.

It is plain that the waters are represented as prevailing above the tops of the loftiest mountains in Armenia—a height which must have been seen to involve the submersion of all the countries with which the Babylonians were acquainted (p. 66).

I may remark, in passing, that many people think the size of Noah's ark "monstrous," considering the probable state of the art of shipbuilding only 1600 years after the origin of man; while others are so unreasonable as to inquire why the translation of Enoch is less an "extravagance" than that of Xisuthros. It is more important, however, to note that the universality of the Deluge is recognised, not merely as a part of the story, but as a necessary consequence of some of its details. The latest exponent of Anglican orthodoxy, as we have seen, insists upon the accuracy of the Pentateuchal history of the Flood in a still more forcible manner. It is cited as one of those very narratives to which the authority of the Founder of Christianity is pledged, and upon the accuracy of which "the trustworthiness of our Lord Jesus Christ" is staked, just as others have staked it upon the truth of the histories of demoniac possession in the Gospels.

Now, when those who put their trust in scientific methods of ascertaining the truth in the province of natural history find themselves confronted and opposed, on their own ground, by ecclesiastical pretensions to better knowledge, it is, undoubtedly, most desirable for them to make sure that their conclusions, whatever they may be, are well founded. And, if they put aside the unauthorised interference with their business and relegate the Pentateuchal history to the region of pure fiction, they are bound to assure themselves that they do so because the plainest teachings of Nature (apart from all doubtful speculations) are irreconcilable with the assertions which they reject.

At the present time, it is difficult to persuade serious scientific inquirers to occupy themselves, in any way, with the Noachian Deluge. They look at you with a smile and a shrug, and say they have more important matters to attend to than mere antiquarianism. But it was not so in my youth. At that time, geologists and biologists could hardly follow to the end any path of inquiry without finding the way blocked by Noah and his ark, or by the first chapter of Genesis; and it was a serious matter, in this country at any rate, for a man to be suspected of doubting the literal truth of the Diluvial or any other Pentateuchal history. The fiftieth anniversary of the foundation of the Geological Club (in 1824), was, if I remember rightly, the last occasion on which the late Sir Charles Lyell spoke to even so small a public as the members of that body. Our veteran leader lighted up once more; and, referring to the difficulties which beset his early efforts to create a rational science of geology, spoke, with his wonted clearness and vigour, of the social ostracism which pursued him after the publication of the *Principles of Geology*, in 1830, on account of the obvious tendency of that noble work to discredit the Pentateuchal accounts of the Creation and the Deluge. If my younger contemporaries find this hard to believe, I may refer them to a grave book, *On the Doctrine of the Deluge*, published eight years later, and dedicated by its author to his father, the then Archbishop of York. The first chapter refers to the treatment of the " Mosaic Deluge," by Dr. Buckland and Mr. Lyell, in the following terms :

> Their respect for revealed religion has prevented them from arraying themselves openly against the Scriptural account of it —much less do they deny its truth—but they are in a great hurry to escape from the consideration of it, and evidently concur in the opinion of Linnæus, that no proofs whatever of the Deluge are to be discovered in the structure of the earth (p. 1).

And after an attempt to reply to some of Lyell's arguments, which it would be cruel to reproduce, the writer continues :—

> When, therefore, upon such slender grounds, it is determined, in answer to those who insist upon its universality, that the Mosaic Deluge must be considered a preternatural event, far beyond the reach of philosophical inquiry ; not only as to the causes employed to produce it, but as to the effects most likely to result from it ; that determination wears an aspect of scepticism, which, however much soever it may be unintentional in the mind of the writer, yet cannot but produce an evil impression on those who are already predisposed to carp and cavil at the evidences of Revelation (pp. 8-9).

The kindly and courteous writer of these curious passages is evidently unwilling to make the geologists the victims of general opprobrium by pressing the obvious consequences of their teaching home. One is therefore pained to think of the feelings with which, if he lived so long as to become acquainted with the *Dictionary of the Bible*, he must have perused the article " Noah," written by a dignitary of the Church for that standard compendium and published in 1863. For the doctrine of the universality of the Deluge is therein altogether given up ; and I

permit myself to hope that a long criticism of the
story from the point of view of natural science, with
which, at the request of the learned theologian who
wrote it, I supplied him, may, in some degree, have
contributed towards this happy result.

Notwithstanding diligent search, I have been unable to discover that the universality of the Deluge
has any defender left, at least among those who have
so far mastered the rudiments of natural knowledge
as to be able to appreciate the weight of evidence
against it.   For example, when I turned to the
Speaker's Bible, published under the sanction of high
Anglican authority, I found the following judicial
and judicious deliverance, the skilful wording of
which may adorn, but does not hide, the complete-
ness of the surrender of the old teaching :—

> Without pronouncing too hastily on any fair inferences
> from the words of Scripture, we may reasonably say that their
> most natural interpretation is, that the whole race of man had
> become grievously corrupted since the faithful had intermingled
> with the ungodly ; that the inhabited world was consequently
> filled with violence, and that God had decreed to destroy all
> mankind except one single family ; that, therefore, all that por-
> tion of the earth, perhaps as yet a very small portion, into which
> mankind had spread was overwhelmed with water.   The ark
> was ordained to save one faithful family ; and lest that family,
> on the subsidence of the waters, should find the whole country
> round them a desert, a pair of all the beasts of the land and of
> the fowls of the air were preserved along with them, and along
> with them went forth to replenish the now desolated continent.
> The words of Scripture (confirmed as they are by universal
> tradition) appear at least to mean as much as this.   They do
> not necessarily mean more.[1]

---

[1] *Commentary on Genesis*, by the Bishop of Ely, p. 77.

In the third edition of Kitto's *Cyclopædia of Biblical Literature* (1876), the article "Deluge," written by my friend, the present distinguished head of the Geological Survey of Great Britain, extinguishes the universality doctrine as thoroughly as might be expected from its authorship; and, since the writer of the article "Noah" refers his readers to that entitled "Deluge," it is to be supposed, notwithstanding his generally orthodox tone, that he does not dissent from its conclusions. Again, the writers in Herzog's *Real-Encyclopädie* (Bd. X. 1882) and in Riehm's *Handwörterbuch* (1884)—both works with a conservative leaning—are on the same side; and Diestel,[1] in his full discussion of the subject, remorselessly rejects the universality doctrine. Even that staunch opponent of scientific rationalism—may I say rationality—Zöckler,[2] flinches from a distinct defence of the thesis, any opposition to which, well within my recollection, was howled down by the orthodox as mere "infidelity." All that, in his sore straits, Dr. Zöckler is able to do, is to pronounce a faint commendation upon a particularly absurd attempt at reconciliation, which would make out the Noachian Deluge to be a catastrophe which occurred at the end of the Glacial Epoch. This hypothesis involves only the trifle of a physical revolution of which geology knows nothing; and which, if it secured the accuracy of the Pentateuchal writer about the fact of the Deluge, would leave the details of his

---

[1] *Die Sintflut*, 1876.
[2] *Theologie und Naturwissenschaft*, ii. 784-791 (1877).

account as irreconcilable with the truths of elementary physical science as ever. Thus I may be permitted to spare myself and my readers the weariness of a recapitulation of the overwhelming arguments against the universality of the Deluge, which they will now find for themselves stated, as fully and forcibly as could be wished, by Anglican and other theologians, whose orthodoxy and conservative tendencies have, hitherto, been above suspicion. Yet many fully admit (and, indeed, nothing can be plainer) that the Pentateuchal narrator means to convey that, as a matter of fact, the whole earth known to him was inundated; nor is it less obvious that, unless all mankind, with the exception of Noah and his family, were actually destroyed, the references to the Flood in the New Testament are unintelligible.

But I am quite aware that the strength of the demonstration that no universal Deluge ever took place has produced a change of front in the army of apologetic writers. They have imagined that the substitution of the adjective " partial " for " universal," will save the credit of the Pentateuch, and permit them, after all, without too many blushes, to declare that the progress of modern science only strengthens the authority of Moses. Nowhere have I found the case of the advocates of this method of escaping from the difficulties of the actual position better put than in the lecture of Professor Diestel to which I have referred. After frankly admitting that the old doctrine of universality involves physical impossibilities, he continues :—

All these difficulties fall away as soon as we give up the universality of the Deluge, and imagine a *partial* flooding of the earth, say in western Asia. But have we a right to do so? The narrative speaks of "the whole earth." But what is the meaning of this expression? Surely not the whole surface of the earth according to the ideas of *modern* geographers, but, at most, according to the conceptions of the Biblical author. This very simple conclusion, however, is never drawn by too many readers of the Bible. But one need only cast one's eyes over the tenth chapter of Genesis in order to become acquainted with the geographical horizon of the Jews. In the north it was bounded by the Black Sea and the mountains of Armenia; extended towards the east very little beyond the Tigris; hardly reached the apex of the Persian Gulf; passed, then, through the middle of Arabia and the Red Sea; went southward through Abyssinia, and then turned westward by the frontiers of Egypt, and inclosed the easternmost islands of the Mediterranean (p. 11).

The justice of this observation must be admitted, no less than the further remark that, in still earlier times, the pastoral Hebrews very probably had yet more restricted notions of what constituted the " whole earth." Moreover, I, for one, fully agree with Professor Diestel that the motive, or generative incident, of the whole story is to be sought in the occasionally excessive and desolating floods of the Euphrates and the Tigris.

Let us, provisionally, accept the theory of a partial deluge, and try to form a clear mental picture of the occurrence. Let us suppose that, for forty days and forty nights, such a vast quantity of water was poured upon the ground that the whole surface of Mesopotamia was covered by water to a depth certainly greater, probably much greater, than fifteen cubits, or twenty feet (Gen. vii. 20). The inundation prevails upon the earth for one hundred and fifty days ; and then the flood gradually decreases, until, on the seventeenth day of the seventh month, the ark, which had previously floated on its surface, grounds upon the " mountains of Ararat "[1] (Gen. viii. 34). Then, as Diestel has acutely pointed out (*Sintflut*, p. 13), we are to imagine the further subsidence of the flood to take place so gradually that it was not until nearly two months and a-half after this time (that is to say, on the first day of the tenth month) that the " tops of the mountains" became visible. Hence it follows that, if the ark drew even as much as twenty feet of water, the level of the inundation fell very slowly— at a rate of only a few inches a day—until the top of the mountain on which it rested became visible. This is an amount of movement which, if it took place in the sea, would be overlooked by ordinary people on the shore. But the Mesopotamian plain slopes gently, from an elevation of 500 or 600 feet at its northern end, to the sea, at its southern end, with hardly so much as a notable ridge to break its uniform flatness, for 300 to 400 miles. These being the conditions of the case, the following inquiry naturally presents itself : not, be it observed, as a recondite problem, generated by modern speculation, but as a plain suggestion flowing out of that very ordinary and archaic piece of knowledge that water cannot be piled up

---

[1] It is very doubtful if this means the region of the Armenian Ararat. More probably it designates some part either of the Kurdish range or of its south-eastern continuation.

in a heap, like sand ; or that it seeks the lowest level. When, after 150 days, " the fountains also of the deep and the windows of heaven were stopped, and the rain from heaven was restrained " (Gen. viii. 2), what pre- vented the mass of water, several, possibly very many, fathoms deep, which covered, say, the present site of Bagdad, from sweeping seaward in a furious torrent ; and, in a very few hours, leaving, not only the " tops of the mountains," but the whole plain, save any minor depressions, bare ? How could its subsidence, by any possibility, be an affair of weeks and months ?

And if this difficulty is not enough, let any one try to imagine how a mass of water several, perhaps very many, fathoms deep, could be accumulated on a flat surface of land rising well above the sea, and separated from it by no sort of barrier. Most people know Lord's Cricket-ground. Would it not be an absurd contradiction to our common knowledge of the properties of water to imagine that, if all the mains of all the waterworks of London were turned on to it, they could maintain a heap of water twenty feet deep over its level surface ? Is it not obvious that the water, whatever momentary accumulation might take place at first, would not stop there, but that it would dash, like a mighty mill-race, southwards down the gentle slope which ends in the Thames ? And is it not further obvious, that whatever depth of water might be maintained over the cricket-ground so long as all the mains poured on to it, anything which floated there would be speedily whirled away by the current, like a cork in a gutter when the rain pours ?

But if this is so, then it is no less certain that Noah's deeply laden, sailless, oarless, and rudderless craft, if by good fortune it escaped capsizing in whirlpools, or having its bottom knocked into holes by snags (like those which prove fatal even to well-built steamers on the Mississippi in our day), would have speedily found itself a good way down the Persian Gulf, and not long after in the Indian Ocean, somewhere between Arabia and Hindostan. Even if, eventually, the ark might have gone ashore, with other jetsam and flotsam, on the coasts of Arabia, or of Hindostan, or of the Maldives, or of Madagascar, its return to the "mountains of Ararat" would have been a miracle more stupendous than all the rest.

Thus, the last state of the would-be reconcilers of the story of the Deluge with fact is worse than the first. All that they have done is to transfer the contradictions to established truth from the region of science proper to that of common information and common sense. For, really, the assertion that the surface of a body of deep water, to which no addition was made, and which there was nothing to stop from running into the sea, sank at the rate of only a few inches or even feet a day, simply outrages the most ordinary and familiar teachings of every man's daily experience. A child may see the folly of it.

In addition, I may remark that the necessary assumption of the "partial Deluge" hypothesis (if it is confined to Mesopotamia) that the Hebrew writer must have meant low hills when he said "high mountains," is quite untenable. On the eastern side of the

Mesopotamian plain, the snowy peaks of the frontier ranges of Persia are visible from Bagdad,[1] and even the most ignorant herdsmen in the neighbourhood of " Ur of the Chaldees," near its western limit, could hardly have been unacquainted with the comparatively elevated plateau of the Syrian desert which lay close at hand. But, surely, we must suppose the Biblical writer to be acquainted with the highlands of Palestine and with the masses of the Sinaitic peninsula, which soar more than 8000 feet above the sea, if he knew of no higher elevations ; and, if so, he could not well have meant to refer to mere hillocks when he said that " all the high mountains which were under the whole heaven were covered " (Genesis vii. 19). Even the hill-country of Galilee reaches an elevation of 4000 feet ; and a flood which covered it could by no possibility have been other than universal in its superficial extent. Water really cannot be got to stand at, say, 4000 feet above the sea-level over Palestine, without covering the rest of the globe to the same height. Even if, in the course of Noah's six hundredth year, some prodigious convulsion had sunk the whole region inclosed within " the horizon of the geographical knowledge " of the Israelites by that much, and another had pushed it up again, just in time to catch the ark upon the " mountains of Ararat," matters are not much mended. I am afraid to think of what would have become of a vessel so little seaworthy as the ark and of its very numerous passengers,

---

[1] So Reclus (*Nouvelle Géographie Universelle*, ix. 386), but I find the statement doubted by an authority of the first rank.

under the peculiar obstacles to quiet flotation which such rapid movements of depression and upheaval would have generated.

Thus, in view, not, I repeat, of the recondite speculations of infidel philosophers, but in the face of the plainest and most commonplace of ascertained physical facts, the story of the Noachian Deluge has no more claim to credit than has that of Deucalion; and whether it was, or was not, suggested by the familiar acquaintance of its originators with the effects of unusually great overflows of the Tigris and Euphrates, it is utterly devoid of historical truth.

That is, in my judgment, the necessary result of the application of criticism, based upon assured physical knowledge, to the story of the Deluge. And it is satisfactory that the criticism which is based, not upon literary and historical speculations, but upon well-ascertained facts in the departments of literature and history, tends to exactly the same conclusion.

For I find this much agreed upon by all Biblical scholars of repute, that the story of the Deluge in Genesis is separable into at least two sets of statements; and that, when the statements thus separated are recombined in their proper order, each set furnishes an account of the event, coherent and complete within itself, but in some respects discordant with that afforded by the other set. This fact, as I understand, is not disputed. Whether one of these is the work of an Elohist, and the other of a Jehovist narrator; whether the two have been pieced together in this

strange fashion because, in the estimation of the compilers and editors of the Pentateuch, they had equal
and independent authority, or not; or whether there
is some other way of accounting for it—are questions
the answers to which do not affect the fact. If possible I avoid *à priori* arguments. But still, I think
it may be urged, without imprudence, that a narrative
having this structure is hardly such as might be
expected from a writer possessed of full and infallibly accurate knowledge. Once more, it would seem
that it is not necessarily the mere inclination of the
sceptical spirit to question everything, or the wilful
blindness of infidels, which prompts grave doubts as
to the value of a narrative thus curiously unlike the
ordinary run of veracious histories.

But the voice of archæological and historical criticism still has to be heard; and it gives forth no
uncertain sound. The marvellous recovery of the
records of an antiquity, far superior to any that can
be ascribed to the Pentateuch, which has been effected
by the decipherers of cuneiform characters, has put
us in possession of a series, once more, not of speculations, but of facts, which have a most remarkable
bearing upon the question of the trustworthiness of
the narrative of the Flood. It is established, that for
centuries before the asserted migration of Terah from
Ur of the Chaldees (which, according to the orthodox
interpreters of the Pentateuch, took place after the
year 2000 B.C.) Lower Mesopotamia was the seat of
a civilisation in which art and science and literature
had attained a development formerly unsuspected, or,

if there were faint reports of it, treated as fabulous. And it is also no matter of speculation, but a fact, that the libraries of these people contain versions of a long epic poem, one of the twelve books of which tells a story of a deluge, which, in a number of its leading features, corresponds with the story attributed to Berosus, no less than with the story given in Genesis, with curious exactness.   Thus, the correctness of Canon Rawlinson's conclusion, cited above, that the story of Berosus was neither drawn from the Hebrew record, nor is the foundation of it, can hardly be questioned.   It is highly probable, if not certain, that Berosus relied upon one of the versions (for there seem to have been several) of the old Babylonian epos, extant in his time ; and, if that is a reasonable conclusion, why is it unreasonable to believe that the two stories, which the Hebrew compiler has put together in such an inartistic fashion, were ultimately derived from the same source ?   I say ultimately, because it does not at all follow that the two versions, possibly trimmed by the Jehovistic writer on the one hand, and by the Elohistic on the other, to suit Hebrew requirements, may not have been current among the Israelites for ages.   And they may have acquired great authority before they were combined in the Pentateuch.

Looking at the convergence of all these lines of evidence to the one conclusion—that the story of the Flood in Genesis is merely a Bowdlerised version of one of the oldest pieces of purely fictitious literature extant; that whether this is, or is not, its origin, the events asserted in it to have taken place assuredly never did take place; further, that, in point of fact, the story, in the plain and logically necessary sense of its words, has long since been given up by orthodox and conservative commentators of the Established Church—I can but admire the courage and clear fore-sight of the Anglican divine who tells us that we must be prepared to choose between the trustworthiness of scientific method and the trustworthiness of that which the Church declares to be Divine authority. For, to my mind, this declaration of war to the knife against secular science, even in its most elementary form; this rejection without a moment's hesitation of any and all evidence which conflicts with theological dogma —is the only position which is logically reconcilable with the axioms of orthodoxy. If the Gospels truly report that which an incarnation of the God of Truth communicated to the world, then it surely is absurd to attend to any other evidence touching matters about which he made any clear statement, or the truth of which is distinctly implied by his words. If the exact historical truth of the Gospels is an axiom of Christianity, it is as just and right for a Christian to say, Let us "close our ears against suggestions" of scientific critics, as it is for the man of science to refuse to waste his time upon circle-squarers and flat-earth fanatics.

It is commonly reported that the manifesto by which the Canon of St. Paul's proclaims that he nails the colours of the straitest Biblical infallibility to the mast of the ship ecclesiastical, was put forth as a

counterblast to *Lux Mundi*; and that the passages
which I have more particularly quoted are directed
against the essay on "The Holy Spirit and Inspiration" in that collection of treatises by Anglican
divines of high standing, who must assuredly be
acquitted of conscious "infidel" proclivities. I fancy
that rumour must, for once, be right, for it is impossible to imagine a more direct and diametrical contradiction than that between the passages from the
sermon cited above and those which follow :—

What is questioned is that our Lord's words foreclose certain
critical positions as to the character of Old Testament literature.
For example, does His use of Jonah's resurrection as a *type* of
His own, depend in any real degree upon whether it is historical
fact or allegory ? . . . Once more, our Lord uses the time before
the Flood, to illustrate the carelessness of men before His own
coming. . . . In referring to the Flood He certainly suggests
that He is treating it as typical, for He introduces circumstances
—"eating and drinking, marrying and giving in marriage"—
which have no counterpart in the original narrative (p. 358-9).

While insisting on the flow of inspiration through
the whole of the Old Testament, the essayist does not
admit its universality. Here, also, the new apologetic
demands a partial flood :

But does the inspiration of the recorder guarantee the exact
historical truth of what he records ? And, in matter of fact,
can the record, with due regard to legitimate historical criticism,
be pronounced true ? Now, to the latter of these two questions
(and they are quite distinct questions) we may reply that there
is nothing to prevent our believing, as our faith strongly disposes us to believe, that the record from Abraham downward is,
in substance, in the strict sense historical (p. 351).

It would appear, therefore, that there is nothing

to prevent our believing that the record, from
Abraham upward, consists of stories in the strict
sense unhistorical, and that the pre-Abrahamic
narratives are mere moral and religious "types" and
parables.

I confess I soon lose my way when I try to follow
those who walk delicately among "types" and alle-
gories. A certain passion for clearness forces me to
ask, bluntly, whether the writer means to say that
Jesus did not believe the stories in question, or that
he did? When Jesus spoke, as of a matter of fact,
that "the Flood came and destroyed them all," did
he believe that the Deluge really took place, or not?
It seems to me that, as the narrative mentions Noah's
wife, and his sons' wives, there is good scriptural
warranty for the statement that the antediluvians
married and were given in marriage; and I should
have thought that their eating and drinking might be
assumed by the firmest believer in the literal truth
of the story. Moreover, I venture to ask what sort
of value, as an illustration of God's methods of dealing
with sin, has an account of an event that never
happened? If no Flood swept the careless people
away, how is the warning of more worth than the cry
of "Wolf" when there is no wolf? If Jonah's three
days' residence in the whale is not an "admitted
reality," how could it "warrant belief" in the "coming resurrection?" If Lot's wife was not turned into
a pillar of salt, the bidding those who turn back from
the narrow path to "remember" it is, morally, about
on a level with telling a naughty child that a bogy is

coming to fetch it away. Suppose that a Conservative orator warns his hearers to beware of great political and social changes, lest they end, as in France, in the domination of a Robespierre; what becomes, not only of his argument, but of his veracity, if he, personally, does not believe that Robespierre existed and did the deeds attributed to him?

Like all other attempts to reconcile the results of scientifically-conducted investigation with the demands of the outworn creeds of ecclesiasticism, the essay on Inspiration is just such a failure as must await mediation, when the mediator is unable properly to appreciate the weight of the evidence for the case of one of the two parties. The question of "Inspiration" really possesses no interest for those who have cast ecclesiasticism and all its works aside, and have no faith in any source of truth save that which is reached by the patient application of scientific methods. Theories of inspiration are speculations as to the means by which the authors of statements, in the Bible or elsewhere, have been led to say what they have said—and it assumes that natural agencies are insufficient for the purpose. I prefer to stop short of this problem, finding it more profitable to undertake the inquiry which naturally precedes it— namely, Are these statements true or false? If they are true, it may be worth while to go into the question of their supernatural generation; if they are false, it certainly is not worth mine.

Now, not only do I hold it to be proven that the story of the Deluge is a pure fiction; but I have no hesitation in affirming the same thing of the story of the Creation.[1] Between these two lies the story of the creation of man and woman and their fall from primitive innocence, which is even more monstrously improbable than either of the other two, though, from the nature of the case, it is not so easily capable of direct refutation. It can be demonstrated that the earth took longer than six days in the making, and that the Deluge, as described, is a physical impossibility; but there is no proving, especially to those who are perfect in the art of closing their ears to that which they do not wish to hear, that a snake did not speak, or that Eve was not made out of one of Adam's ribs.

The compiler of Genesis, in its present form, evidently had a definite plan in his mind. His countrymen, like all other men, were doubtless curious to know how the world began; how men, and especially wicked men, came into being, and how existing nations and races arose among the descendants of one stock; and, finally, what was the history of their own particular tribe. They, like ourselves, desired to solve the four great problems of cosmogeny, anthropogeny, ethnogeny, and geneogeny. The Pentateuch

---

[1] So far as I know, the narrative of the Creation is not now held to be true, in the sense in which I have defined historical truth, by any of the reconcilers. As for the attempts to stretch the Pentateuchal days into periods of thousands or millions of years, the verdict of the eminent biblical scholar, Dr. Riehm (*Der biblische Schöpfungsbericht*, 1881, pp. 15, 16), on such pranks of "Auslegungskunst" should be final. Why do the reconcilers take Goethe's advice seriously?—

"Im Auslegen seyd frisch und munter!
Legt ihr's nicht aus, so legt was unter."

furnishes the solutions which appeared satisfactory to its author. One of these, as we have seen, was borrowed from a Babylonian fable; and I know of no reason to suspect any different origin for the rest. Now, I would ask, is the story of the fabrication of Eve to be regarded as one of those pre-Abrahamic narratives, the historical truth of which is an open question, in face of the reference to it in a speech unhappily famous for the legal oppression to which it has been wrongfully forced to lend itself?

Have ye not read, that he which made them from the beginning made them male and female, and said, For this cause shall a man leave his father and mother, and cleave to his wife; and the twain shall become one flesh? (Matt. xix. 5).

If divine authority is not here claimed for the twenty-fourth verse of the second chapter of Genesis, what is the value of language? And again, I ask, if one may play fast and loose with the story of the Fall as a "type" or "allegory," what becomes of the foundation of Pauline theology?—

For since by man came death, by man came also the resurrection of the dead. For as in Adam all die, so also in Christ shall all be made alive (1 Corinthians xv. 21, 22).

If Adam may be held to be no more real a personage than Prometheus, and if the story of the Fall is merely an instructive "type," comparable to the profound Promethean mythus, what value has Paul's dialectic?

While, therefore, every right-minded man must sympathise with the efforts of those theologians, who have not been able altogether to close their ears to the still, small voice of reason, to escape from the fetters which ecclesiasticism has forged, the melancholy fact remains, that the position they have taken up is hopelessly untenable. It is raked alike by the old-fashioned artillery of the Churches and by the fatal weapons of precision with which the *enfants perdus* of the advancing forces of science are armed. They must surrender, or fall back into a more sheltered position. And it is possible that they may long find safety in such retreat.

It is, indeed, probable that the proportional number of those who will distinctly profess their belief in the transubstantiation of Lot's wife, and the anticipatory experience of submarine navigation by Jonah; in water standing fathoms deep on the side of a declivity without anything to hold it up; and in devils who enter swine—will not increase. But neither is there ground for much hope that the proportion of those who cast aside these fictions and adopt the consequence of that repudiation, are, for some generations, likely to constitute a majority. Our age is a day of compromises. The present and the near future seem given over to those happily, if curiously, constituted people who see as little difficulty in throwing aside any amount of post-Abrahamic Scriptural narrative, as the authors of *Lux Mundi* see in sacrificing the pre-Abrahamic stories; and, having distilled away every inconvenient matter of fact in Christian history, continue to pay divine honours to the residue. There really seems to be no reason why the next generation

should not listen to a Bampton Lecture modelled
upon that addressed to the last :—

Time was—and that not very long ago—when all the relations of Biblical authors concerning the old world were received
with a ready belief; and an unreasoning and uncritical faith
accepted with equal satisfaction the narrative of the Captivity
and the doings of Moses at the court of Pharaoh, the account of
the Apostolic meeting in the Epistle to the Galatians, and that
of the fabrication of Eve.   We can most of us remember when,
in this country, the whole story of the Exodus, and even the
legend of Jonah, were seriously placed before boys as history,
and discoursed of in as dogmatic a tone as the tale of Agincourt
or the history of the Norman Conquest.

But all this is now changed.   The last century has seen the
growth of scientific criticism to its full strength.   The whole
world of history has been revolutionised and the mythology
which embarrassed earnest Christians has vanished as an evil
mist, the lifting of which has only more fully revealed the
lineaments of infallible Truth.   No longer in contact with fact
of any kind, Faith stands now and for ever proudly inaccessible
to the attacks of the infidel.

So far the apologist of the future.   Why not?
Cantabit vacuus.

XIV

THE KEEPERS OF THE HERD OF SWINE

I HAD fondly hoped that Mr. Gladstone and I had come to an end of disputation, and that the hatchet of war was finally superseded by the calumet, which, as Mr. Gladstone, I believe, objects to tobacco, I was quite willing to smoke for both. But I have had, once again, to discover that the adage that whoso seeks peace will ensue it, is a somewhat hasty generalisation. The renowned warrior with whom it is my misfortune to be opposed in most things has dug up the axe and is on the war-path once more. The weapon has been wielded with all the dexterity which long practice has conferred on a past master in craft, whether of wood or state. And I have reason to believe that the simpler sort of the great tribe which he heads imagine that my scalp is already on its way to adorn their big chief's wigwam. I am glad therefore to be able to relieve any anxieties which my friends may entertain without delay. I assure them that my skull retains its normal covering, and that though, naturally, I may have felt alarmed, nothing serious has happened. My doughty adver-

sary has merely performed a war dance, and his blows
have for the most part cut the air. I regret to add,
however, that by misadventure, and I am afraid I
must say carelessness, he has inflicted one or two
severe contusions on himself.

When the noise of approaching battle roused me
from the dreams of peace which occupy my retire-
ment, I was glad to observe (since I must fight)
that the campaign was to be opened upon a new
field. When the contest raged over the Pentateuchal
myth of the creation, Mr. Gladstone's manifest want
of acquaintance with the facts and principles involved
in the discussion, no less than with the best literature
on his own side of the subject, gave me the uncomfortable feeling that I had my adversary at a disadvantage. The sun of science, at my back, was in
his eyes. But, on the present occasion, we are
happily on an equality. History and Biblical
criticism are as much, or as little, my vocation
as they are that of Mr. Gladstone ; the blinding
from too much light, or the blindness from too
little, may be presumed to be equally shared by
both of us.

Mr. Gladstone takes up his new position in the
country of the Gadarenes. His strategic sense
justly leads him to see that the authority of the
teachings of the synoptic Gospels, touching the
nature of the spiritual world, turns upon the
acceptance or the rejection of the Gadarene and
other like stories. As we accept or repudiate such
histories as that of the possessed pigs, so shall we

accept or reject the witness of the synoptics to such miraculous interventions.

It is exactly because these stories constitute the key-stone of the orthodox arch, that I originally drew attention to them ; and, in spite of my longing for peace, I am truly obliged to Mr. Gladstone for compelling me to place my case before the public once more. It may be thought that this is a work of supererogation by those who are aware that my essay is the subject of attack in a work so largely circulated as the *Impregnable Rock of Holy Scripture*; and who may possibly, in their simplicity, assume that it must be truthfully set forth in that work. But the warmest admirers of Mr. Gladstone will hardly be prepared to maintain that mathematical accuracy in stating the opinions of an opponent is the most prominent feature of his controversial method. And what follows will show that, in the present case, the desire to be fair and accurate, the existence of which I am bound to assume, has not borne as much fruit as might have been expected.

In referring to the statement of the narrators that the herd of swine perished in consequence of the entrance into them of the demons by the permission, or order, of Jesus of Nazareth, I said :

" Everything that I know of law and justice convinces me that the wanton destruction of other people's property is a misdemeanour of evil example " (*Nineteenth Century*, February 1889, p. 172).

Mr. Gladstone has not found it convenient to cite this passage ; and, in view of various considerations,

I dare not assume that he would assent to it, without sundry subtle modifications which, for me, might possibly rob it of its argumentative value. But, until the proposition is seriously controverted, I shall assume it to be true, and content myself with warning the reader that neither he nor I have any grounds for assuming Mr. Gladstone's concurrence. With this caution, I proceed to remark that I think it may be granted that the people whose herd of 2000 swine (more or fewer) was suddenly destroyed suffered great loss and damage. And it is quite certain that the narrators of the Gadarene story do not, in any way, refer to the point of morality and legality thus raised ; as I said, they show no inkling of the moral and legal difficulties which arise.

Such being the facts of the case, I submit that for those who admit the principle laid down, the conclusion which I have drawn necessarily follows ; though I repeat that, since Mr. Gladstone does not explicitly admit the principle, I am far from suggesting that he is bound by its logical consequences. However, I distinctly repeat the opinion that any one who acted in the way described in the story would, in my judgment, be guilty of " a misdemeanour of evil example." About that point I desire to leave no ambiguity whatever ; and it follows that, if I believed the story, I should have no hesitation in applying this judgment to the chief actor in it.

But if any one will do me the favour to turn to the paper in which these passages occur, he will find that a considerable part of it is devoted to the exposure of the familiar trick of the "counsel for creeds," who, when they wish to profit by the easily stirred *odium theologicum*, are careful to confuse disbelief in a narrative of a man's act, or disapproval of the acts as narrated, with disbelieving and vilipending the man himself. If I say that "according to paragraphs in several newspapers, my valued Separatist friend A. B. has houghed a lot of cattle which he considered to be unlawfully in the possession of an Irish land-grabber; that in my opinion any such act is a misdemeanour of evil example; but that I utterly disbelieve the whole story and have no doubt that it is a mere fabrication:" it really appears to me that, if any one charges me with calling A. B. an immoral misdemeanant, I should be justified in using very strong language respecting either his sanity or his veracity. And, if an analogous charge has been brought in reference to the Gadarene story, there is certainly no excuse producible on account of any lack of plain speech on my part. Surely no language can be more explicit than that which follows:

"I can discern no escape from this dilemma; either Jesus said what he is reported to have said, or he did not. In the former case, it is inevitable that his authority on matters connected with the 'unseen world' should be roughly shaken; in the latter, the blow falls upon the authority of the synoptic Gospels" (p. 173). "The choice then lies between discrediting those who compiled the gospel biographies and disbelieving the Master, whom

they, simple souls, thought to honour by preserving such traditions of the exercise of his authority
over Satan's invisible world " (p. 174).   And I leave
no shadow of doubt as to my own choice : " After
what has been said, I do not think that any sensible
man, unless he happen to be angry, will accuse of
' contradicting the Lord and his Apostles ' if I
reiterate my total disbelief in the whole Gadarene
story " (p. 178).

I am afraid, therefore, that Mr. Gladstone must
have been exceedingly angry when he committed
himself to such a statement as follows :

So, then, after eighteen centuries of worship offered to our
Lord by the most cultivated, the most developed, and the most
progressive portion of the human race, it has been reserved to a
scientific inquirer to discover that He was no better than a law-
breaker and an evil-doer. . . . How, in such a matter, came the
honours of originality to be reserved to our time and to Professor Huxley ? (pp. 269, 270.)

Truly, the hatchet is hardly a weapon of precision, but would seem to have rather more the
character of the boomerang, which returns to damage
the reckless thrower.   Doubtless such incidents are
somewhat ludicrous.   But they have a very serious
side ; and, if I rated the opinion of those who
blindly follow Mr. Gladstone's leading, but not light,
in these matters, much higher than the great Duke
of Wellington's famous standard of minimum value,
I think I might fairly beg them to reflect upon
the general bearings of this particular example of
his controversial method.   I imagine it can hardly
commend itself to their cool judgment.

After this tragi-comical ending to what an old historian calls a " robustious and rough coming on " ; and after some praises of the provisions of the Mosaic law in the matter of not eating pork—in which, as pork disagrees with me and for some other reasons, I am much disposed to concur, though I do not see what they have to do with the matter in hand— comes the serious onslaught.

> Mr. Huxley, exercising his rapid judgment on the text, does not appear to have encumbered himself with the labour of in- quiring what anybody else had known or said about it. He has thus missed a point which might have been set up in support of his accusation against our Lord (p. 273).

Unhappily for my comfort, I have been much exercised in controversy during the past thirty years ; and the only compensation for the loss of time and the trials of temper which it has inflicted upon me, is that I have come to regard it as a branch of the fine arts, and to take an impartial and æsthetic interest in the way it is conducted, even by those whose efforts are directed against myself. Now, from the purely artistic point of view (which, as we are all being told, has nothing to do with morals), I consider it an axiom, that one should never appear to doubt that the other side has per- formed the elementary duty of acquiring proper elementary information, unless there is demon- strative evidence to the contrary. And I think, though I admit that this may be a purely sub- jective appreciation, that (unless you are quite certain) there is a " want of finish," as a great

master of disputation once put it, about the suggestion that your opponent has missed a point on
his own side.   Because it may happen that he has
not missed it at all, but only thought it unworthy of
serious notice.   And if he proves that, the suggestion
looks foolish.

Merely noting the careful repetition of a charge,
the absurdity of which has been sufficiently exposed
above, I now ask my readers to accompany me on a
little voyage of discovery in search of the side on
which the rapid judgment and the ignorance of the
literature of the subject lie.   I think I may promise
them very little trouble, and a good deal of entertainment.

Mr. Gladstone is of opinion that the Gadarene
swinefolk were " Hebrews bound by the Mosaic
law" (p. 274), and he conceives that it has not
occurred to me to learn what may be said in favour
of and against this view.   He tells us that

> Some commentators have alleged the authority of Josephus
> for stating that Gadara was a city of Greeks rather than of
> Jews, from whence it might be inferred that to keep swine was
> innocent and lawful (p. 273).

Mr. Gladstone then goes on to inform his readers
that in his painstaking search after truth he has
submitted to the labour of personally examining the
writings of Josephus.   Moreover, in a note, he positively exhibits an acquaintance, in addition, with the
works of Bishop Wordsworth and of Archbishop
Trench ; and even shows that he has read Hudson's
commentary on Josephus.   And yet people say that

our Biblical critics do not equal the Germans in research! But Mr. Gladstone's citation of Cuvier and Sir John Herschel about the Creation myth, and his ignorance of all the best modern writings on his own side, produced a great impression on my mind. I have had the audacity to suspect that his acquaintance with what has been done in biblical history might stand at no higher level than his information about the natural sciences. However unwillingly, I have felt bound to consider the possibility that Mr. Gladstone's labours in this matter may have carried him no further than Josephus and the worthy, but somewhat antique, episcopal and other authorities to whom he refers; that even his reading of Josephus may have been of the most cursory nature, directed not to the understanding of his author, but to the discovery of useful controversial matter; and that, in view of the not inconsiderable misrepresentation of my statements to which I have drawn attention, it might be that Mr. Gladstone's exposition of the evidence of Josephus was not more trustworthy. I proceed to show that my previsions have been fully justified. I doubt if controversial literature contains anything more *piquant* than the story I have to unfold.

That I should be reproved for rapidity of judgment is very just: however quaint the situation of Mr. Gladstone, as the reprover, may seem to people blessed with a sense of humour. But it is a quality, the defects of which have been painfully obvious to me all my life; and I try to keep my Pegasus—at best a poor Shetland variety of that species of quadruped—at a respectable jog-trot, by loading him heavily with bales of reading. Those who took the trouble to study my paper in good faith, and not for mere controversial purposes, have a right to know, that something more than a hasty glimpse of two or three passages of Josephus (even with as many episcopal works thrown in) lay at the back of the few paragraphs I devoted to the Gadarene story. I proceed to set forth, as briefly as I can, some results of that preparatory work. My artistic principles do not permit me, at present, to express a doubt that Mr. Gladstone was acquainted with the facts I am about to mention when he undertook to write. But, if he did know them, then both what he has said and what he has not said, his assertions and his omissions alike, will require a paragraph to themselves.

The common consent of the synoptic Gospels affirms that the miraculous transference of devils from a man, or men, to sundry pigs took place somewhere on the eastern shore of the Lake of Tiberias ; " on the other side of the sea over against Galilee," the western shore being, without doubt, included in the latter province. But there is no such concord when we come to the name of the part of the eastern shore on which, according to the story, Jesus and his disciples landed. In the revised version Matthew calls it the " country of the Gadarenes :" Luke and Mark have " Gerasenes." In sundry very ancient manuscripts " Gergesenes " occurs.

The existence of any place called Gergesa, however, is declared by the weightiest authorities whom I have consulted to be very questionable; and no such town is mentioned in the list of the cities of the Decapolis, in the territory of which (as it would seem from Mark v. 20) the transaction was supposed to take place. About Gerasa, on the other hand, there hangs no such doubt. It was a large and important member of the group of the Decapolitan cities. But Gerasa is more than thirty miles distant from the nearest part of the Lake of Tiberias, while the city mentioned in the narrative could not have been very far off the scene of the event. However, as Gerasa was a very important Hellenic city, not much more than a score of miles from Gadara, it is easily imaginable that a locality which was part of Decapolitan territory may have been spoken of as belonging to one of the two cities, when it really appertained to the other. After weighing all the arguments, no doubt remains on my mind that "Gadarene" is the proper reading. At the period under consideration, Gadara appears to have been a good-sized fortified town, about two miles in circumference. It was a place of considerable strategic importance, inasmuch as it lay on a high ridge at the point of intersection of the roads from Tiberias, Scythopolis, Damascus, and Gerasa. Three miles north from it, where the Tiberias road descended into the valley of the Hieromices, lay the famous hot springs and the fashionable baths of Amatha. On the north-east side, the remains of the extensive necropolis of Gadara are still to be seen.

2 N

Innumerable sepulchral chambers are excavated in the limestone cliffs, and many of them still contain sarcophaguses of basalt ; while not a few are converted into dwellings by the inhabitants of the present village of Um Keis.   The distance of Gadara from the south-eastern shore of the Lake of Tiberias is less than seven miles.   The nearest of the other cities of the Decapolis, to the north, is Hippos, which also lay some seven miles off on the south-eastern corner of the shore of the lake.   In accordance with the ancient Hellenic practice that each city should be surrounded by a certain amount of territory amenable to its jurisdiction,[1] and on the other grounds, it may be taken for certain that the intermediate country was divided between Gadara and Hippos, and that the citizens of Gadara had free access to a port on the lake.   Hence the title of "country of the Gadarenes" applied to the locality of the porcine catastrophe becomes easily intelligible.   The swine may well be imagined to have been feeding (as they do now in the adjacent region) on the hillsides, which slope somewhat steeply down to the lake from the northern boundary wall of the valley of the Hieromices (*Nahr Yarmuk*), about half-way between the city and the shore, and doubtless lay well within the territory of the *polis* of Gadara.

The proof that Gadara was, to all intents and purposes, a Gentile and not a Jewish city is complete.

---

[1] Thus Josephus (lib. ix.) says that his rival, Justus, persuaded the citizens of Tiberias to "set the villages that belonged to Gadara and Hippos on fire ; which villages were situated on the borders of Tiberias and of the region of Scythopolis."

The date and the occasion of its foundation are un-
known; but it certainly existed in the third century
B.C. Antiochus the Great annexed it to his dominions
in B.C. 198. After this, during the brief revival of
Jewish autonomy, Alexander Jannæus took it; and
for the first time, so far as the records go, it fell under
Jewish rule.[1] From this it was rescued by Pompey
(B.C. 63), who rebuilt the city and incorporated it
with the province of Syria. In gratitude to the
Romans for the dissolution of a hated union, the
Gadarenes adopted the Pompeian era on their coinage. Gadara was a commercial centre of some importance, and therefore, it may be assumed, Jews
settled in it, as they settled in almost all considerable
Gentile cities. But a wholly mistaken estimate of the
magnitude of the Jewish colony has been based upon
the notion that Gabinius, proconsul of Syria in
57-55 B.C., seated one of the five sanhedrims in
Gadara. Schürer has pointed out that what he
really did was to lodge one of them in Gazara, far
away on the other side of the Jordan. This is one
of the many errors which have arisen out of the confusion of the names Gadara, Gazara, and Gabara.

Augustus made a present of Gadara to Herod the
Great, as an appanage personal to himself; and, upon
Herod's death, recognising it to be a "Grecian city like
Hippos and Gaza," [2] he transferred it back to its former

---

[1] It is said to have been destroyed by its captors.

[2] "But as to the Grecian cities Gaza and Gadara and Hippos, he
cut them off from the kingdom and added them to Syria."—Josephus,
Wars, II. vi. 3. See also *Antiquities*, XVII. xi. 4.

place in the province of Syria. That Herod made no
effort to judaise his temporary possession, but rather
the contrary, is obvious from the fact that the coins
of Gadara, while under his rule, bear the image of
Augustus with the superscription Σεβαστός—a flying
in the face of Jewish prejudices which even he did
not dare to venture upon in Judæa. And I may
remark that, if my co-trustee of the British Museum
had taken the trouble to visit the splendid numis-
matic collection under our charge, he might have
seen two coins of Gadara, one of the time of Tiberius
and the other of that of Titus, each bearing the
effigies of the emperor on the obverse: while the
personified genius of the city is on the reverse of the
former. Further, the well-known works of De Saulcy
and of Ekhel would have supplied the information
that, from the time of Augustus to that of Gordian,
the Gadarene coinage had the same thoroughly Gentile character. Curious that a city of " Hebrews bound
by the Mosaic law " should tolerate such a mint!

Whatever increase in population the Ghetto of
Gadara may have undergone between B.C. 4 and A.D.
66, it nowise affected the Gentile and anti-judaic
character of the city at the outbreak of the great war;
for Josephus tells us that immediately after the great
massacre at Cæsarea, the revolted Jews "laid waste
the villages of the Syrians and their neighbouring
cities, Philadelphia and Sebonitis and Gerasa and
Pella and Scythopolis, and after them Gadara and
Hippos" (Wars, II. xviii. 1). I submit that if
Gadara had been a city of "Hebrews bound by the

Mosaic law," the ravaging of their territory by their brother Jews in revenge for the massacre of the Cæsarean Jews by the Gentile population of that place, would surely have been a somewhat unaccountable proceeding. But when we proceed a little further, to the fifth section of the chapter in which this statement occurs, the whole affair becomes intelligible enough.

> Besides this murder at Scythopolis, the other cities rose up against the Jews that were among them : those of Askelon slew two thousand five hundred, and those of Ptolemais two thousand, and put not a few into bonds; those of Tyre also put a great number to death, but kept a greater number in prison; moreover, those of Hippos and those of Gadara did the like, while they put to death the boldest of the Jews, but kept those of whom they were most afraid in custody; as did the rest of the cities of Syria according as they every one either hated them or were afraid of them.

Josephus is not always trustworthy, but he has no conceivable motive for altering facts here; he speaks of contemporary events, in which he himself took an active part, and he characterises the cities in the way familiar to him. For Josephus, Gadara is just as much a Gentile city as Ptolemais; it was reserved for his latest commentator, either ignoring, or ignorant of, all this, to tell us that Gadara had a Hebrew population bound by the Mosaic law.

In the face of all this evidence, most of which has been put before serious students, with full reference to the needful authorities and in a thoroughly judicial manner, by Schürer in his classical work,[1] one reads

---

[1] *Geschichte des jüdischen Volkes im Zeitalter Christi*, 1886-90.

with stupefaction the statement which Mr. Gladstone
has thought fit to put before the uninstructed
public :

> Some commentators have alleged the authority of Josephus
> for stating that Gadara was a city of Greeks rather than of Jews,
> from whence it might be inferred that to keep swine was inno-
> cent and lawful.   This is not quite the place for a critical
> examination of the matter ; but I have examined it, and have
> satisfied myself that Josephus gives no reason whatever to
> suppose that the population of Gadara, and still less (if less may
> be) the population of the neighbourhood, and least of all the
> swine-herding or lower portion of that population, were other
> than Hebrews bound by the Mosaic law.   (Pp. 373-4.)

Even " rapid judgment" cannot be pleaded in excuse
for this surprising statement, because a " Note on the
Gadarene miracle" is added (in a special appendix), in
which the references are given to the passages of
Josephus, by the improved interpretation of which
Mr. Gladstone has thus contrived to satisfy himself
of the thing which is not.   One of these is *Antiquities*,
XVII. xiii. 4, in which section I regret to say I can
find no mention of Gadara.   In *Antiquities*, XVII. xi.
4, however, there is a passage which would appear to
be that which Mr. Gladstone means, and I will give
it in full, although I have already cited part of it :

> There were also certain of the cities which paid tribute to
> Archelaus ; Strato's tower, and Sebaste, with Joppa and Jeru-
> salem ; for, as to Gaza, Gadara, and Hippos, they were Grecian
> cities, which Cæsar separated from his government, and added
> them to the province of Syria.

That is to say, Augustus simply restored the state of
things which existed before he gave Gadara, then
certainly a Gentile city, lying outside Judæa, to

Herod as a mark of great personal favour. Yet Mr. Gladstone can gravely tell those who are not in a position to check his statements :

The sense seems to be not that these cities were inhabited by a Greek population, but that they had politically been taken out of Judæa and added to Syria, which I presume was classified as simply Hellenic, a portion of the great Greek empire erected by Alexander. (Pp. 295-6.)

Mr. Gladstone's next reference is to the *Wars*, III. vii. 1 :

So Vespasian marched to the city Gadara, and took it upon the first onset, because he found it destitute of a considerable number of men grown up fit for war. He then came into it, and slew all the youth, the Romans having no mercy on any age whatsoever; and this was done out of the hatred they bore the nation, and because of the iniquity they had been guilty of in the affair of Cestius.

Obviously, then, Gadara was an ultra-Jewish city. Q.E.D. But a student trained in the use of weapons of precision, rather than in that of rhetorical tomahawks, has had many and painful warnings to look well about him before trusting an argument to the mercies of a passage, the context of which he has not carefully considered. If Mr. Gladstone had not been too much in a hurry to turn his imaginary prize to account—if he had paused just to look at the preceding chapter of Josephus—he would have discovered that his much haste meant very little speed. He would have found (*Wars*, III. vi. 2) that Vespasian marched from his base, the port of Ptolemais (Acre), on the shores of the Mediterranean, into Galilee ; and, having dealt with the so-called " Gadara," was minded to finish with Jotapata, a strong place about fourteen miles south-east of Ptolemais, into which Josephus, who at first had fled to Tiberias, eventually threw himself—Vespasian arriving before Jotapata "the very next day."   Now, if any one will take a decent map of Ancient Palestine in hand, he will see that Jotapata, as I have said, lies about fourteen miles in a straight line east-south-east of Ptolemais, while a certain town, "Gabara" (which was also held by the Jews), is situated about the same distance to the east of that port.   Nothing can be more obvious than that Vespasian, wishing to advance from Ptolemais into Galilee, could not afford to leave these strongholds in the possession of the enemy ; and as Gabara would lie on his left flank when he moved to Jotapata, he took that city, whence his communications with his base could easily be threatened, first.   It might really have been fair evidence of demoniac possession, if the best general of Rome had marched forty odd miles, as the crow flies, through hostile Galilee, to take a city (which, moreover, had just tried to abolish its Jewish population) on the other side of the Jordan ; and then marched back again to a place fourteen miles off his starting-point.[1]   One would think that the most careless of readers must be startled by this incongruity into inquiring whether there might not be something wrong with the text ; and if he had done so he would

[1] If William the Conqueror, after fighting the battle of Hastings had marched to capture Chichester and then returned to assault Rye, being all the while anxious to reach London, his proceedings would not have been more eccentric than Mr. Gladstone must imagine those of Vespasian were.

have easily discovered that since the time of Reland, a century and a half ago, careful scholars have read Gabara for Gadara.[1]

Once more, I venture to point out that training in the use of the weapons of precision of science may have its value in historical studies, if only in preventing the occurrence of droll blunders in geography.

In the third citation (*Wars*, IV. vii.) Josephus tells us that Vespasian marched against " Gadara," which he calls the metropolis of Peræa (it was possibly the seat of a common festival of the Decapolitan cities), and entered it without opposition, the wealthy and powerful citizens having opened negotiations with him without the knowledge of an opposite party, who, " as being inferior in number to their enemies who were within the city, and seeing the Romans very near the city," resolved to fly.   Before doing so, however, they, after a fashion unfortunately too common among the Zealots, murdered and shockingly mutilated Dolesus, a man of the first rank, who had promoted the embassy to Vespasian, and then " ran out of the city."   Hereupon " the people of Gadara " (surely not this time " Hebrews bound by the Mosaic law ") received Vespasian with joyful acclamations, voluntarily pulled down their wall, so that the city could not in future be used as a fortress by the Jews, and accepted a Roman garrison for their future protection.   Granting that this Gadara really is the city of the Gadarenes, the reference, without citation, to

[1] See Reland, *Palestina* (1714), t. ii. p. 771. Also Robinson, *Later Biblical Researches* (1856), p. 87 *note*.

the passage in support of Mr. Gladstone's contention seems rather remarkable. Taken in conjunction with the shortly antecedent ravaging of the Gadarene territory by the Jews, in fact, better proof could hardly be expected of the real state of the case ; namely, that the population of Gadara (and notably the wealthy and respectable part of it) was thoroughly Hellenic ; though, as in Cæsarea and elsewhere among the Palestinian cities, the rabble contained a considerable body of fanatical Jews, whose reckless ferocity made them, even though a mere minority of the population, a standing danger to the city.

Thus Mr. Gladstone's conclusion from his study of Josephus, that the population of Gadara were "Hebrews bound by the Mosaic law," turns out to depend upon nothing better than a marvellously complete misinterpretation of what that author says, combined with equally marvellous geographical misunderstandings, long since exposed and rectified ; while the positive evidence that Gadara, like other cities of the Decapolis, was thoroughly Hellenic in organisation and essentially Gentile in population is overwhelming.

And, that being the fact of the matter, patent to all who will take the trouble to inquire about what has been said about it, however obscure to those who merely talk of so doing, the thesis that the Gadarene swineherds, or owners, were Jews violating the Mosaic law shows itself to be an empty and most unfortunate guess. But really, whether they that kept the swine were Jews, or whether they were Gentiles, is a con-

sideration which has no relevance whatever to my
case. The legal provisions which alone had authority
over an inhabitant of the country of the Gadarenes
were the Gentile laws sanctioned by the Roman
suzerain of the province of Syria, just as the only
law which has authority in England is that recognised
by the sovereign Legislature. Jewish communities in
England may have their private code, as they doubtless had in Gadara. But an English magistrate, if
called upon to enforce their peculiar laws, would
dismiss the complainants from the judgment seat, let
us hope with more politeness than Gallio did in a like
case, but quite as firmly. Moreover, in the matter of
keeping pigs, we may be quite certain that Gadarene
law left everybody free to do as he pleased, indeed
encouraged the practice rather than otherwise. Not
only was pork one of the commonest and one of the
most favourite articles of Roman diet; but, to both
Greeks and Romans, the pig was a sacrificial animal
of high importance. Sucking pigs played an important part in Hellenic purificatory rites; and everybody
knows the significance of the Roman suovetaurilia,
depicted on so many bas-reliefs.

Under these circumstances, only the extreme need
of a despairing "reconciler" drowning in a sea of
adverse facts, can explain the catching at such a poor
straw as the reckless guess that the swineherds of the
" country of the Gadarenes " were erring Jews, doing
a little clandestine business on their own account.
The endeavour to justify the asserted destruction of
the swine by the analogy of breaking open a cask of

smuggled spirits, and wasting their contents on the ground, is curiously unfortunate. Does Mr. Gladstone mean to suggest that a Frenchman landing at Dover, and coming upon a cask of smuggled brandy in the course of a stroll along the cliffs, has the right to break it open and waste its contents on the ground? Yet the party of Galileans who, according to the narrative, landed and took a walk on the Gadarene territory, were as much foreigners in the Decapolis as Frenchmen would be at Dover. Herod Antipas, their sovereign, had no jurisdiction in the Decapolis—they were strangers and aliens, with no more right to interfere with a pig-keeping Hebrew than I have a right to interfere with an English professor of the Israelitic faith, if I see a slice of ham on his plate. According to the law of the country in which these Galilean foreigners found themselves, men might keep pigs if they pleased. If the men who kept them were Jews, it might be permissible for the strangers to inform the religious authority acknowledged by the Jews of Gadara, but to interfere themselves in such a matter was a step devoid of either moral or legal justification.

Suppose a modern English Sabbatarian fanatic, who believes, on the strength of his interpretation of the fourth commandment, that it is a deadly sin to work on the "Lord's Day," sees a fellow Puritan yielding to the temptation of getting in his harvest on a fine Sunday morning—is the former justified in setting fire to the latter's corn? Would not an English court of justice speedily teach him better?

In truth, the government which permits private persons, on any pretext (especially pious and patriotic pretexts), to take the law into their own hands, fails in the performance of the primary duties of all governments; while those who set the example of such acts, or who approve them, or who fail to disapprove them, are doing their best to dissolve civil society —they are compassers of illegality and fautors of immorality.

I fully understand that Mr. Gladstone may not see the matter in this light. He may possibly consider that the union of Gadara with the Decapolis by Augustus was a "blackguard" transaction, which deprived Hellenic Gadarene law of all moral force; and that it was quite proper for a Jewish Galilean, going back to the time when the land of the Girgashites was given to his ancestors, some 1500 years before, to act as if the state of things which ought to obtain in territory which traditionally, at any rate, belonged to his forefathers, did really exist. And, that being so, I can only say I do not agree with him, but leave the matter to the appreciation of those of our countrymen, happily not yet the minority, who believe that the first condition of enduring liberty is obedience to the law of the land.

The end of the month drawing nigh, I thought it well to send away the manuscript of the foregoing pages yesterday, leaving open, in my own mind, the possibility of adding a succinct characterisation of Mr. Gladstone's controversial methods as illustrated therein. This morning, however, I had the pleasure of reading a speech which I think must satisfy the requirements of the most fastidious of controversial artists ; and there occurs in it so concise, yet so complete, a delineation of Mr. Gladstone's way of dealing with disputed questions of another kind, that no poor effort of mine could better it as a description of the aspect which his treatment of scientific, historical, and critical questions presents to me.

The smallest examination would have told a man of his capacity and of his experience that he was uttering the grossest exaggerations, that he was basing arguments upon the slightest hypotheses, and that his discussions only had to be critically examined by the most careless critic in order to show their intrinsic hollowness.

Those who have followed me through this paper will hardly dispute the justice of this judgment, severe as it is. But the Chief Secretary for Ireland has science in the blood ; and has the advantage of a natural, as well as a highly cultivated, aptitude for the use of methods of precision in investigation, and for the exact enunciation of the results thereby obtained.

ILLUSTRATIONS OF MR. GLADSTONE'S
CONTROVERSIAL METHODS

THE series of essays in defence of the historical accuracy of the Jewish and Christian Scriptures contributed by Mr. Gladstone to *Good Words*, having been revised and enlarged by their author, appeared last year as a separate volume, under the somewhat defiant title of *The Impregnable Rock of Holy Scripture*.

The last of these essays, entitled "Conclusion," contains an attack, or rather several attacks, couched in language which certainly does not err upon the side of moderation or of courtesy, upon statements and opinions of mine. One of these assaults is a deliberately devised attempt, not merely to rouse the theological prejudices ingrained in the majority of Mr. Gladstone's readers, but to hold me up as a person who has endeavoured to besmirch the personal character of the object of their veneration. For Mr. Gladstone asserts that I have undertaken to try "the character of our Lord" (p. 268); and he tells the many who are, as I think unfortunately, predisposed to place implicit credit in his assertions, that it has been reserved for me to discover that Jesus " was no better than a law-breaker and an evil-doer ! " (p. 269).

It was extremely easy for me to prove, as I did in the pages of this Review last December, that, under the most favourable interpretation, this amazing declaration must be ascribed to extreme confusion of thought. And, by bringing an abundance of good-will to the consideration of the subject, I have now convinced myself that it is right for me to admit that a person of Mr. Gladstone's intellectual acuteness really did mistake the reprobation of the course of conduct ascribed to Jesus, in a story of which I expressly say I do not believe a word, for an attack on his character and a declaration that he was " no better than a law-breaker and evil-doer." At any rate, so far as I can see, this is what Mr. Gladstone wished to be believed when he wrote the following passage :—

I must, however, in passing, make the confession that I did not state with accuracy, as I ought to have done, the precise form of the accusation. I treated it as an imputation on the action of our Lord ; he replies that it is only an imputation on the narrative of three evangelists respecting Him. The difference, from his point of view, is probably material, and I therefore regret that I overlooked it.[1]

Considering the gravity of the error which is here admitted, the fashion of the withdrawal appears more singular than admirable. From my " point of view " —not from Mr. Gladstone's apparently—the little discrepancy between the facts and Mr. Gladstone's

[1] *Nineteenth Century*, February 1891, pp. 339-40.

carefully offensive travesty of them is "probably" (only "probably") material. However, as Mr. Gladstone concludes with an official expression of regret for his error, it is my business to return an equally official expression of gratitude for the attenuated reparation with which I am favoured.

Having cleared this specimen of Mr. Gladstone's controversial method out of the way, I may proceed to the next assault, that on a passage in an article on Agnosticism (*Nineteenth Century*, February 1889), published two years ago. I there said, in referring to the Gadarene story, "Everything I know of law and justice convinces me that the wanton destruction of other people's property is a misdemeanour of evil example." On this, Mr. Gladstone, continuing his candid and urbane observations, remarks (*Impregnable Rock*, p. 273) that, "Exercising his rapid judgment on the text," and "not inquiring what anybody else had known or said about it," I had missed a point in support of that "accusation against our Lord" which he has now been constrained to admit I never made.

The "point" in question is that "Gadara was a city of Greeks rather than of Jews, from whence it might be inferred that to keep swine was innocent and lawful." I conceive that I have abundantly proved that Gadara answered exactly to the description here given of it; and I shall show, by-and-by, that Mr. Gladstone has used language which, to my mind, involves the admission that the authorities of the city were not Jews. But I have also taken a good deal of pains to show that the question thus raised is of no importance in relation to the main issue.[1] If Gadara was, as I maintain it was, a city of the Decapolis, Hellenistic in constitution and containing a predominantly Gentile population, my case is superabundantly fortified. On the other hand, if the hypothesis that Gadara was under Jewish government, which Mr. Gladstone seems sometimes to defend and sometimes to give up, were accepted, my case would be nowise weakened. At any rate, Gadara was not included within the jurisdiction of the tetrarch of Galilee; if it had been, the Galileans who crossed over the lake to Gadara had no official status; and they had no more civil right to punish law-breakers than any other strangers.

In my turn, however, I may remark that there is a "point" which appears to have escaped Mr. Gladstone's notice. And that is somewhat unfortunate, because his whole argument turns upon it. Mr. Gladstone assumes, as a matter of course, that pig-keeping was an offence against the "Law of Moses"; and, therefore, that Jews who kept pigs were as much liable to legal pains and penalties

---

[1] Neither is it of any consequence whether the locality of the supposed miracle was Gadara, or Gerasa, or Gergesa. But I may say that I was well acquainted with Origen's opinion respecting Gergesa. It is fully discussed and rejected in Riehm's *Handwörterbuch*. In Kitto's *Biblical Cyclopædia* (ii. p. 51) Professor Porter remarks that Origen merely "*conjectures*" that Gergesa was indicated; and he adds, "Now, in a question of this kind, conjectures cannot be admitted. We must implicitly follow the most ancient and creditable testimony, which clearly pronounces in favour of Γαδαρηνῶν. This reading is adopted by Tischendorf, Alford, and Tregelles."

as Englishmen who smuggle brandy (*Impregnable Rock*, p. 274).

There can be no doubt that, according to the Law, as it is defined in the Pentateuch, the pig was an "unclean" animal, and that pork was a forbidden article of diet. Moreover, since pigs are hardly likely to be kept for the mere love of those unsavoury animals, pig-owning, or swine-herding, must have been, and evidently was, regarded as a suspicious and degrading occupation by strict Jews, in the first century A.D. But I should like to know on what provision of the Mosaic Law, as it is laid down in the Pentateuch, Mr. Gladstone bases the assumption, which is essential to his case, that the possession of pigs and the calling of a swineherd were actually illegal. The inquiry was put to me the other day; and, as I could not answer it, I turned up the article "Schwein" in Riehm's standard *Handwörterbuch*, for help out of my difficulty; but unfortunately without success. After speaking of the martyrdom which the Jews, under Antiochus Epiphanes, preferred to eating pork, the writer proceeds :—

It may be, nevertheless, that the practice of keeping pigs may have found its way into Palestine in the Græco-Roman time, in consequence of the great increase of the non-Jewish population; yet there is no evidence of it in the New Testament; the great herd of swine, 2000 in number, mentioned in the narrative of the possessed, was feeding in the territory of Gadara, which belonged to the Decapolis; and the prodigal son became a swineherd with the native of a far country into which he had wandered; in neither of these cases is there reason for thinking that the possessors of these herds were Jews.[1]

Having failed in my search, so far, I took up the next work of reference at hand, Kitto's *Cyclopædia* (vol. iii. 1876). There, under " Swine," the writer, Colonel Hamilton Smith, seemed at first to give me what I wanted, as he says that swine " appear to have been repeatedly introduced and reared by the Hebrew people,[2] notwithstanding the strong prohibition in the Law of Moses (Is. lxv. 4)." But, in the first place, Isaiah's writings form no part of the " Law of Moses "; and, in the second place, the people denounced by the prophet in this passage are neither the possessors of pigs, nor swineherds, but those " which eat swine's flesh and broth of abominable things is in their vessels." And when, in despair, I turned to the provisions of the Law itself, my difficulty was not cleared up. Leviticus xi. 8 (Revised Version) says, in reference to the pig and other unclean animals : " Of their flesh ye shall not eat, and their carcases ye shall not touch." In the revised version of Deuteronomy xiv. 8 the words of the prohibition are identical, and a skilful refiner might possibly satisfy himself, even if he satisfied nobody else, that " carcase " means the body of a live

---

[1] I may call attention, in passing, to the fact that this authority, at any rate, has no sort of doubt of the fact that Jewish Law did not rule in Gadara (indeed, under the head of " Gadara," in the same work, it is expressly stated that the population of the place consisted " predominantly of heathens "), and that he scouts the notion that the Gadarene swineherds were Jews.

[2] The evidence adduced, so far as post-exile times are concerned, appears to me insufficient to prove this assertion.

animal as well as of a dead one; and that, since swineherds could hardly avoid contact with their charges, their calling was implicitly forbidden.[1] Unfortunately, the authorised version expressly says "dead carcase"; and thus the most rabbinically minded of reconcilers might find his casuistry foiled by that great source of surprises, the "original Hebrew." That such check is at any rate possible, is clear from the fact that the legal uncleanness of some animals, as food, did not interfere with their being lawfully possessed, cared for, and sold by Jews. The provisions for the ransoming of unclean beasts (Lev. xxvii. 27) and for the redemption of their sucklings (Numbers xviii. 15) sufficiently prove this. As the late Dr. Kalisch has observed in his *Commentary on Leviticus*, part ii. p. 129, note :—

> Though asses and horses, camels and dogs, were kept by the Israelites, they were, to a certain extent, associated with the notion of impurity; they might be turned to profitable account by their labour or otherwise, but in respect to food they were an abomination.

The same learned commentator (*loc. cit.* p. 88) proves that the Talmudists forbade the rearing of pigs by Jews, unconditionally and everywhere; and even included it under the same ban as the study of Greek philosophy, "since both alike were considered to lead to the desertion of the Jewish faith." It is very possible, indeed probable, that the Pharisees of the fourth decade of our first century took as

---

[1] Even Leviticus xi. 26, cited without reference to the context, will not serve the purpose; because the swine *is* "cloven footed" (Lev. xi. 7).

strong a view of pig-keeping as did their spiritual descendants. But, for all that, it does not follow that the practice was illegal. The stricter Jews could not have despised and hated swineherds more than they did publicans; but, so far as I know, there is no provision in the Law against the practice of the calling of a tax-gatherer by a Jew. The publican was in fact very much in the position of an Irish process-server at the present day—more, rather than less, despised and hated on account of the perfect legality of his occupation. Except for certain sacrificial purposes, pigs were held in such abhorrence by the ancient Egyptians that swineherds were not permitted to enter a temple, or to intermarry with other castes; and any one who had touched a pig, even accidentally, was unclean. But these very regulations prove that pig-keeping was not illegal; it merely involved certain civil and religious disabilities. For the Jews, dogs were typically "unclean" animals; but, when that eminently pious Hebrew, Tobit, "went forth" with the angel "the young man's dog" went "with them" (Tobit v. 16) without apparent remonstrance from the celestial guide. I really do not see how an appeal to the Law could have justified any one in drowning Tobit's dog, on the ground that his master was keeping and feeding an animal quite as "unclean" as any pig. Certainly the excellent Raguel must have failed to see the harm of dog-keeping, for we are told that, on the travellers' return homewards, "the dog went after them" (xi. 4).

XV MR. GLADSTONE'S CONTROVERSIAL METHODS 567

Until better light than I have been able to obtain is thrown upon the subject, therefore, it is obvious that Mr. Gladstone's argumentative house has been built upon an extremely slippery quicksand; perhaps even has no foundation at all.

Yet another "point" does not seem to have occurred to Mr. Gladstone, who is so much shocked that I attach no overwhelming weight to the assertions contained in the synoptic Gospels, even when all three concur. These Gospels agree in stating, in the most express, and to some extent verbally identical, terms, that the devils entered the pigs at their own request,[1] and the third Gospel (viii. 31) tells us what the motive of the demons was in asking the singular boon: "They intreated him that he would not command them to depart into the abyss." From this, it would seem that the devils thought to exchange the heavy punishment of transportation to the abyss for the lighter penalty of imprisonment in swine. And some commentators, more ingenious than respectful to the supposed chief actor in this extraordinary fable, have dwelt, with satisfaction, upon the very unpleasant quarter of an hour which the evil spirits must have had, when the headlong rush of their maddened tenements convinced them how completely they were taken in. In the whole story there is not one solitary hint that the destruction of

---

[1] 1st Gospel: "And the devils *besought him*, saying, If Thou cast us out send us away *into* the herd of swine." 2d Gospel: "They *besought him*, saying, Send us *into* the swine." 3d Gospel: "They *intreated him* that he would give them leave to enter *into* them."

the pigs was intended as a punishment of their owners, or of the swineherds. On the contrary, the concurrent testimony of the three narratives is to the effect that the catastrophe was the consequence of diabolic suggestion. And, indeed, no source could be more appropriate for an act of such manifest injustice and illegality.

I can but marvel that modern defenders of the faith should not be glad of any reasonable excuse for getting rid of a story which, if it had been invented by Voltaire, would have justly let loose floods of orthodox indignation.

Thus, the hypothesis to which Mr. Gladstone so fondly clings finds no support in the provisions of the " Law of Moses " as that law is defined in the Pentateuch ; while it is wholly inconsistent with the concurrent testimony of the synoptic Gospels, to which Mr. Gladstone attaches so much weight. In my judgment, it is directly contrary to everything which profane history tells us about the constitution and the population of the city of Gadara ; and it commits those who accept it to a story which, if it were true, would implicate the founder of Christianity in an illegal and inequitable act.

Such being the case, I consider myself excused from following Mr. Gladstone through all the meanderings of his late attempt to extricate himself from the maze of historical and exegetical difficulties in which he is entangled. I content myself with assuring those who, with my paper (not Mr. Gladstone's version of my arguments) in hand, consult the original authorities, that they will find full justification for every statement I have made. But in order to dispose those who cannot, or will not, take that trouble, to believe that the proverbial blindness of one that judges his own cause plays no part in inducing me to speak thus decidedly, I beg their attention to the following examination, which shall be as brief as I can make it, of the seven propositions in which Mr. Gladstone professes to give a faithful summary of my " errors."

When, in the middle of the seventeenth century, the Holy See declared that certain propositions contained in the works of Bishop Jansen were heretical, the Jansenists of Port Royal replied that, while they were ready to defer to the Papal authority about questions of faith and morals, they must be permitted to judge about questions of fact for themselves; and that, really, the condemned propositions were not to be found in Jansen's writings. As everybody knows, His Holiness and the Grand Monarque replied to this, surely not unreasonable, plea after the manner of Lord Peter in the *Tale of a Tub*. It is, therefore, not without some apprehension of meeting with a similar fate, that I put in a like plea against Mr. Gladstone's Bull. The seven propositions declared to be false and condemnable, in that kindly and gentle way which so pleasantly compares with the authoritative style of the Vatican (No. 5 more particularly), may or may not be true. But they are not to be found in anything I have written.

And some of them diametrically contravene that which I have written. I proceed to prove my assertions.

PROP. 1. *Throughout the paper he confounds together what I had distinguished, namely, the city of Gadara and the vicinage attached to it, not as a mere pomœrium, but as a rural district.*

In my judgment, this statement is devoid of foundation. In my paper on " The Keepers of the Herd of Swine" I point out, at some length, that, " in accordance with the ancient Hellenic practice," each city of the Decapolis must have been "surrounded by a certain amount of territory amenable to its jurisdiction :" and, to enforce this conclusion, I quote what Josephus says about the " villages that belonged to Gadara and Hippos." As I understand the term *pomerium* or *pomœrium*,[1] it means the space which, according to Roman custom, was kept free from buildings, immediately within and without the walls of a city ; and which defined the range of the *auspicia urbana*. The conception of a *pomœrium* as a "vicinage attached to" a city, appears to be something quite novel and original. But then, to be sure, I do not know how many senses Mr. Gladstone may attach to the word " vicinage."

Whether Gadara had a *pomœrium*, in the proper technical sense, or not, is a point on which I offer no opinion. But that the city had a very considerable " rural district" attached to it and, notwithstanding

---

[1] See Marquardt, *Römische Staatsverwaltung*, Bd. III. p. 408.

its distinctness, amenable to the jurisdiction of the Gentile municipal authorities, is one of the main points of my case.

PROP. 2. *He more fatally confounds the local civil government and its following, including, perhaps, the whole wealthy class and those attached to it, with the ethnical character of the general population.*

Having survived confusion No. 1, which turns out not to be on my side, I am now confronted in No. 2 with a "more fatal" error—and so it is, if there be degrees of fatality; but, again, it is Mr. Gladstone's and not mine. It would appear, from this proposition (about the grammatical interpretation of which, however, I admit there are difficulties), that Mr. Gladstone holds that the "local civil government and its following among the wealthy," were ethnically different from the "general population." On p. 348 he further admits that the "wealthy and the local governing power" were friendly to the Romans. Are we then to suppose that it was the persons of Jewish "ethnical character" who favoured the Romans, while those of Gentile "ethnical character" were opposed to them? But if that supposition is absurd, the only alternative is that the local civil government was ethnically Gentile. This is exactly my contention.

At pp. 547 and 553 of the Essay on "The Keepers of the Herd of Swine" I have fully discussed the question of the ethnical character of the general population. I have shown that, according to Josephus, who surely ought to have known, Gadara was as much a Gentile

city as Ptolemais; I have proved that he includes
Gadara amongst the cities "that rose up against the
Jews that were amongst them," which is a pretty definite expression of his belief that the " ethnical character
of the general population" was Gentile. There is no
question here of Jews of the Roman party fighting
with Jews of the Zealot party, as Mr. Gladstone
suggests. It is the non-Jewish and anti-Jewish
general population which rises up against the Jews
who had settled "among them."

PROP. 3. *His one item of direct evidence as to the
Gentile character of the city refers only to the former
and not to the latter.*

More fatal still. But, once more, not to me. I
adduce not one, but a variety of "items" in proof of
the non-Judaic character of the population of Gadara :
the evidence of history ; that of the coinage of the
city ; the direct testimony of Josephus, just cited—to
mention no others. I repeat, if the wealthy people
and those connected with them—the " classes" and
the "hangers on" of Mr. Gladstone's well-known
taxonomy—were, as he appears to admit they were,
Gentiles ; if the " civil government" of the city was
in their hands, as the coinage proves it was ; what
becomes of Mr. Gladstone's original proposition in
The Impregnable Rock of Scripture that " the population of Gadara, and still less (if less may be) the
population of the neighbourhood," were " Hebrews
bound by the Mosaic law" ? And what is the importance of estimating the precise proportion of Hebrews
who may have resided, either in the city of Gadara

or in its dependant territory, when, as Mr. Gladstone now seems to admit (I am careful to say " seems "), the government, and consequently the law, which ruled in that territory and defined civil right and wrong was Gentile and not Judaic? But perhaps Mr. Gladstone is prepared to maintain that the Gentile " local civil government" of a city of the Decapolis administered Jewish Law; and showed their respect for it, more particularly, by stamping their coinage with effigies of the Emperors.

In point of fact, in his haste to attribute to me errors which I have not committed, Mr. Gladstone has given away his case.

PROP. 4. *He fatally confounds the question of political party with those of nationality and of religion, and assumes that those who took the side of Rome in the factions that prevailed could not be subject to the Mosaic Law.*

It would seem that I have a feline tenacity of life; once more, a "fatal error." But Mr. Gladstone has forgotten an excellent rule of controversy; say what is true, of course, but mind that it is decently probable. Now it is not decently probable, hardly indeed conceivable, that any one who has read Josephus, or any other historian of the Jewish war, should be unaware that there were Jews (of whom Josephus himself was one) who " Romanised" and, more or less openly, opposed the war party. But, however that may be, I assert that Mr. Gladstone neither has produced, nor can produce, a passage of my writing which affords the slightest

foundation for this particular article of his indict-
ment.

PROP. 5. *His examination of the text of Josephus
is alike one-sided, inadequate, and erroneous.*

Easy to say, hard to prove. So long as the authorities whom I have cited are on my side, I do not
know why this singularly temperate and convincing
dictum should trouble me. I have yet to become
acquainted with Mr. Gladstone's claims to speak with
an authority equal to that of scholars of the rank
of Schürer, whose obviously just and necessary emen-
dations he so unceremoniously pooh-poohs.

PROP. 6. *Finally, he sets aside, on grounds not
critical or historical, but partly subjective, the
primary historical testimony on the subject, namely,
that of the three Synoptic Evangelists, who write as
contemporaries and deal directly with the subject,
neither of which is done by any other authority.*

Really this is too much! The fact is, as anybody
can see who will turn to my article of February 1889,
out of which all this discussion has arisen, that the
arguments upon which I rest the strength of my case
touching the swine-miracle, are exactly "historical"
and "critical." Expressly, and in words that cannot
be misunderstood, I refuse to rest on what Mr. Gladstone calls "subjective" evidence. I abstain from
denying the possibility of the Gadarene occurrence,
and I even go so far as to speak of some physical
analogies to possession. In fact, my quondam opponent, Dr. Wace, shrewdly, but quite fairly, made the
most of these admissions, and stated that I had

removed the only "consideration which would have been a serious obstacle" in the way of his belief in the Gadarene story.[1]

So far from setting aside the authority of the synoptics on "subjective" grounds, I have taken a great deal of trouble to show that my non-belief in the story is based upon what appears to me to be evident ; firstly, that the accounts of the three synoptic Gospels are not independent, but are founded upon a common source ; secondly, that, even if the story of the common tradition proceeded from a contemporary, it would still be worthy of very little credit, seeing the manner in which the legends about mediæval miracles have been propounded by contemporaries. And in illustration of this position I wrote a special essay about the miracles reported by Eginhard.[2]

In truth, one need go no further than Mr. Gladstone's sixth proposition to be convinced that contemporary testimony, even of well-known and distinguished persons, may be but a very frail reed for the support of the historian, when theological prepossession blinds the witness.[3]

[1] *Nineteenth Century*, March 1889 (p. 362).

[2] "The Value of Witness to the Miraculous." *Nineteenth Century*, March 1889.

[3] I cannot ask the Editor of this Review to reprint pages of an old article,—but the following passages sufficiently illustrate the extent and the character of the discrepancy between the facts of the case and Mr. Gladstone's account of them :—

"Now, in the Gadarene affair, I do not think I am unreasonably sceptical if I say that the existence of demons who can be transferred from a man to a pig does thus contravene probability. Let me be

PROP. 7. *And he treats the entire question, in the narrowed form in which it arises upon secular testimony, as if it were capable of a solution so clear and summary as to warrant the use of the extremest weapons of controversy against those who presume to differ from him.*

The six heretical propositions which have gone before are enunciated with sufficient clearness to enable me to prove without any difficulty that, whosesoever they are, they are not mine. But number seven, I confess, is too hard for me. I cannot undertake to contradict that which I do not understand.

What is the " entire question " which " arises " in a " narrowed form " upon " secular testimony "? After much guessing, I am fain to give up the conundrum. The " question " may be the ownership of the pigs; or the ethnological character of the

perfectly candid. I admit I have no *à priori* objection to offer. . . . I declare, as plainly as I can, that I am unable to show cause why these transferable devils should not exist." . . . (" Agnosticism," *Nineteenth Century*, 1889, p. 177).

" What then do we know about the originator, or originators, of this groundwork—of that threefold tradition which all three witnesses (in Paley's phrase) agree upon—that we should allow their mere statements to outweigh the counter arguments of humanity, of common sense, of exact science, and to imperil the respect which all would be glad to be able to render to their Master ?" (*ibid.* p. 175).

I then go on through a couple of pages to discuss the value of the evidence of the synoptics on critical and historical grounds. Mr. Gladstone cites the essay from which these passages are taken, whence I suppose he has read it; though it may be that he shares the impatience of Cardinal Manning where my writings are concerned. Such impatience will account for, though it will not excuse, his sixth proposition.

Gadarenes; or the propriety of meddling with other people's property without legal warrant. And each of these questions might be so "narrowed" when it arose on "secular testimony" that I should not know where I was. So I am silent on this part of the proposition.

But I do dimly discern in the latter moiety of this mysterious paragraph a reproof of that use of "the extremest weapons of controversy" which is attributed to me. Upon which I have to observe that I guide myself in such matters very much by the maxim of a great statesman, "Do ut des." If Mr. Gladstone objects to the employment of such weapons in defence, he would do well to abstain from them in attack. He should not frame charges which he has, afterwards, to admit are erroneous, in language of carefully calculated offensiveness (*Impregnable Rock*, pp. 269-70); he should not assume that persons with whom he disagrees are so recklessly unconscientious as to evade the trouble of inquiring what has been said or known about a grave question (*Impregnable Rock*, p. 273); he should not qualify the results of careful thought as "hand-over-head reasoning" (*Impregnable Rock*, p. 274); he should not, as in the extraordinary propositions which I have just analysed, make assertions respecting his opponent's position and arguments which are contradicted by the plainest facts.

Persons who, like myself, having spent their lives outside the political world, yet take a mild and philosophical concern in what goes on in it, often find

it difficult to understand what our neighbours call
the psychological moment of this or that party leader
and are, occasionally, loth to believe in the seeming
conditions of certain kinds of success. And when
some chieftain, famous in political warfare, adventures
into the region of letters or of science, in full confidence that the methods which have brought fame and
honour in his own province will answer there, he is
apt to forget that he will be judged by these people,
on whom rhetorical artifices have long ceased to take
effect; and to whom mere dexterity in putting
together cleverly ambiguous phrases, and even the
great art of offensive misrepresentation, are unspeak-
ably wearisome. And, if that weariness finds its
expression in sarcasm, the offender really has no right
to cry out. Assuredly, ridicule is no test of truth,
but it is the righteous meed of some kinds of error.
Nor ought the attempt to confound the expression of
a revolted sense of fair dealing with arrogant impatience of contradiction, to restrain those to whom
' the extreme weapons of controversy" come handy
from using them. The function of police in the
intellectual, if not in the civil, economy may sometimes be legitimately discharged by volunteers.

Some time ago, in one of the many criticisms with
which I am favoured, I met with the remark that, at
our time of life, Mr. Gladstone and I might be better
occupied than in fighting over the Gadarene pigs.
And, if these too famous swine were the only parties
to the suit, I, for my part, should fully admit the

justice of the rebuke. But, under the beneficent rule
of the Court of Chancery, in former times, it was not
uncommon that a quarrel about a few perches of
worthless land ended in the ruin of ancient families
and the engulfing of great estates ; and I think that
our admonisher failed to observe the analogy—to
note the momentous consequences of the judgment
which may be awarded in the present apparently insignificant action *in re* the swineherds of Gadara.

The immediate effect of such judgment will be the
decision of the question whether the men of the
nineteenth century are to adopt the demonology of
the men of the first century as divinely revealed
truth, or to reject it as degrading falsity. The reverend Principal of King's College has delivered his
judgment in perfectly clear and candid terms. Two
years since, Dr. Wace said that he believed the story
as it stands ; and consequently he holds, as a part of
divine revelation, that the spiritual world comprises
devils, who, under certain circumstances, may enter
men and be transferred from them to four-footed
beasts. For the distinguished Anglican divine and
Biblical scholar that is part and parcel of the teachings respecting the spiritual world which we owe to
the founder of Christianity. It is an inseparable part
of that Christian orthodoxy which, if a man rejects,
he is to be considered and called an "infidel."
According to the ordinary rules of interpretation of
language, Mr. Gladstone must hold the same view.

If antiquity and universality are valid tests of the
truth of any belief, no doubt this is one of the beliefs

so certified. There are no known savages, nor people sunk in the ignorance of partial civilisation, who do not hold them. The great majority of Christians have held them and still hold them. Moreover, the oldest records we possess of the early conceptions of mankind in Egypt and in Mesopotamia prove that exactly such demonology, as is implied in the Gadarene story, formed the substratum, and, among the early Accadians, apparently the greater part, of their supposed knowledge of the spiritual world. M. Lenormant's profoundly interesting work on Babylonian magic and the magical texts given in the Appendix to Professor Sayce's *Hibbert Lectures* leave no doubt on this head. They prove that the doctrine of possession, and even the particular case of pig possession,[1] were firmly believed in by the Egyptians and the Mesopotamians before the tribes of Israel invaded Palestine. And it is evident that these beliefs, from some time after the exile and probably much earlier, completely interpenetrated the Jewish mind, and thus became inseparably interwoven with the fabric of the synoptic Gospels.

Therefore, behind the question of the acceptance of the doctrines of the oldest heathen demonology as part of the fundamental beliefs of Christianity, there lies the question of the credibility of the Gospels, and of their claim to act as our instructors, outside that ethical

[1] The wicked, before being annihilated, returned to the world to disturb men; they entered into the body of unclean animals, "often that of a pig, as on the Sarcophagus of Seti I. in the Soane Museum." —Lenormant, *Chaldean Magic*, p. 88, Editorial Note.

province in which they appeal to the consciousness of all thoughtful men. And still, behind this problem, there lies another—how far do these ancient records give a sure foundation to the prodigious fabric of Christian dogma which has been built upon them by the continuous labours of speculative theologians during eighteen centuries?

I submit that there are few questions before the men of the rising generation on the answer to which the future hangs more fatally than this. We are at the parting of the ways. Whether the twentieth century shall see a recrudescence of the superstitions of mediæval papistry, or whether it shall witness the severance of the living body of the ethical ideal of prophetic Israel from the carcase, foul with savage superstitions and cankered with false philosophy, to which the theologians have bound it, turns upon their final judgment of the Gadarene tale.

The gravity of the problems ultimately involved in the discussion of the legend of Gadara will, I hope, excuse a persistence in returning to the subject, to which I should not have been moved by merely personal considerations.

With respect to the diluvial invective which over-flowed thirty-three pages of this Review last January, I doubt not that it has a catastrophic importance in the estimation of its author. I, on the other hand, may be permitted to regard it as a mere spate; noisy and threatening while it lasted, but forgotten almost as soon as it was over. Without my help, it will be judged by every instructed and clear-headed reader; and that is fortunate, because, were aid neeessary, I have cogent reasons for withholding it.

In an article characterised by the same qualities of thought and diction, entitled " A Great Lesson," which appeared in this Review for September 1887, the Duke of Argyll, firstly, charged the whole body of men of science interested in the question with having conspired to ignore certain criticisms of Mr. Darwin's theory of the origin and coral reefs; and, secondly, he asserted that some person unnamed had " actually induced " Mr. John Murray to delay the publication of his views on that subject " for two years."

It was easy for me and for others to prove that the first statement was not only, to use the Duke of Argyll's favourite expression, " contrary to fact," but that it was without any foundation whatever. The second statement rested on the Duke of Argyll's personal authority. All I could do was to demand the production of the evidence for it. Up to the present time, so far as I know, that evidence has not made its appearance; nor has there been any withdrawal of, or apology for, the erroneous charge.

Under these circumstances, most people will understand why the Duke of Argyll may feel quite secure of having the battle all to himself, whenever it pleases him to attack me.

XVI

HASISADRA'S ADVENTURE

SOME thousands of years ago, there was a city in Mesopotamia called Surippak. One night a strange dream came to a dweller therein, whose name, if rightly reported, was Hasisadra. The dream foretold the speedy coming of a great flood; and it warned Hasisadra to lose no time in building a ship, in which, when notice was given, he, his family and friends, with their domestic animals and a collection of the wild creatures and seed of plants of the land, might take refuge and be rescued from destruction. Hasisadra awoke, and at once acted upon the warning. A strong decked ship was built, and her sides were paid, inside and out, with the mineral pitch, or bitumen, with which the country abounded; the vessel's seaworthiness was tested, the cargo was stowed away, and a trusty pilot or steersman appointed.

The promised signal arrived. Wife and friends embarked; Hasisadra, following, prudently "shut the door," or, as we should say, put on the hatches; and Nes-Hea, the pilot, was left alone on deck to do

his best for the ship.   Thereupon a hurricane began
to rage ; rain fell in torrents ; the subterranean waters
burst forth; a deluge swept over the land, and the
wind lashed it into waves sky high ; heaven and
earth became mingled in chaotic gloom.   For six
days and seven nights the gale raged, but the good
ship held out until, on the seventh day, the storm
lulled.   Hasisadra ventured on deck ; and, seeing
nothing but a waste of waters strewed with floating
corpses and wreck, wept over the destruction of his
land and people.   Far away, the mountains of Nizir
were visible ; the ship was steered for them and ran
aground upon the higher land.   Yet another seven
days passed by.   On the seventh, Hasisadra sent
forth a dove, which found no resting place and
returned ; then he liberated a swallow, which also
came back; finally, a raven was let loose, and that
sagacious bird, when it found that the water had
abated, came near the ship, but refused to return
to it.   Upon this, Hasisadra liberated the rest of
the wild animals, which immediately dispersed in
all directions, while he, with his family and friends,
ascending a mountain hard by, offered sacrifices upon
its summit to the gods.

The story thus given in summary abstract, told in
an ancient Semitic dialect, is inscribed in cuneiform
characters upon a tablet of burnt clay.   Many
thousands of such tablets, collected by Assurbanipal,
King of Assyria in the middle of the seventh century
B.C., were stored in the library of his palace at

Nineveh; and, though in a sadly broken and mutilated condition, they have yielded a marvellous amount of information to the patient and sagacious labour which modern scholars have bestowed upon them. Among the multitude of documents of various kinds, this narrative of Hasisadra's adventure has been found in a tolerably complete state. But Assyriologists agree that it is only a copy of a much more ancient work; and there are weighty reasons for believing that the story of Hasisadra's flood was well known in Mesopotamia before the year 2000 B.C.

No doubt, then, we are in presence of a narrative which has all the authority which antiquity can confer; and it is proper to deal respectfully with it, even though it is quite as proper, and indeed necessary, to act no less respectfully towards ourselves; and, before professing to put implicit faith in it, to inquire what claim it has to be regarded as a serious account of an historical event.

It is of no use to appeal to contemporary history, although the annals of Babylonia, no less than those of Egypt, go much further back than 2000 B.C. All that can be said is, that the former are hardly consistent with the supposition that any catastrophe, competent to destroy all the population, has befallen the land since civilisation began, and that the latter are notoriously silent about deluges. In such a case as this, however, the silence of history does not leave the inquirer wholly at fault. Natural science has something to say when the phenomena of nature

are in question. Natural science may be able to
show, from the nature of the country, either that
such an event as that described in the story is
impossible, or at any rate highly improbable; or,
on the other hand, that it is consonant with probability. In the former case, the narrative must be
suspected or rejected; in the latter, no such summary
verdict can be given: on the contrary, it must be
admitted that the story may be true. And then,
if certain strangely prevalent canons of criticism are
accepted, and if the evidence that an event might
have happened is to be accepted as proof that it
did happen, Assyriologists will be at liberty to congratulate one another on the " confirmation by modern
science " of the authority of their ancient books.

It will be interesting, therefore, to inquire how
far the physical structure and the other conditions of
the region in which Surippak was situated are compatible with such a flood as is described in the
Assyrian record.

The scene of Hasisadra's adventure is laid in the
broad valley, six or seven hundred miles long, and
hardly anywhere less than a hundred miles in width,
which is traversed by the lower courses of the rivers
Euphrates and Tigris, and which is commonly known
as the "Euphrates valley." Rising, at the one end,
into a hill country, which gradually passes into the
Alpine heights of Armenia; and, at the other, dipping
beneath the shallow waters of the head of the Persian
Gulf, which continues in the same direction, from
north-west to south-east, for some eight hundred

miles farther, the floor of the valley presents a gradual
slope, from eight hundred feet above the sea level to
the depths of the southern end of the Persian Gulf.
The boundary between sea and land, formed by the
extremest mudflats of the delta of the two rivers, is
but vaguely defined; and, year by year, it advances
seaward. On the north-eastern side, the western
frontier ranges of Persia rise abruptly to great
heights; on the south-western side, a more gradual
ascent leads to a table-land of less elevation, which,
very broad in the south, where it is occupied by the
deserts of Arabia and of Southern Syria, narrows,
northwards, into the highlands of Palestine, and is
continued by the ranges of the Lebanon, the Antilebanon, and the Taurus, into the highlands of Armenia.

The wide and gently inclined plain, thus inclosed
between the gulf and the highlands, on each side and
at its upper extremity, is distinguishable into two
regions of very different character, one of which lies
north, and the other south of the parallel of Hit, on
the Euphrates. Except in the immediate vicinity of
the river, the northern division is stony and scantily
covered with vegetation, except in spring. Over the
southern division, on the contrary, spreads a deep
alluvial soil, in which even a pebble is rare; and
which, though, under the existing misrule, mainly a
waste of marsh and wilderness, needs only intelligent
attention to become, as it was of old, the granary of
western Asia. Except in the extreme south, the
rainfall is small and the air dry. The heat in
summer is intense, while bitterly cold northern blasts

sweep the plain in winter. Whirlwinds are not uncommon; and, in the intervals of the periodical
inundations, the fine, dry, powdery soil is swept,
even by moderate breezes, into stifling clouds, or
rather fogs, of dust. Low inequalities, elevations
here and depressions there, diversify the surface of
the alluvial region. The latter are occupied by
enormous marshes, while the former support the
permanent dwellings of the present scanty and
miserable population.

In antiquity, so long as the canalisation of the
country was properly carried out, the fertility of
the alluvial plain enabled great and prosperous
nations to have their home in the Euphrates
valley. Its abundant clay furnished the materials
for the masses of sun-dried and burnt bricks, the
remains of which, in the shape of huge artificial
mounds, still testify to both the magnitude and the
industry of the population, thousands of years ago.
Good cement is plentiful, while the bitumen, which
wells from the rocks at Hit and elsewhere, not only
answers the same purpose, but is used to this day, as
it was in Hasisadra's time, to pay the inside and the
outside of boats.

In the broad lower course of the Euphrates, the
stream rarely acquires a velocity of more than three
miles an hour, while the lower Tigris attains double
that rate in times of flood. The water of both great
rivers is mainly derived from the northern and eastern
highlands in Armenia and in Kurdistan, and stands
at its lowest level in early autumn and in January.

But when the snows accumulated in the upper basins of the great rivers, during the winter, melt under the hot sunshine of spring, they rapidly rise,[1] and at length overflow their banks, covering the alluvial plain with a vast inland sea, interrupted only by the higher ridges and hummocks which form islands in a seemingly boundless expanse of water.

In the occurrence of these annual inundations lies one of several resemblances between the valley of the Euphrates and that of the Nile. But there are important differences. The time of the annual flood is reversed, the Nile being highest in autumn and winter, and lowest in spring and early summer. The periodical overflows of the Nile, regulated by the great lake basins in the south, are usually punctual in arrival, gradual in growth, and beneficial in operation. No lakes are interposed between the mountain torrents of the upper basis of the Tigris and the Euphrates and their lower courses. Hence, heavy rain, or an unusually rapid thaw in the uplands, gives rise to the sudden irruption of a vast volume of water which not even the rapid Tigris, still less its more sluggish companion, can carry off in time to prevent violent and dangerous overflows. Without an elaborate system of canalisation, providing an escape for such sudden excesses of the supply of water, the annual floods of the Euphrates, and especially of the

---

[1] In May 1849 the Tigris at Bagdad rose 22½ feet—5 feet above its usual rise—and nearly swept away the town. In 1831 a similarly exceptional flood did immense damage, destroying 7000 houses. See Loftus, *Chaldea and Susiana*, p. 7.

Tigris, must always be attended with risk, and often prove harmful.

There are other peculiarities of the Euphrates valley which may occasionally tend to exacerbate the evils attendant on the inundations. It is very subject to seismic disturbances; and the ordinary consequences of a sharp earthquake shock might be seriously complicated by its effect on a broad sheet of water. Moreover, the Indian Ocean lies within the region of typhoons; and if, at the height of an inundation, a hurricane from the south-east swept up the Persian Gulf, driving its shallow waters upon the delta and damming back the outflow, perhaps for hundreds of miles up-stream, a diluvial catastrophe, fairly up to the mark of Hasisadra's, might easily result.[1]

Thus there seems to be no valid reason for rejecting Hasisadra's story on physical grounds. I do not gather from the narrative that the "mountains of Nizir" were supposed to be submerged, but merely that they came into view above the distant horizon of the waters, as the vessel drove in that direction. Certainly the ship is not supposed to ground on any of their higher summits, for Hasisadra has to ascend a peak in order to offer his sacrifice. The country of Nizir lay on the north-eastern side of the Euphrates

---

[1] See the instructive chapter on Hasisadra's flood in Suess, *Das Antlitz der Erde*, Abth. I. Only fifteen years ago a cyclone in the Bay of Bengal gave rise to a flood which covered 3000 square miles of the delta of the Ganges, 3 to 45 feet deep, destroying 100,000 people, innumerable cattle, houses, and trees. It broke inland, on the rising ground of Tipperah, and may have swept a vessel from the sea that far, though I do not know that it did.

valley, about the courses of the two rivers Zab, which
enter the Tigris where it traverses the plain of
Assyria some eight or nine hundred feet above the
sea ; and, so far as I can judge from maps [1] and other
sources of information, it is possible, under the circumstances supposed, that such a ship as Hasisadra's
might drive before a southerly gale, over a continuously flooded country, until it grounded on some
of the low hills between which both the lower and the
upper Zab enter upon the Assyrian plain.

The tablet which contains the story under consideration is the eleventh of a series of twelve. Each
of these answers to a month, and to the corresponding
sign of the Zodiac. The Assyrian year began with
the spring equinox; consequently, the eleventh
month, called "the rainy," answers to our January-
February, and to the sign which corresponds with
our Aquarius. The aquatic adventure of Hasisadra,
therefore, is not inappropriately placed. It is curious,
however, that the season thus indirectly assigned to
the flood is not that of the present highest level of
the rivers. It is too late for the winter rise and too
early for the spring floods.

I think it must be admitted that, so far, the
physical cross-examination to which Hasisadra has
been subjected does not break down his story. On the
contrary, he proves to have kept it in all essential
respects [2] within the bounds of probability or possi-

---

[1] See Cernik's maps in *Petermanns Mittheilungen*, Ergänzungshefte
44 and 45, 1875-76.

[2] I have not cited the dimensions given to the ship in most trans-

bility.   However, we have not yet done with him.
For the conditions which obtained in the Euphrates
valley, four or five thousand years ago, may have
differed to such an extent from those which now
exist that we should be able to convict him of having
made up his tale.   But here again everything is in
favour of his credibility.   Indeed, he may claim very
powerful support, for it does not lie in the mouths of
those who accept the authority of the Pentateuch to
deny that the Euphrates valley was what it is, even
six thousand years back.   According to the book of
Genesis, Phrat and Hiddekel—the Euphrates and the
Tigris—are coeval with Paradise.   An edition of the
Scriptures, recently published under high authority,
with an elaborate apparatus of " Helps " for the use
of students—and therefore, as I am bound to suppose,
purged of all statements that could by any possibility
mislead the young—assigns the year B.C. 4004 as the
date of Adam's too brief residence in that locality.

But I am far from depending on this authority for
the age of the Mesopotamian plain.   On the contrary,
I venture to rely, with much more confidence, on
another kind of evidence, which tends to show that
the age of the great rivers must be carried back to a
date earlier than that at which our ingenuous youth
is instructed that the earth came into existence.   For,
the alluvial deposit having been brought down by the
rivers, they must needs be older than the plain it

lations of the story, because there appears to be a doubt about them.
Haupt (*Keilinschriftliche Sindfluth-Bericht*, p. 13) says that the figures
are illegible.

forms, as navvies must needs antecede the embank-
ment painfully built up by the contents of their wheel-
barrows. For thousands of years, heat and cold,
rain, snow, and frost, the scrubbing of glaciers, and
the scouring of torrents laden with sand and gravel,
have been wearing down the rocks of the upper basins
of the rivers, over an area of many thousand square
miles; and these materials, ground to fine powder in
the course of their long journey, have slowly sub-
sided, as the water which carried them spread out and
lost its velocity in the sea. It is because this process
is still going on that the shore of the delta constantly
encroaches on the head of the gulf[1] into which the
two rivers are constantly throwing the waste of
Armenia and of Kurdistan. Hence, as might be expected, fluviatile and marine shells are common in
the alluvial deposit; and Loftus found strata, containing subfossil marine shells of species now living, in
the Persian Gulf, at Warka, two hundred miles in a
straight line from the shore of the delta.[2] It follows
that, if a trustworthy estimate of the average rate of
growth of the alluvial can be formed, the lowest limit
(by no means the highest limit) of age of the rivers
can be determined. All such estimates are beset

---

[1] It is probable that a slow movement of elevation of the land at
one time contributed to the result—perhaps does so still.

[2] At a comparatively recent period, the littoral margin of the
Persian Gulf extended certainly 250 miles farther to the north-west
than the present embouchure of the Shatt-el Arab. (Loftus, *Quarterly
Journal of the Geological Society*, 1853, p. 251.) The actual extent of
the marine deposit inland cannot be defined, as it is covered by later
fluviatile deposits.

with sources of error of very various kinds; and the best of them can only be regarded as approximations to the truth. But I think it will be quite safe to assume a maximum rate of growth of four miles in a century for the lower half of the alluvial plain.

Now, the cycle of narratives of which Hasisadra's adventure forms a part contains allusions not only to Surippak, the exact position of which is doubtful, but to other cities, such as Erech. The vast ruins at the present village of Warka have been carefully explored and determined to be all that remains of that once great and flourishing city, "Erech the lofty." Supposing that the two hundred miles of alluvial country, which separates them from the head of the Persian Gulf at present, have been deposited at the very high rate of four miles in a century, it will follow that 4000 years ago, or about the year 2100 B.C., the city of Erech still lay forty miles inland. Indeed, the city might have been built a thousand years earlier. Moreover, there is plenty of independent archæological and other evidence that in the whole thousand years, 2000 to 3000 B.C., the alluvial plain was inhabited by a numerous people, among whom industry, art, and literature had attained a very considerable development. And it can be shown that the physical conditions and the climate of the Euphrates valley, at that time, must have been extremely similar to what they are now.

Thus, once more, we reach the conclusion that, as a question of physical probability, there is no ground for objecting to the reality of Hasisadra's adventure.

It would be unreasonable to doubt that such a flood might have happened, and that such a person might have escaped in the way described, any time during the last 5000 years. And if the postulate of loose thinkers in search of scientific "confirmations" of questionable narratives—proof that an event may have happened is evidence that it did happen—is to be accepted, surely Hasisadra's story is "confirmed by modern scientific investigation" beyond all cavil. However, it may be well to pause before adopting this conclusion, because the original story, of which I have set forth only the broad outlines, contains a great many statements which rest upon just the same foundation as those cited, and yet are hardly likely to meet with general acceptance. The account of the circumstances which led up to the flood, of those under which Hasisadra's adventure was made known to his descendant, of certain remarkable incidents before and after the flood, are inseparably bound up with the details already given. And I am unable to discover any justification for arbitrarily picking out some of these and dubbing them historical verities, while rejecting the rest as legendary fictions. They stand or fall together.

Before proceeding to the consideration of these less satisfactory details, it is needful to remark that Hasisadra's adventure is a mere episode in a cycle of stories of which a personage, whose name is provisionally read "Izdubar," is the centre. The nature of Izdubar hovers vaguely between the heroic and the divine; sometimes he seems a mere man, sometimes

approaches so closely to the divinities of fire and of
the sun as to be hardly distinguishable from them.
As I have already mentioned, the tablet which sets
forth Hasisadra's perils is one of twelve; and, since
each of these represents a month and bears a story
appropriate to the corresponding sign of the Zodiac,
great weight must be attached to Sir Henry Rawlinson's suggestion that the epos of Izdubar is a poetical
embodiment of solar mythology.

In the earlier books of the epos, the hero, not content with rejecting the proffered love of the Chaldæan
Aphrodite, Istar, freely expresses his very low estimate
of her character; and it is interesting to observe that,
even in this early stage of human experience, men
had reached a conception of that law of nature which
expresses the inevitable consequences of an imperfect
appreciation of feminine charms. The injured goddess
makes Izdubar's life a burden to him, until at last,
sick in body and sorry in mind, he is driven to seek
aid and comfort from his forbears in the world of
spirits. So this antitype of Odysseus journeys to the
shore of the waters of death, and there takes ship with a
Chaldæan Charon, who carries him within hail of his
ancestor Hasisadra. That venerable personage not
only gives Izdubar instructions how to regain his
health, but tells him, somewhat *à propos des bottes*
(after the manner of venerable personages), the long
story of his perilous adventure; and how it befell
that he, his wife, and his steersman came to dwell
among the blessed gods, without passing through the
portals of death like ordinary mortals.

According to the full story, the sins of mankind had become grievous; and, at a council of the gods, it was resolved to extirpate the whole race by a great flood. And, once more, let us note the uniformity of human experience. It would appear that, four thousand years ago, the obligations of confidential intercourse about matters of state were sometimes violated—of course from the best of motives. Ea, one of the three chiefs of the Chaldæan Pantheon, the god of justice and of practical wisdom, was also the god of the sea; and, yielding to the temptation to do a friend a good turn, irresistible to kindly seafaring folks of all ranks, he warned Hasisadra of what was coming. When Bel subsequently reproached him for this breach of confidence, Ea defended himself by declaring that he did not tell Hasisadra anything; he only sent him a dream. This was undoubtedly sailing very near the wind; but the attribution of a little benevolent obliquity of conduct to one of the highest of the gods is a trifle compared with the truly Homeric anthropomorphism which characterises other parts of the epos.

The Chaldæan deities are, in truth, extremely human; and, occasionally, the narrator does not scruple to represent them in a manner which is not only inconsistent with our idea of reverence, but is sometimes distinctly humorous.[1] When the storm is at its height, he exhibits them flying in a state of panic to Anu, the god of heaven, and crouching

[1] Tiele (*Babylonisch-Assyrische Geschichte*, pp. 572-3) has some very just remarks on this aspect of the epos.

before his portal like frightened dogs.   As the smoke of Hasisadra's sacrifice arises, the gods, attracted by the sweet savour, are compared to swarms of flies.   I have already remarked that the lady Istar's reputation is torn to shreds; while she and Ea scold Bel handsomely for his ferocity and injustice in destroying the innocent along with the guilty.   One is reminded of Here hung up with weighted heels; of misleading dreams sent by Zeus; of Ares howling as he flies from the Trojan battlefield; and of the very questionable dealings of Aphrodite with Helen and Paris.

But to return to the story.   Bel was, at first, excluded from the sacrifice as the author of all the mischief; which really was somewhat hard upon him, since the other gods agreed to his proposal.   But eventually a reconciliation takes place; the great bow of Anu is displayed in the heavens; Bel agrees that he will be satisfied with what war, pestilence, famine, and wild beasts can do in the way of destroying men ; and that, henceforward, he will not have recourse to extraordinary measures.   Finally, it is Bel himself who, by way of making amends, transports Hasisadra, his wife, and the faithful Nes-Hea to the abode of the gods.

It is as indubitable as it is incomprehensible to most of us, that, for thousands of years, a great people, quite as intelligent as we are, and living in as high a state of civilisation as that which had been attained in the greater part of Europe a few centuries ago, entertained not the slightest doubt that Anu,

Bel, Ea, Istar, and the rest, were real personages, possessed of boundless powers for good and evil. The sincerity of the monarchs whose inscriptions gratefully attribute their victories to Merodach, or to Assur, is as little to be questioned as that of the authors of the hymns and penitential psalms which give full expression to the heights and depths of religious devotion. An "infidel" bold enough to deny the existence, or to doubt the influence, of these deities probably did not exist in all Mesopotamia; and even constructive rebellion against their authority was apt to end in the deprivation, not merely of the good name, but of the skin of the offender. The adherents of modern theological systems dismiss these objects of the love and fear of a hundred generations of their equals, offhand, as "gods of the heathen," mere creations of a wicked and idolatrous imagination; and, along with them, they disown, as senseless, the crude theology, with its gross anthropomorphism and its low ethical conception of the divinity, which satisfied the pious souls of Chaldæa.

I imagine, though I do not presume to be sure, that any endeavour to save the intellectual and moral credit of Chaldæan religion, by suggesting the application to it of that universal solvent of absurdities, the allegorical method, would be scouted; I will not even suggest that any ingenuity can be equal to the discovery of the antitypes of the personifications effected by the religious imagination of later ages, in the triad Anu, Ea, and Bel, still less in Istar. Therefore, unless some plausible reconciliatory scheme should be propounded by a Neo-Chaldæan devotee (and, with Neo-Buddhists to the fore, this supposition is not so wild as it looks), I suppose the moderns will continue to smile, in a superior way, at the grievous absurdity of the polytheistic idolatry of these ancient people.

It is probably a congenital absence of some faculty which I ought to possess which withholds me from adopting this summary procedure. But I am not ashamed to share David Hume's want of ability to discover that polytheism is, in itself, altogether absurd. If we are bound, or permitted, to judge the government of the world by human standards, it appears to me that directorates are proved, by familar experience, to conduct the largest and the most complicated concerns quite as well as solitary despots. I have never been able to see why the hypothesis of a divine syndicate should be found guilty of innate absurdity. Those Assyrians, in particular, who held Assur to be the one supreme and creative deity, to whom all the other supernal powers were subordinate, might fairly ask that the essential difference between their system and that which obtains among the great majority of their modern theological critics should be demonstrated. In my apprehension, it is not the quantity, but the quality, of the persons, among whom the attributes of divinity are distributed, which is the serious matter. If the divine might is associated with no higher ethical attributes than those which obtain among ordinary men; if the divine intelligence is supposed to be so imperfect that it cannot foresee the consequences of its own contrivances; if the supernal powers can become furiously angry with the creatures of their omnipotence and, in their senseless wrath, destroy the innocent along with the guilty; or if they can show themselves to be as easily placated by presents and gross flattery as any oriental or occidental despot; if, in short, they are only stronger than mortal men and no better, as it must be admitted Hasisadra's deities proved themselves to be—then, surely, it is time for us to look somewhat closely into their credentials, and to accept none but conclusive evidence of their existence.

To the majority of my respected contemporaries this reasoning will doubtless appear feeble, if not worse. However, to my mind, such are the only arguments by which the Chaldæan theology can be satisfactorily upset. So far from there being any ground for the belief that Ea, Anu, and Bel are, or ever were, real entities, it seems to me quite infinitely more probable that they are products of the religious imagination, such as are to be found everywhere and in all ages, so long as that imagination riots uncontrolled by scientific criticism.

It is on these grounds that I venture, at the risk of being called an atheist by the ghosts of all the principals of all the colleges of Babylonia, or by their living successors among the Neo-Chaldæans, if that sect should arise, to express my utter disbelief in the gods of Hasisadra. Hence, it follows, that I find Hasisadra's account of their share in his adventure

incredible; and, as the physical details of the flood
are inseparable from its theophanic accompaniments,
and are guaranteed by the same authority, I must let
them go with the rest. The consistency of such
details with probability counts for nothing. The
inhabitants of Chaldæa must always have been
familiar with inundations; probably no generation
failed to witness an inundation which rose unusually
high, or was rendered serious by coincident atmospheric, or other, disturbances. And the memory of
the general features of any exceptionally severe and
devastating flood, would be preserved by popular
tradition for long ages. What, then, could be more
natural than that a Chaldæan poet should seek for
the incidents of a great catastrophe among such
phenomena? In what other way than by such an
appeal to their experience could he so surely
awaken in his audience the tragic pity and terror?
What possible ground is there for insisting that he
must have had some individual flood in view, and
that his history is historical, in the sense that the
account of the effects of a hurricane in the Bay of
Bengal, in the year 1875, is historical?

More than three centuries after the time of
Assurbanipal, Berosus of Babylon, born in the reign
of Alexander the Great, wrote an account of the
history of his country in Greek. The work of
Berosus has vanished; but extracts from it—how
far faithful is uncertain—have been preserved by
later writers. Among these occurs the well-known

story of the Deluge of Xisuthros, which is evidently
built upon the same foundation as that of Hasisadra.
The incidents of the divine warning, the building of
the ship, the sending out of birds, the ascension of
the hero, betray their common origin.   But stories,
like Madeira, acquire a heightened flavour with time
and travel; and the version of Berosus is characterised by those circumstantial improbabilities which
habitually gather round the legend of a legend.   The
later narrator knows the exact day of the month on
which the flood began.   The dimensions of the ship
are stated with Munchausenian precision at five stadia
by two—say, half by one-fifth of an English mile.
The ship runs aground among the " Gordæan mountains " to the south of Lake Van, in Armenia, beyond
the limits of any imaginable real inundation of the
Euphrates valley; and, by way of climax, we have
the assertion, worthy of the sailor who said that he
had brought up one of Pharaoh's chariot wheels on
the fluke of his anchor in the Red Sea, that pilgrims
visited the locality and made amulets of the bitumen
which they scraped off from the still extant remains
of the mighty ship of Xisuthros.

Suppose that some later polyhistor, as devoid of
critical faculty as most of his tribe, had found the
version of Berosus, as well as another much nearer
the original story; that, having too much respect
for his authorities to make up a *tertium quid* of
his own, out of the materials offered, he followed
a practice, common enough among ancient and,
particularly, among Semitic historians, of dividing

both into fragments and piecing them together, without troubling himself very much about the resulting repetitions and inconsistencies; the product of such a primitive editorial operation would be a narrative analogous to that which treats of the Noachian deluge in the book of Genesis. For the Pentateuchal story is indubitably a patchwork, composed of fragments of at least two, different and partly discrepant, narratives, quilted together in such an inartistic fashion that the seams remain conspicuous. And, in the matter of circumstantial exaggeration, it in some respects excels even the second-hand legend of Berosus.

There is a certain practicality about the notion of taking refuge from floods and storms in a ship provided with a steersman; but, surely, no one who had ever seen more water than he could wade through would dream of facing even a moderate breeze, in a huge three-storied coffer, or box, three hundred cubits long, fifty wide and thirty high, left to drift without rudder or pilot.[1]   Not content with giving the exact year of Noah's age in which the flood began, the Pentateuchal story adds the month and the day of

[1] In the second volume of the *History of the Euphrates Expedition*, p. 637, Col. Chesney gives a very interesting account of the simple and rapid manner in which the people about Tekrit and in the marshes of Lemlum construct large barges, and make them watertight with bitumen.   Doubtless the practice is extremely ancient ; and as Colonel Chesney suggests, may possibly have furnished the conception of Noah's ark.   But it is one thing to build a barge 44 ft. long by 11ft. wide and 4 ft. deep in the way described ; and another to get a vessel of ten times the dimensions, so constructed, to hold together.

the month. It is the Deity himself who "shuts in" Noah. The modest week assigned to the full deluge in Hasisadra's story becomes forty days, in one of the Pentateuchal accounts, and a hundred and fifty in the other. The flood, which, in the version of Berosus, has grown so high as to cast the ship among the mountains of Armenia, is improved upon in the Hebrew account until it covers "all the high hills that were under the whole heaven"; and, when it begins to subside, the ark is left stranded on the summit of the highest peak, commonly identified with Ararat itself.

While the details of Hasisadra's adventure are, at least, compatible with the physical conditions of the Euphrates valley, and, as we have seen, involve no catastrophe greater than such as might be brought under those conditions, many of the very precisely stated details of Noah's flood contradict some of the best established results of scientific inquiry.

If it is certain that the alluvium of the Mesopotamian plain has been brought down by the Tigris and the Euphrates, then it is no less certain that the physical structure of the whole valley has persisted, without material modification, for many thousand years before the date assigned to the flood. If the summits, even of the moderately elevated ridges which immediately bound the valley, still more those of the Kurdish and Armenian mountains, were ever covered by water, for even forty days, that water must have extended over the whole earth. If the earth was thus covered, anywhere between 4000 and 5000 years ago, or, at any other time, since the

higher terrestrial animals came into existence, they
must have been destroyed from the whole face of
it, as the Pentateuchal account declares they were
three several times (Genesis vii. 21, 22, 23), in
language which cannot be made more emphatic, or
more solemn, than it is; and the present population
must consist of the descendants of emigrants from
the ark. And, if that is the case, then, as has often
been pointed out, the sloths of the Brazilian forests,
the kangaroos of Australia, the great tortoises of the
Galapagos islands, must have respectively hobbled,
hopped, and crawled over many thousand miles of
land and sea from " Ararat" to their present habita-
tions. Thus, the unquestionable facts of the geographical distribution of recent land animals, alone, form an
insuperable obstacle to the acceptance of the assertion
that the kinds of animals composing the present
terrestrial fauna have been, at any time, universally
destroyed in the way described in the Pentateuch.

It is upon this and other unimpeachable grounds,
that, as I ventured to say some time ago, persons
who are duly conversant with even the elements of
natural science decline to take the Noachian deluge
seriously; and that, as I also pointed out, candid
theologians, who, without special scientific knowledge,
have appreciated the weight of scientific arguments,
have long since given it up. But, as Goethe has
remarked, there is nothing more terrible than
energetic ignorance [1]; and there are, even yet, very

---

[1] " Es ist nichts schrecklicher als eine thätige Unwissenheit."
Maximen und Reflexionen, iii.

energetic people, who are neither candid, nor clear-
headed, nor theologians, still less properly instructed
in the elements of natural science, who make prodigious efforts to obscure the effect of these plain
truths, and to conceal their real surrender of the
historical character of Noah's deluge under cover of
the smoke of a great discharge of pseudoscientific
artillery.   They seem to imagine that the proofs
which abound in all parts of the world, of large
oscillations of the relative level of land and sea,
combined with the probability that, when the sea-
level was rising, sudden incursions of the sea, like
that which broke in over Holland and formed the
Zuyder Zee, may have often occurred, can be made
to look like evidence that something that, by courtesy,
might be called a general Deluge has really taken
place.   Their discursive energy drags misunderstood
truth into their service; and " the glacial epoch " is
as sure to crop up among them as King Charles's
head in a famous memorial—with about as much
appropriateness.   The old story of the raised
beach on Moel Tryfaen is trotted out; though,
even if the facts are as yet rightly interpreted,
there is not a shadow of evidence that the change
of sea-level in that locality was sudden, or that
glacial Welshmen would have known it was taking
place.[1]   Surely it is difficult to perceive the relevancy
of bringing in something that happened in the glacial

[1] The well-known difficulties connected with this case have recently
been carefully discussed by Mr. Bell in the *Transactions* of the
Geological Society of Glasgow.

epoch (if it did happen) to account for the tradition
of a flood in the Euphrates valley between 2000 and
3000 B.C. But the date of the Noachian flood is
solidly fixed by the sole authority for it ; no shuffling
of the chronological data will carry it so far back as
3000 B.C. ; and the Hebrew epos agrees with the
Chaldæan in placing it after the development of a
somewhat advanced civilisation. The only authority
for the Noachian deluge assures us that, before it
visited the earth, Cain had built cities ; Jubal had
invented harps and organs ; while mankind had
advanced so far beyond the neolithic, nay even
the bronze, stage that Tubal-cain was a worker in
iron. Therefore, if the Noachian legend is to be
taken for the history of an event which happened in
the glacial epoch, we must revise our notions of
pleistocene civilisation. On the other hand, if the
Pentateuchal story only means something quite
different, that happened somewhere else, thousands
of years earlier, dressed up, what becomes of its
credit as history ? I wonder what would be said
to a modern historian who asserted that Pekin was
burnt down in 1886, and then tried to justify the
assertion by adducing evidence of the Great Fire
of London in 1666. Yet the attempt to save the
credit of the Noachian story by reference to something which is supposed to have happened in the
far north, in the glacial epoch, is far more preposterous.

Moreover, these dust-raising dialecticians ignore
some of the most important and well-known facts

which bear upon the question. Anything more than a parochial acquaintance with physical geography and geology would suffice to remind its possessor that the Holy Land itself offers a standing protest against bringing such a deluge as that of Noah anywhere near it, either in historical times or in the course of that pleistocene period, of which the " great ice age " formed a part.

Judæa and Galilee, Moab and Gilead, occupy part of that extensive tableland at the summit of the western boundary of the Euphrates valley, to which I have already referred. If that valley had ever been filled with water to a height sufficient, not indeed to cover a third of Ararat, in the north, or half some of the mountains of the Persian frontier in the east, but to reach even four or five thousand feet, it must have stood over the Palestinian hog's-back, and have filled, up to the brim, every depression on its surface. Therefore it could not have failed to fill that remarkable trench in which the Dead Sea, the Jordan, and the Sea of Galilee lie, and which is known as the " Jordan-Arabah " valley.

This long and deep hollow extends more than 200 miles, from near the site of ancient Dan in the north, to the water-parting at the head of the Wady Arabah in the south ; and its deepest part, at the bottom of the basin of the Dead Sea, lies 2500 feet below the surface of the adjacent Mediterranean. The lowest portion of the rim of the Jordan-Arabah valley is situated at the village of El Fuleh, 257 feet above the Mediterranean. Everywhere else the circumjacent heights rise to a very much greater altitude. Hence, of the water which stood over the Syrian tableland, when as much drained off as could run away, enough would remain to form a "Mere" without an outlet, 2757 feet deep, over the present site of the Dead Sea. From this time forth, the level of the Palestinian mere could be lowered only by evaporation. It is an extremely interesting fact, which has happily escaped capture for the purposes of the energetic misunderstanding, that the valley, at one time, was filled, certainly within 150 feet of this height—probably higher. And it is almost equally certain, that the time at which this great Jordan-Arabah mere reached its highest level coincides with the glacial epoch. But then the evidence which goes to prove this, also leads to the conclusion that this state of things obtained at a period considerably older than even 4004 B.C., when the world, according to the "Helps" (or shall we say "Hindrances") provided for the simple student of the Bible, was created; that it was not brought about by any diluvial catastrophe, but was the result of a change in the relative activities of certain natural operations which are quietly going on now; and that, since the level of the mere began to sink, many thousand years ago, no serious catastrophe of any description has affected the valley.

The evidence that the Jordan-Arabah valley really was once filled with water, the surface of which reached within 160 feet of the level of the pass of Jezrael, and possibly stood higher, is this: Remains of alluvial strata, containing shells of the freshwater mollusks which still inhabit the valley, worn down into terraces by waves which long rippled at the same level, and furrowed by the channels excavated by modern rainfalls, have been found at the former height; and they are repeated, at intervals, lower down, until the Ghor, or plain of the Jordan, itself an alluvial deposit, is reached. These strata attain a considerable thickness; and they indicate that the epoch at which the freshwater mere of Palestine reached its highest level is extremely remote; that its diminution has taken place very slowly, and with periods of rest, during which the first formed deposits were cut down into terraces. This conclusion is strikingly borne out by other facts. A volcanic region stretches from Galilee to Gilead and the Hauran, on each side of the northern end of the valley. Some of the streams of basaltic lava which have been thrown out from its craters and clefts in times of which history has no record, have run athwart the course of the Jordan itself, or of that of some of its tributary streams. The lava streams, therefore, must be of later date than the depressions they fill. And yet, where they have thus temporarily dammed the Jordan and the Jermuk, these streams have had time to cut through the hard basalts and lay bare the beds, over which, before the lava streams invaded them, they flowed.

In fact, the antiquity of the present Jordan-Arabah valley, as a hollow in a tableland, out of reach of the sea, and troubled by no diluvial or other disturbances, beyond the volcanic eruptions of Gilead and of Galilee, is vast, even as estimated by a geological standard. No marine deposits of later than miocene age occur in or about it; and there is every reason to believe that the Syro-Arabian plateau has been dry land, throughout the pliocene and later epochs, down to the present time. Raised beaches, containing recent shells, on the Levantine shores of the Mediterranean and on those of the Red Sea, testify to a geologically recent change of the sea level to the extent of 250 or 300 feet, probably produced by the slow elevation of the land; and, as I have already remarked, the alluvial plain of the Euphrates and Tigris appears to have been affected in the same way though seemingly to a less extent. But of violent, or catastrophic, change there is no trace. Even the volcanic outbursts have flowed in even sheets over the old land surface; and the long lines of the horizontal terraces which remain, testify to the geological insignificance of such earthquakes as have taken place. It is, indeed, possible that the original formation of the valley may have been determined by the well-known fault, along which the western rocks are relatively depressed and the eastern elevated. But, whether that fault was effected slowly or quickly, and whenever it came into existence, the excavation of the valley to its present width, no less than the sculpturing of its steep walls and of the innumerable deep ravines which score them down to the very bottom, are indubitably due to the operation of rain and streams, during an enormous length of time,

without interruption or disturbance of any magnitude.
The alluvial deposits which have been mentioned are
continued into the lateral ravines, and have more or less
filled them. But, since the waters have been lowered,
these deposits have been cut down to great depths,
and are still being excavated by the present temporary, or permanent, streams. Hence, it follows, that
all these ravines must have existed before the time at
which the valley was occupied by the great mere.
This fact acquires a peculiar importance when we
proceed to consider the grounds for the conclusion
that the old Palestinian mere attained its highest
level in the cold period of the pleistocene epoch. It
is well known that glaciers formerly came low down
on the flanks of Lebanon and Antilebanon; indeed,
the old moraines are the haunts of the few survivors
of the famous cedars. This implies a perennial snow-
cap of great extent on Hermon; therefore, a vastly
greater supply of water to the sources of the Jordan
which rise on its flanks; and, in addition, such a
total change in the general climate, that the innumerable Wadys, now traversed only by occasional storm
torrents, must have been occupied by perennial
streams. All this involves a lower annual tempera-
ture and a moist and rainy atmosphere. If such a
change of meteorological conditions could be effected
now, when the loss by evaporation from the surface
of the Dead Sea salt-pan balances all the gain from
the Jordan and other streams, the scale would be
turned in the other direction. The waters of the
Dead Sea would become diluted; its level would rise;

it would cover, first the plain of the Jordan, then the lake of Galilee, then the middle Jordan between this lake and that of Huleh (the ancient Merom); and, finally, it would encroach, northwards, along the course of the upper Jordan, and, southwards, up the Wady Arabah, until it reached some 260 feet above the level of the Mediterranean, when it would attain a permanent level, by sending any superfluity through the pass of Jezrael to swell the waters of the Kishon, and flow thence into the Mediterranean.

Reverse the process, in consequence of the excess of loss by evaporation over gain by inflow, which must have set in as the climate of Syria changed after the end of the pleistocene epoch, and (without taking into consideration any other circumstances) the present state of things must eventually be reached—a concentrated saline solution in the deepest part of the valley—water, rather more charged with saline matter than ordinary fresh water, in the lower Jordan and the lake of Galilee—fresh waters, still largely derived from the snows of Hermon, in the upper Jordan and in Lake Huleh. But, if the full state of Jordan valley marks the glacial epoch, then it follows that the excavation of that valley by atmospheric agencies must have occupied an immense antecedent time—a large part, perhaps the whole, of the pliocene epoch ; and we are thus forced to the conclusion that, since the miocene epoch, the physical conformation of the Holy Land has been substantially what it is now. It has been more or less rained upon, searched by earthquakes here and there, partially overflowed by lava streams, slowly raised (relatively to the sea-level) a few hundred feet. But there is not a shadow of ground for supposing that, throughout all this time, terrestrial animals have ceased to inhabit a large part of its surface; or that, in many parts, they have been, in any respect, incommoded by the changes which have taken place.

The evidence of the general stability of the physical conditions of Western Asia, which is furnished by Palestine and by the Euphrates Valley, is only fortified if we extend our view northwards to the Black Sea and the Caspian. The Caspian is a sort of magnified replica of the Dead Sea. The bottom of the deepest part of this vast inland mere is 3000 feet below the level of the Mediterranean, while its surface is lower by 85 feet. At present, it is separated, on the west, by wide spaces of dry land from the Black Sea, which has the same height as the Mediterranean, and, on the east, from the Aral, 138 feet above that level. The waters of the Black Sea, now in communication with the Mediterranean by the Dardanelles and the Bosphorus, are salt, but become brackish northwards, where the rivers of the steppes pour in a great volume of fresh water. Those of the shallower northern half of the Caspian are similarly affected by the Volga and the Ural, while, in the shallow bays of the southern division, they become extremely saline in consequence of the intense evaporation. The Aral Sea, though supplied by the Jaxartes and the Oxus, has brackish water. There is evidence

that, in the pliocene and pleistocene periods, to go no
farther back, the strait of the Dardanelles did not
exist, and that the vast area, from the valley of the
Danube to that of the Jaxartes, was covered by
brackish or, in some parts, fresh water to a height of
at least 200 feet above the level of the Mediterranean.
At the present time, the water-parting which separates
the northern part of the basin of the Caspian from
the vast plains traversed by the Tobol and the Obi,
in their course to the Arctic Ocean, appears to be
less than 200 feet above the latter.   It would seem,
therefore, to be very probable that, under the climatal
conditions of part of the pleistocene period, the valley
of the Obi played the same part in relation to the
Ponto-Aralian sea, as that of the Kishon may have
done to the great mere of the Jordan valley ; and
that the outflow formed the channel by which the
well-known Arctic elements of the fauna of the Caspian entered it.   For the fossil remains imbedded in
the strata continuously deposited in the Aralo-Caspian
area, since the latter end of the miocene epoch, show
no sign that, from that time onward, it has ever been
covered by sea water.   Therefore, the supposition
of a free inflow of the Arctic Ocean, which at one
time was generally received, as well as that of various
hypothetical deluges from that quarter, must be
seriously questioned.

The Caspian and the Aral stand in somewhat the
same relation to the vast basin of dry land in which
they lie, as the Dead Sea and the lake of Galilee to
the Jordan valley.   They are the remains of a vast,

mostly brackish, mere, which has dried up in consequence of the excess of evaporation over supply, since the cold and damp climate of the pleistocene epoch gave place to the increasing dryness and great summer heats of Central Asia in more modern times. The desiccation of the Aralo-Caspian basin, which communicated with the Black Sea only by a comparatively narrow and shallow strait along the present valley of Manytsch, the bottom of which was less than 100 feet above the Mediterranean, must have been vastly aided by the erosion of the strait of the Dardanelles towards the end of the pleistocene epoch, or perhaps later. For the result of thus opening a passage for the waters of the Black Sea into the Mediterranean must have been the gradual lowering of its level to that of the latter sea. When this process had gone so far as to bring down the Black Sea water to within less than a hundred feet of its present level, the strait of Manytsch ceased to exist; and the vast body of fresh water brought down by the Danube, the Dnieper, the Don, and other South Russian rivers was cut off from the Caspian, and eventually delivered into the Mediterranean. Thus, there is as conclusive evidence as one can well hope to obtain in these matters, that, north of the Euphrates valley, the physical geography of an area as large as all Central Europe has remained essentially unchanged, from the miocene period down to our time; just as, to the west of the Euphrates valley, Palestine has exhibited a similar persistence of geographical type. To the south, the valley of the Nile tells exactly the same story. The holes bored by miocene mollusks in the cliffs east and west of Cairo bear witness that, in the miocene epoch, it contained an arm of the sea, the bottom of which has since been gradually filled up by the alluvium of the Nile, and elevated to its present position. But the higher parts of the Mokattam and of the desert about Ghizeh, have been dry land from that time to this. Too little is known of the geology of Persia, at present, to allow any positive conclusion to be enunciated. But, taking the name to indicate the whole continental mass of Iran, between the valleys of the Indus and the Euphrates, the supposition that its physical geography has remained unchanged for an immensely long period is hardly rash. The country is, in fact, an enormous basin, surrounded on all sides by a mountainous rim, and subdivided within by ridges into plateaus and hollows, the bottom of the deepest of which, in the province of Seistan, probably descends to the level of the Indian Ocean. These depressions are occupied by salt marshes and deserts, in which the waters of the streams which flow down the sides of the basin are now dissipated by evaporation. I am acquainted with no evidence that the present Iranian basin was ever occupied by the sea; but the accumulations of gravel over a great extent of its surface indicate long-continued water action. It is, therefore, a fair presumption that large lakes have covered much of its present deserts, and that they have dried up by the operation of the same changed climatal conditions as those which have reduced the Caspian and the Dead Sea to their present dimensions.[1]

Thus it would seem that the Euphrates valley, the centre of the fabled Noachian deluge, is also the centre of a region covering some millions of square miles of the present continents of Europe, Asia, and Africa, in which all the facts, relevant to the argument, at present known, converge to the conclusion that, since the miocene epoch, the essential features of its physical geography have remained unchanged; that it has neither been depressed below the sea, nor swept by diluvial waters since that time; and that the Chaldean version of the legend of a flood in the Euphrates valley is, of all those which are extant, the only one which is even consistent with probability, since it depicts a local inundation not more severe than one which might be brought about by a concurrence of favourable conditions at the present day, and which might probably have been more easily effected when the Persian Gulf extended farther north. Hence, the recourse to the " glacial epoch " for some event which might colourably represent a flood, distinctly asserted by the only authority for it to have occurred in historical times, is peculiarly unfortunate. Even a Welsh antiquarian might hesitate over the supposition that a tradition of the fate of Moel Tryfaen, in the glacial epoch, had furnished the basis of fact for a legend

---

[1] An instructive parallel is exhibited by the " Great Basin " of North America. See the remarkable memoir on " Lake Bonneville " by Mr. G. K. Gilbert, of the United States Geological Survey, just published.

which arose among people whose own experience abundantly supplied them with the needful precedents. Moreover, if evidence of interchanges of land and sea are to be accepted as "confirmations" of Noah's deluge, there are plenty of sources for the tradition to be had much nearer than Wales.

The depression now filled by the Red Sea, for example, appears to be, geologically, of very recent origin. The later deposits found on its shores, two or three hundred feet above the sea level, contain no remains older than those of the present fauna; while, as I have already mentioned, the valley of the adjacent delta of the Nile was a gulf of the sea in miocene times. But there is not a particle of evidence that the change of relative level which admitted the waters of the Indian Ocean between Arabia and Africa, took place any faster than that which is now going on in Greenland and Scandinavia, and which has left their inhabitants undisturbed. Even more remarkable changes were effected, towards the end of, or since, the glacial epoch, over the region now occupied by the Levantine Mediterranean and the Ægean Sea. The eastern coast region of Asia Minor, the western of Greece, and many of the intermediate islands, exhibit thick masses of stratified deposits of later tertiary age and of purely lacustrine characters; and it is remarkable that, on the south side of the island of Crete, such masses present steep cliffs facing the sea, so that the southern boundary of the lake in which they were formed must have been situated where the sea now flows. Indeed, there are valid reasons for the supposition that the dry land once extended far to the west of the present Levantine coast, and not improbably forced the Nile to seek an outlet to the north-east of its present delta—a possibility of no small importance in relation to certain puzzling facts in the geographical distribution of animals in this region.   At any rate, continuous land joined Asia Minor with the Balkan peninsula; and its surface bore deep freshwater lakes, apparently disconnected with the Ponto-Aralian sea. This state of things lasted long enough to allow of the formation of the thick lacustrine strata to which I have referred.   I am not aware that there is the smallest ground for the assumption that the Ægean land was broken up in consequence of any of the "catastrophes" which are so commonly invoked.[1] For anything that appears to the contrary, the narrow, steep-sided, straits between the islands of the Ægean archipelago may have been originally brought about by ordinary atmospheric and stream action; and then filled from the Mediterranean, during a slow submergence proceeding from the south northwards.   The strait of the Dardanelles is bounded by undisturbed pleistocene strata forty feet thick, through which, to all appearance, the present passage has been quietly cut.

That Olympus and Ossa were torn asunder and the waters of the Thessalian basin poured forth, is a very ancient notion, and an often cited " confirmation " of Deucalion's flood.   It has not yet ceased to

---

[1] It is true that earthquakes are common enough, but they are incompetent to produce such changes as those which have taken place.

be in vogue, apparently because those who entertain it are not aware that modern geological investigation has conclusively proved that the gorge of the Peneus is as typical an example of a valley of erosion as any to be seen in Auvergne or in Colorado.[1]

Thus, in the immediate vicinity of the vast expanse of country which can be proved to have been untouched by any catastrophe before, during, and since the " glacial epoch," lie the great areas of the Ægean and the Red Sea, in which, during or since the glacial epoch, changes of the relative positions of land and sea have taken place, in comparison with which the submergence of Moel Tryfaen, with all Wales and Scotland to boot, does not come to much.

What, then, is the relevancy of talk about the " glacial epoch " to the question of the historical veracity of the narrator of the story of the Noachian deluge ? So far as my knowledge goes, there is not a particle of evidence that destructive inundations were more common over the general surface of the earth in the glacial epoch than they have been before or since.   No doubt the fringe of an ice-covered region must be always liable to them ; but, if we examine the records of such catastrophes in historical times, those produced in the deltas of great rivers, or in lowlands like Holland, by sudden floods, combined with gales of wind or with unusual tides, far excel all others.

With respect to such inundations as are the con-

---

[1] See Teller, *Geologische Beschreibung des sud-östlichen Thessalien* : Denkschriften d. Akademie der Wissenschaften, Wien, Bd. xl. p. 199.

sequences of earthquakes, and other slight movements of the crust of the earth, I have never heard of anything to show that they were more frequent and severer in the quaternary or tertiary epochs than they are now.   In the discussion of these, as of all other geological problems, the appeal to needless catastrophes is born of that impatience of the slow and painful search after sufficient causes in the ordinary course of nature which is a temptation to all, though only energetic ignorance nowadays completely succumbs to it.

POSTSCRIPT.

My best thanks are due to Mr. Gladstone for his courteous withdrawal of one of the statements to which I have thought it needful to take exception.   The familiarity with controversy, to which Mr. Gladstone alludes, will have accustomed him to the misadventures which arise when, as sometimes will happen in the heat of fence, the buttons come off the foils.   I trust that any scratch which he may have received will heal as quickly as my own flesh wounds have done.

A contribution to the last number of this Review of a different order would be left unnoticed, were it not that my silence would convert me into an accessory to misrepresentations of a very grave character.   However, I shall restrict myself to the barest possible statement of facts, leaving my readers to draw their own conclusions.

In an article entitled " A Great Lesson," published in this Review for September 1887 :

(1) The Duke of Argyll says the "overthrow of Darwin's speculations " (p. 301) concerning the origin of coral reefs, which he fancied had taken place, had been received by men of science " with a grudging silence as far as public discussion is concerned " (p. 301).

The truth is that, as every one acquainted with the literature of the subject was well aware, the views supposed to have effected this overthrow had been fully and publicly discussed by Dana in the United States; by Geikie, Green, and Prestwich in this country; by Lapparent in France; and by Credner in Germany.

(2) The Duke of Argyll says "that no serious reply has ever been attempted" (p. 305).

The truth is that the highest living authority on the subject, Professor Dana, published a most weighty reply, two years before the Duke of Argyll committed himself to this statement.

(3) The Duke of Argyll uses the preceding products of defective knowledge, multiplied by excessive imagination, to illustrate the manner in which "certain accepted opinions" established "a sort of Reign of Terror in their own behalf" (p. 307).

The truth is that no plea, except that of total ignorance of the literature of the subject, can excuse the errors cited, and that the "Reign of Terror" is a purely subjective phenomenon.

(4) The letter in *Nature* for the 17th of November 1887, to which I am referred, contains neither substantiation, nor retractation, of statements 1 and 2. Nevertheless, it repeats number 3. The Duke of Argyll says of his article that it "has done what I intended it to do. It has called wide attention to the influence of mere authority in establishing erroneous theories and in retarding the progress of scientific truth."

(5) The Duke of Argyll illustrates the influence of his fictitious "Reign of Terror" by the statement that Mr. John Murray "was strongly advised against the publication of his views in derogation of Darwin's long-accepted theory of the coral islands, and was actually induced to delay it for two years" (p. 307). And in *Nature* for the 17th November 1887, the Duke of Argyll states that he has seen a letter from Sir Wyville Thomson in which he "urged and almost insisted that Mr. Murray should withdraw the reading of his papers on the subject from the Royal Society of Edinburgh. This was in February 1877." The next paragraph, however, contains the confession: "No special reason was assigned." The Duke of Argyll proceeds to give a speculative opinion that "Sir Wyville

dreaded some injury to the scientific reputation of the body of which he was the chief." Truly, a very probable supposition; but as Sir Wyville Thomson's tendencies were notoriously anti-Darwinian, it does not appear to me to lend the slightest justification to the Duke of Argyll's insinuation that the Darwinian "terror" influenced him. However, the question was finally set at rest by a letter which appeared in *Nature* (29th of December 1887), in which the writer says that:

talking with Sir Wyville about "Murray's new theory," I asked what objection he had to its being brought before the public? The answer simply was : he considered that the grounds of the theory had not, as yet, been sufficiently investigated or sufficiently corroborated, and that therefore any immature, dogmatic publication of it would do less than little service either to science or to the author of the paper.

Sir Wyville Thomson was an intimate friend of mine, and I am glad to have been afforded one more opportunity of clearing his character from the aspersions which have been so recklessly cast upon his good sense and his scientific honour.

(6) As to the "overthrow" of Darwin's theory, which, according to the Duke of Argyll, was patent to every unprejudiced person four years ago, I have recently become acquainted with a work, in which a really competent authority,[1] thoroughly acquainted with all the new lights which have been thrown upon the subject during the last ten years, pronounces the judgment; firstly, that some of the facts brought forward by Messrs. Murray and Guppy against Darwin's theory are not facts; secondly, that the others are reconcilable with Darwin's theory; and, thirdly, that the theories of Messrs. Murray and Guppy "are contradicted by a series of important facts" (p. 13).

Perhaps I had better draw attention to the circumstance that Dr. Langenbeck writes under shelter of the guns of the fortress of Strassburg; and may therefore be presumed to be unaffected by those dreams of a "Reign of Terror" which seem to disturb the peace of some of us in these islands (April 1891).

---

[1] Dr. Langenbeck, *Die Theorien über die Entstehung der Korallen-Inseln und Korallen-Riffe* (p. 13), 1890.

Printed in the United States
By Bookmasters